Modern Gas Kinetics

To Gwen and Sue

Modern Gas Kinetics

THEORY, EXPERIMENT AND
APPLICATION

EDITED BY

M.J. PILLING MA, PhD
Lecturer in Physical Chemistry
University of Oxford

I.W.M. SMITH MA, PhD
Professor of Physical Chemistry
University of Birmingham

Blackwell Scientific Publications

OXFORD LONDON EDINBURGH

BOSTON PALO ALTO MELBOURNE

© 1987 by
Blackwell Scientific Publications
Editorial offices:
Osney Mead, Oxford OX2 0EL
8 John Street, London WC1N 2ES
23 Ainslie Place, Edinburgh EH3 6AJ
52 Beacon Street, Boston
    Massachusetts 02108, USA
667 Lytton Avenue, Palo Alto
    California 94301, USA
107 Barry Street, Carlton
    Victoria 3053, Australia

First published 1987

Set by Datapage International Ltd.,
Unit 12A, IDA Enterprise Centre,
East Wall Road, Dublin 3, Ireland.

Printed and bound in Great Britain.

DISTRIBUTORS

USA and Canada
    Blackwell Scientific Publications Inc
    P O Box 50009, Palo Alto
    California 94303

Australia
    Blackwell Scientific Publications
    (Australia) Pty Ltd
    107 Barry Street
    Carlton, Victoria 3053

British Library
Cataloguing in Publication Data

Pilling, M.J.
    Modern gas kinetics: theory, experiment
    and application.
    1. Gases, Kinetic theory of   2. Chemical
    reaction, Rate of
    I. Title   II. Smith, I.W.M.
    541.3′94   QD502

    ISBN 0-632-01615-9

Library of Congress
Cataloging-in-Publication Data

Modern gas kinetics.

    Bibliography: p.
    Includes index.
    1. Chemical reaction, Rate of. 2. Gases,
Kinetic theory of.   I. Pilling, M. J.   II. Smith,
Ian W. M.
    QD502.M63   1987   541.3′94   86-20725
    ISBN 0-632-01615-9

Contents

C   Kinetics of complex reactions

*Figures in parentheses after each chapter title refer to the number of lectures on which that chapter is based.

Contributors

MARK S. CHILD MA, PhD, Lecturer in Theoretical Chemistry, University of Oxford

R. ANTHONY COX BSc, PhD, Environmental and Medical Sciences Division, Atomic Energy Research Establishment, Harwell, Oxfordshire

GRAHAM HANCOCK MA, PhD, Lecturer in Physical Chemistry, University of Oxford

MICHAEL J. PILLING MA, PhD, Lecturer in Physical Chemistry, University of Oxford

JOHN P. SIMONS MA, PhD, ScD, FRSChem, Professor of Chemistry, University of Nottingham

IAN W. M. SMITH MA, PhD, FRSChem, Professor of Physical Chemistry, University of Birmingham

Preface

The origins of this book are in a Summer School on Gas-Phase Kinetics which was held in Cambridge between 26th June and 3rd July, 1985. The Summer School was sponsored by the Science and Engineering Research Council and was attended by 43 students whom they support with research studentships. In addition, 21 other British students and 18 students from abroad attended the course.

The intention of the organisers of the Summer School was to provide a course which would acquaint post-graduate students with the wide range of research activity in gas-phase kinetics, whatever was their own, necessarily narrow, research topic. The major—but certainly not the only—mode of instruction was a course of lectures. They were accompanied by a fairly extensive set of lecture notes which have served as the progenitor of this book. In addition to the lectures, there was a daily class session and two evenings when students presented short seminars on their research. The participants were divided into three groups, each with two lecturers, for the classes. The main focus for these sessions was the extensive problem sets which accompanied the lectures. These problems and their worked answers are reproduced here and are an unusual, and we hope valuable, feature of the book. We have also indicated in the list of contents how many lectures were devoted to the topics covered in each chapter. We believe that this feature along with the problems should make our text especially attractive to those planning, giving or taking a course of postgraduate lectures in chemical kinetics.

This book, like the course on which it is based, addresses the three main themes suggested in the title. Part A considers theories of elementary chemical reactions, both bimolecular and unimolecular. Although the calculation of potential energy surfaces for reaction systems is not treated explicitly, the role of the potential energy surface in controlling the dynamics by which reagent molecules are transformed to products is an underlying theme throughout this chapter. Space is also found for some comparisons of theoretical predictions with experimental data. The methods by which both the kinetics and dynamics of elementary reactions are studied experimentally are described in Part B, together with some case histories and the important, and so often neglected, matter of error analysis. Finally, in Part C, a number of chapters describe the kinetics of complex reactions particularly in commercially or environmentally important situations. Some stress is placed on the need for a good data base for the crucial elementary reactions in complex systems, and on the identification of which reactions are indeed crucial.

The first chapter of the book stands on its own. Written by Professor John Simons, it is based on the first lecture given at the Summer School. Its aim was to whet the appetite of the audience by describing the aims of the course, by providing a map of the terrain to be covered, and by highlighting some of the splendours to be encountered. In addition, the editors have written a short introduction to each part of the book. These paragraphs set the contents of that part in context and connect them with the topics covered in the other parts of the book. An attempt is also made to define the objectives of the chapters that follow and to specify what the serious student (and problem solver) should learn from them.

We conclude this preface with thanks to a number of people. First, to our four fellow-authors—who remain our very good friends even after this project! Secondly, to several people who helped in a variety of ways with the Summer School: Mrs S. Clements, whose office dealt efficiently with the arrangements in Christ's College; Mr Martin Woodman, whose cheerful assistance with the accounts was most welcome; and Drs A. B. Callear (Cambridge) and Dr. J. A. Kerr (Birmingham) who helped with tutorial sessions. Thirdly, Mr Navin Sullivan of Blackwell's. He has shown great enthusiasm for this book since the idea was first put to him. His firmness has done much to ensure that we have managed to keep to the timetable that we originally put forward.

Our final and most heartfelt thanks must be reserved for Mrs Margot Long. Her efforts in typing all the material for the Summer School were quite astonishing. They might have been described as unbelievable, except they have been repeated during the preparation of the manuscript for this book. For this Herculean effort, all six authors are extremely grateful.

June 1986   Michael Pilling
                                                                      Ian Smith

Introduction
The Interplay Between Theory and Experiment in the Kinetics of Gas-Phase Reactions

J. P. SIMONS

The patterns of enquiry in gas kinetics range from (i) the microscopic study of the dynamics of elementary reactive collisions, seen as individual scattering events at selected collision energies involving reagents in selected quantum states, to (ii) the study of elementary reaction rates at selected bulk temperatures, averaged over all populated internal quantum states and collision energies, to (iii) the study of complex reactions—often at high temperatures—the anatomy of which involves extended sequences of elementary reactions. Examples of complex reactions include pyrolysis (e.g., hydrocarbon cracking, which is dealt with later in Chapter C5), combustion (Chapter C2), atmospheric chemistry (Chapter C1) at altitudes ranging from the troposphere to the stratosphere, or, if complex sequences of ion–molecule reactions are also included, into the ionosphere, comet tails, inter-stellar clouds and—closer to home—gas laser media (Chapter C6). The levels of enquiry implicit in these patterns form a hierachy which links many talents, ranging from those of the quantum mechanic to those of the chemical engineer, and this diversity is reflected in the design of this book which addresses three interrelated topics:

Section A: theories of elementary gas-phase reactions;
Section B: experimental methods for the study of the kinetics and dynamics of elementary reactions;
Section C: the kinetics of complex gas-phase reactions.

1 Interplay between theory and experiment

In scientific enquiry one may choose to follow, at one extreme, the Baconian inductive approach: that is, collect as much observational data as possible and hope that a pattern of comprehension and thence understanding will eventually emerge. Experimental data are valueless, however, unless set against some work-ing theoretical model or hypothesis. This leads to the other extreme, the deductive (inspirational) approach of Descartes, where a basic theoretical model is devel-oped to be tested by subsequent experiment (see Fig. 1). Experimental activity in the kinetics of gas-phase reactions lies somewhere between these two extremes, whilst theoretical activity tends towards the approach of Descartes.

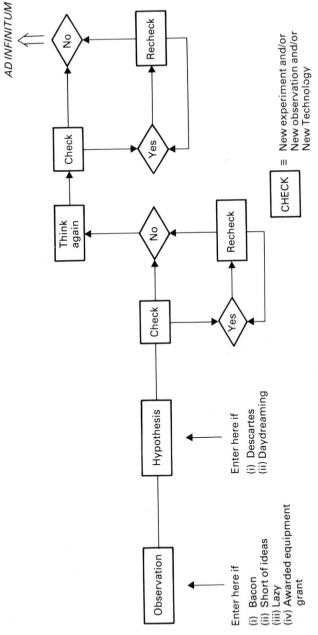

The Scientific Method (Ideal)

Observation

Enter here if

(i) Bacon
(ii) Short of ideas
(iii) Lazy
(iv) Awarded equipment grant

Hypothesis

Enter here if

(i) Descartes
(ii) Daydreaming

Check

Yes → Recheck

No → Think again

Check

Yes → Recheck

No → AD INFINITUM

CHECK ≡ New experiment and/or
         New observation and/or
         New Technology

Fig. 1. The scientific approach: intuition versus exploration.

At its most basic, theory attempts the *ab initio* calculation of the potential energy hypersurface for the reactive system and the quantum mechanical description of the reactive scattering events that may occur under its influence. This approach is at present constrained, by its very difficulty, to a few simple, model systems. At a more tractable and practical level, molecular events can be simulated using the equations of classical, rather than quantum, mechanics to describe the scattering. The computation of such trajectories (see Chapter A2) can provide an adequate description of the reaction dynamics in systems of only a few atoms, but where molecular complexity or dynamical ignorance prevails (i.e., in most of chemistry) statistically based models, such as transition state theory (Chapters A1 and A3), must complement experimental measurement. The main aim of such models is the prediction of elementary reaction rates in advance or in place of direct experimental measurement. The realistic predictive precision of theory was provocatively (and characteristically) questioned at a conference a few years ago by Professor Fred Kaufman.[1] If elementary reaction rate theory is not quantitatively predictive, so putting the experimenters out of business, is it of any practical use as a tool for the analysis of more complex reacting systems? The conclusion, at least for unimolecular reactions (Chapter A4), seems to be a guarded 'Yes!', though qualified by uncertainties in critical parameters such as the efficiency, quantum state dependence, and temperature dependence of collisional energy transfer processes, and by ignorance of the details of the potential energy hypersurface.[2]

Even in the absence of quantitative reliability there is a deeper motive for 'theoretical understanding' that is based on simple curiosity. At its most fundamental level the theory of elementary gas-phase reactons seeks to understand the way chemical bonds are made and broken during energetic atomic, molecular or ionic collisions or following single or multiple photon absorption. There is a continuum between the 'core' science of the chemical physicists and its applications to the 'strategic' science of the chemical engineers. Gas-phase kineticists lie somewhere in the middle.

2 Bimolecular reactions

In the macroscopic world of reactions in bulk between reagents at thermal equilibrium (Chapter B1), the crucial parameters are the temperature-dependent rate constant $k(T)$ and the activation energy E_{act} defined by

$$E_{act} = -d \ln k(T)/d (1/\mathbf{R}T). \tag{1}$$

When the internal quantum states of the reagents and products, denoted collectively by the quantum numbers n and n', are selected and observed (in bulk systems), we enter the realm of state-to-state kinetics (Chapter B2) in which the rates are described by sets of detailed rate constants $k(n'|n;T)$. If we increase the resolution of the experiment still further and enter the microscopic world of individual reactive collisions at energy E_t, usually by studying reactive scattering under molecular beam conditions, the counterparts to the bulk world of rate constants and activation energies become the set of energy-dependent reactive cross-sections $S_{reac}(n'|n;E_t)$ and the threshold energy E_t^0. Averaging over the thermal distributions of relative velocity (or collision energy) at temperature T, links the two worlds through the Laplace transform:

$$k(n'|n;T) \propto \int_{E_t^0}^{\infty} S_{reac}(n'|n;E_t) \, E_t \exp(-E_t/k_B T) \, \mathrm{d}E_t. \tag{2}$$

A knowledge of $S_{reac}(n'|n;E_t)$, the state-to-state reactive excitation functions, would allow, in principle, the calculation of the detailed rate constants $k(n'|n;T)$ and, after appropriate averaging, the evaluation of the thermal rate constant $k(T)$.

$$k(T) = \sum_{n'} \sum_{n} g(n) \, k(n'|n;T) \tag{3}$$

where n defines the internal quantum states of the reagents and $g(n)$ is the initial reagent state Boltzmann distribution for the temperature T.

In practice, the procedure just outlined is not yet technically viable and there is no microscopic substitute for the direct experimental measurement of macroscopic rate constants. (Similarly, the inverse transform from the macroscopic rate constant and its temperature dependence to the microscopic cross-sections places such demands on the precision of the measurements of $k(T)$ as to be also non-viable.) The real value of excitation function measurements lies in their direct probing of the detailed reaction dynamics, which are controlled by the potential energy surface over which the elementary reactions proceed. At a higher level of averaging this control is also reflected in the temperature dependence $k(n'|n;T)$ or ultimately, $k(T)$. If one wants to understand the temperature dependence one is inevitably led back to the potential energy hypersurface and the nature of the transition state(s) of the reactive collisions proceeding under its influence. The surface lies at the heart of reaction kinetics and is the meeting ground of the experimentalists and the theoreticians. Figure 2 summarises these converging patterns of enquiry and also some of the interactions between the various aspects of gas-phase reaction kinetics discussed in this book. We turn now to the question of the transition state.

If one assumes that the motion of atoms can be described by classical equations of motion, then the transition state of an elementary reaction can be identified with a 'dividing surface' (within the complete potential energy hypersurface) which separates the reagent and product regions of the configuration space and through which all reactive trajectories must pass for reagents to be transformed to products. Classical transition state theory chooses the dividing surface through which the total calculated flux is least, in order to minimise the number of trajectories which pass through the dividing surface but then return through it

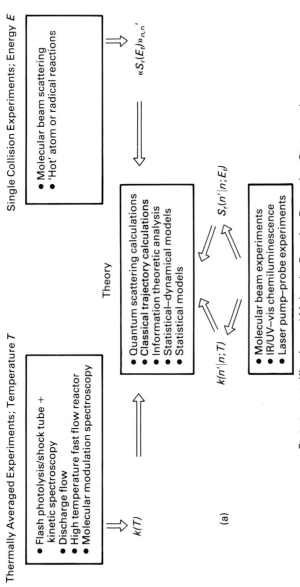

Thermally Averaged Experiments; Temperature T

Single Collision Experiments; Energy E

- Flash photolysis/shock tube + kinetic spectroscopy
- Discharge flow
- High temperature fast flow reactor
- Molecular modulation spectroscopy

- Molecular beam scattering
- 'Hot' atom or radical reactions

Theory

- Quantum scattering calculations
- Classical trajectory calculations
- Information theoretic analysis
- Statistical–dynamical models
- Statistical models

- Molecular beam experiments
- IR/UV–vis chemiluminescence
- Laser pump–probe experiments

$k(T)$

$\langle\!\langle S_r(E_t)\rangle\!\rangle_{n,n'}$

$S_r(n'|n;E_t)$

$k(n'|n;T)$

(a)

State-to-state Kinetics and Molecular Reaction Dynamics; States, n,n'

Fig. 2(a).

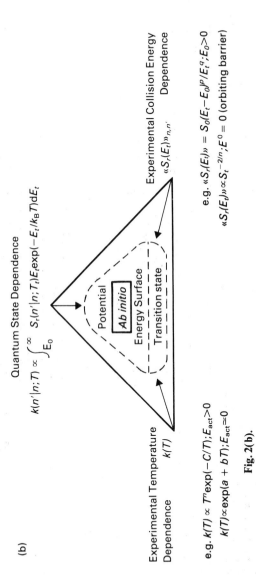

Fig. 2(b).

Fig. 2. The interface between theory and experiment: bimolecular reactions.

rather than continue on and yield separated products. If the product region of the potential energy hypersurface lies behind an energy barrier, then at energies near the reaction threshold the conventional choice of a dividing surface which passes through the saddle point of the potential energy surface is likely to be the best—that is, to be the surface through which the flux is a minimum. However, at higher collision energies, this may not remain the optimum choice. At least for collinear collisions, the best dividing surfaces can be defined by sections through the potential energy surface associated with 'trapped' classical trajectories. These trajectories oscillate continuously, crossing the reaction coordinate at right angles, and veering towards neither the reagent nor the product region of the potential energy surface. They are known as PODS—periodic orbit dividing surfaces. Examples are shown in Fig. 3 and they are discussed in detail in Chapter A1. Variational transition state theory chooses the 'best' transition state to estimate an upper bound to the reaction probability (even with the best choice of dividing surface some trajectories may 'return' through it); a lower bound is obtained by eliminating all trajectories which cross the surface through the saddle point more than twice (see Chapter A1).

Once the notion of trapped trajectories has taken root it is a short step to the notion of resonant periodic orbits—classical trajectories which correspond to quasi-bound quantum states in the collision complex and which can be related to dynamical scattering resonances—and then to the notion of a 'spectroscopy of the transition state'.[3] This could be an optical spectroscopy such as resonance Raman or fluorescence emission from bimolecular collision complexes or from molecules that are in the process of dissociating from an electronically excited state.[3,4] Alternatively, it could be scattering spectroscopy, as in the marvellously detailed angle and state-resolved crossed molecular beam scattering studies of the HF/DF products generated through the reaction of $F + H_2/HD/D_2$.[5]

Unimolecular reactions

The average rate of decomposition or isomerisation of a critically energised molecule (i.e. one that has an internal energy $E > E^0$, where E^0 is a threshold or critical energy) depends upon its internal energy distribution, its environment and possibly on its method of preparation. If this is collisional, as in a thermal unimolecular reaction, the distribution is controlled by the dynamical balance of intermolecular energy transfer from highly excited vibrational–rotational levels, and reaction. As Luther and Troe point out[6] 'the (principal) dilemma of thermal unimolecular rate theory is the lack of understanding of this elementary process', i.e. the intermolecular energy transfer. At sufficiently high pressures, of course, the Boltzmann distribution is maintained, but at lower pressures, in the absence of the strong collision assumption (section A4.3), we are in trouble. New techniques of activation have been developed which avoid this problem (as well as new experimental methods which address the problem of relaxation) though the question of

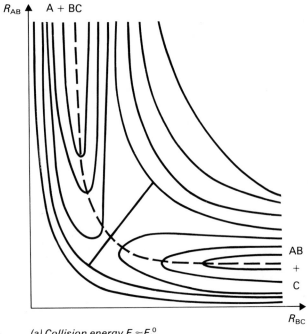

(a) Collision energy $E_t \approx E_t^0$

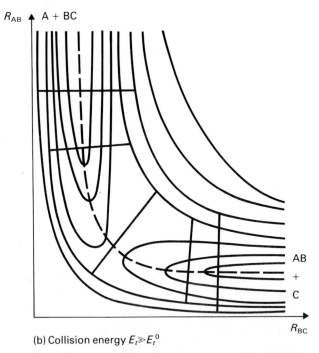

(b) Collision energy $E_t \gg E_t^0$

Fig. 3. Periodic orbit dividing surfaces (PODS) defining transition states for a bimolecular reaction at successively higher collision energies.

the internal energy distribution in the activated molecules still remains. Among the new techniques are 'chemical' activation (via bimolecular atomic or free radical addition/insertion reactions), 'photochemical' activation (via internal conversion following single photon electronic excitation), 'overtone' activation (via visible laser single photon absorption into high overtone vibrational bands) and infrared multiple photon absorption (IRMPA, where the absorbed intensity and fluence from a pulsed laser replace pressure and temperature as the rate controlling parameters).

At the microscopic level, both theory and experiment focus on the set of energy and quantum state-selected rate constants $k(E, J)$ for the unimolecular reaction of molecules with specific internal energy $E > E^0$ (section A4.5), where J represents the totality of quantum numbers that are conserved during the elementary reaction step (e.g., the total angular momentum). The microcanonical formulation of transition state theory (section A4.5) gives the key relationship[6]

$$k(E, J) = N^*(E, J)/h\rho(E, J) \tag{4}$$

where $N^*(E, J)$ is the number of 'open channels' or internal states in the activated molecule at the transition state, $1/h\rho(E, J)$ is the rate constant for reaction through a single, open channel or internal state, and $\rho(E, J)$ is the density of rovibronic states in the reagent at energy E and angular momentum J. The key assumption in deriving equation 4 is that all open channels contribute democractically, with the total rate as a statistical sum and complete mixing of energy among active internal states prior to reaction (the RRKM assumption, see section A4.6). In practice, this may not necessarily be the case, and the study of intramolecular energy transfer and kinetics remains a fertile field of research:[7] significant dynamical constraints on the flow of energy within highly energised molecules cannot be identified readily when the reaction proceeds over timescales much longer than 10^{-12} s. Calculations of state-resolved rate constants depend on evaluating the sum $N^*(E, J)$, which in turn depends on the identification of the transition state. One approach, the 'statistical adiabatic channel' model of Quack and Troe (section A4.7) resolves the net reaction into a set of individual contributions, each one of which corresponds to the temporary conservation of a set of quantum numbers during reaction—hence the adiabatic assumption. In each individual channel, a transition state can be defined as the configuration at the maximum of the adiabatic channel potential. Another, simpler approach chooses the configuration at which the number of states $N^*(E, J)$ is minimised, an approach which is equivalent to that employed in variational transition state calculations on bimolecular reactions.

The macroscopic, canonical rate constant $k(T)$ for a unimolecular reaction occurring at high enough pressures to maintain effectively a Boltzmann distribution among the internal quantum states, is obtained (section A4.6) by
(a) taking the Boltzmann probability for an internal energy in the range

$E \rightarrow E + dE$ and angular momentum J, i.e.

$$P(E, J)dE = \rho(E, J) \exp(-E/k_B T)dE/Q_{int}^{\dagger}$$

(b) multiplying by the specific rate constant $k(E, J)$
(c) summing over J and integrating over E

$$k(T) = (k_B T/h)(1/Q_{reag}) \sum_J \int_{E^0}^{\infty} N^*(E, J) \exp(-E/k_B T) \, d(E/k_B T). \qquad (5)$$

The result is the familiar expression of statistical transition state theory

$$k^{\infty}(T) = (k_B T/h)(Q^{\neq}/Q_{reag}) \exp(-E^0/k_B T) \qquad (6)$$

where Q_{reag}, $Q^{\neq}$ are the partition functions for the reagent and transition state, respectively, and E^0, the threshold energy for reaction over the zero-point potential, is equal to the difference between the zero-point energies in the transition state and in the reagent. A laboratory measurement of $k(T)$ gives an activation energy which is close to the threshold energy E^0 at the high pressure limit, but the measured values of E_{act} become considerably lower as the pressure drops and the reaction rate approaches second-order behaviour. The decline is due to the imbalance between the rate of collisional up-pumping of the highly activated molecules and their depletion by reaction. As the pumping rate falls, so too does the average energy of the reacting molecules. As Tolman first realised, the experimental activation energy is defined by the difference between this average energy and the average energy of the entire ensemble, i.e.

$$E_{act} = \langle E^* \rangle - \langle E \rangle. \qquad (7)$$

At high pressures, the rapid up-pumping sustains the Boltzmann distribution f_i, in levels i, through and beyond the threshold energy region, and a relatively broad range of energy dependent unimolecular rate constants $k(E_i)$, is sampled (see Fig. 4).

The activation energy is thus given as

$$E_{act}^{\infty} = \left(\frac{\Sigma k(E_i)f_i E_i}{\Sigma k(E_i)f_i} \right)_{E_i > E^0} - \left(\Sigma f_i E_i \right)_{E_i > 0}$$

$$\simeq (E^0 + \langle E_{th} \rangle) - \langle E_{th} \rangle$$

$$\simeq E^0. \qquad (8)$$

At low pressures, very few molecules will be activated to levels much beyond the threshold energy E^0; only a very narrow range of rate constants $k(E_i)$ are now

---

†For a *bimolecular* reaction $\rho(E, J)$ would also include the density of relative translational states per unit volume.

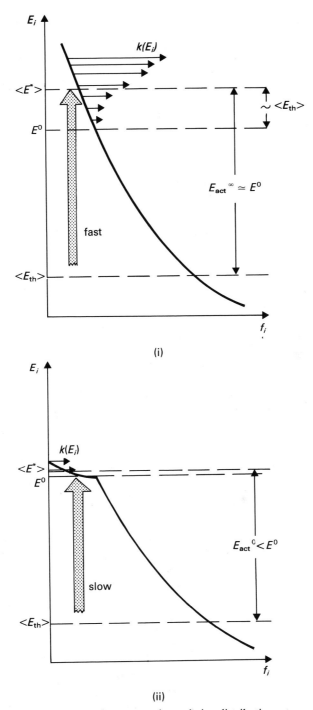

Fig. 4. Thermal unimolecular reaction rates and population distributions at very high pressures (i) and very low pressures (ii). In case (ii) depletion of the molecular population at $E > E^0$ is severe, but in case (i) reaction imposes a negligible perturbation on the Boltzmann population distribution.

sampled and the activation energy becomes

$$E^0_{\text{act}} = \left(\frac{\Sigma f_i E_i}{\Sigma f_i}\right)_{E_i > E^0} - \langle E_{\text{th}} \rangle \simeq E^0 - \langle E_{\text{th}} \rangle. \tag{9}$$

The quantitative understanding of the change in the experimental activation energy depends on our knowledge or assumptions about the efficiencies of the collisional up-pumping and deactivation processes at high energy levels and their dependence on the temperature or collision energy. Ignorance of this is conventionally 'accommodated' by the strong collision assumption (section A4.3). If the method of activation is *non*-thermal, e.g. IRMPA, knowledge of the energy state distribution in the ensemble with $E > E^0$ is even more necessary. The search for a proper understanding coupled with the advent of laser pump–probe techniques has led to the new class of activation techniques mentioned earlier, which are specifically aimed at the measurement of the unknown energy transfer rates. Figure 5 summarises the experimental/theoretical interface of unimolecular processes.

Complex reactions

'Practical' gas-phase reaction systems are hardly likely to involve just a single elementary reaction: rather the elementary reactions will be the components of a complex network of sequential or coupled steps which together comprise the 'mechanism' of the overall reaction. The study of gas-phase reaction kinetics began with complex systems such as flames, explosions, combustion and pyrolysis (Chapters C2, C3, C5). The isolation of individual elementary reactions and the resolution of state-to-state reaction rates came about much later, developing rapidly in the 1950's and 1960's with the advent of flash photolysis, discharge flow and laser pump–probe techniques. Now, the pendulum swings back and forth. There are anxieties about atmospheric pollution; there is a demand for improved efficiency in combustion chambers and the optimisation of thermal 'cracking' processes in industry. These involve the development of complex computer codes incorporating many elementary reaction rates to model, for example, atmospheric and combustion systems. There is a consequent demand for accurate and precise elementary rate data for the quantitative development of the computer models but the precise data required, often at elevated temperatures, are simply not available. Here the interface between theory and experiment is clearly defined. If we cannot measure the elementary rate constants under extreme conditions, can we calculate them; or extrapolate them reliably from data obtained at lower temperatures; or at least estimate their orders of magnitude, to assess their importance via computer modelling using sensitivity analysis techniques (Chapter C4)? If they prove to be of major importance, that will justify the investment of time and effort in the laboratory devising methods for their direct measurement; for example, in self-

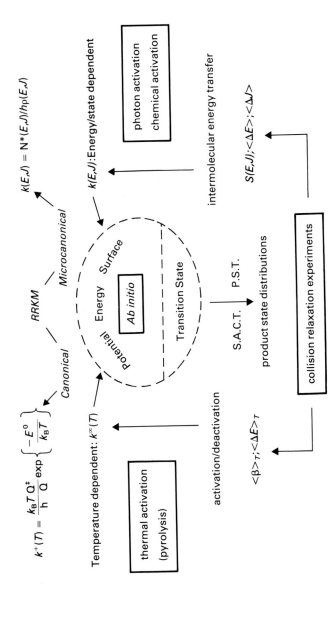

Fig. 5. The interface between theory and experiment: unimolecular reaction. S.A.C.T. ≡ statistical adiabatic channel theory. P.S.T. ≡ phase space theory.

sustaining flames, in furnaces, in high temperature fast flow reactors, or in shock tubes.

To illustrate the philosophy, consider a complex hydrocarbon flame, which may be sampled or probed via a molecular beam inlet system coupled to a mass spectrometer. The concentration–time profiles of the major reaction products may be predicted on the basis of a plausible reaction scheme and a computer code for the numerical analysis of the coupled differential equations incorporating known or assumed rate constant data. Suppose agreement with observation is encouraging but far from perfect. Sensitivity analysis is then introduced to highlight those elementary reactions whose rate constants need to be known accurately (Chapter C4). Their experimental determination can be approached via iterative procedures, in which the composition of the flame is systematically varied, while the sets of observed concentration profiles are compared with those predicted by the computer modelling until an acceptable degree of self-consistency is achieved. If the self-consistency is not acceptable, direct experimental measurements of those elementary reaction rates highlighted by the analysis must be undertaken to establish the reliability of the critical rate constants.

All this may still not be enough, since the overall behaviour of a complex reaction in a 'practical' system such as a flame or in the chemistry of the earth's atmosphere may be as sensitive to energy and mass transport as it is to the chemistry. The introduction of physical variables such as gas flows, convection, radiation and conduction, in addition to the chemical steps, calls for additional demands on the numerical modelling and an increase in the computing power. All of these matters are expanded in Part C of this book: the interface with theory involves the two-way flow between chemical understanding and numerical or analytical skill in the modelling of non-steady state chemically reacting systems, but . . . the chemistry comes first!

Suggestions for further reading

The remainder of this book!

References

1 See *J. Phys. Chem.*, 1979, **83**, 1.
2 J. Troe, *J. Phys. Chem.*, 1979, **83**, 114.
3 H.-J. Foth, J. C. Polanyi and H. E. Telle, *J. Phys. Chem.*, 1982, **86**, 5027.
4 D. Imre, J. L. Kinsey, A. Sinha and J. Krenos, *J. Phys. Chem.*, 1984, **88**, 3956.
5 D. M. Neumark, A. M. Wodtke, G. N. Robinson, C. C. Hayden and Y. T. Lee, in ACS Symposium Series, *Resonances*, 1984, **263**, 479.
6 K. Luther and J. Troe, in *Reactions of Small Transient Species*, ed. A. Fontijn and M. A. A. Clyne, Academic Press (New York, 1983), p. 63.
7 Intramolecular Kinetics, *Faraday Discuss. Chem. Soc.*, 1983, **75**.

Section A
Theories of elementary reactions

As in many branches of experimentally based science, theory in chemical kinetics has two main purposes. First, it provides an intellectual framework which stimulates new pathways of experimental enquiry and prevents the subject from merely becoming, in Rutherford's phrase, a higher form of 'stamp collecting'. Secondly, the application of theory can extend our knowledge of chemical reactions into regions—for example, of temperature and pressure—where it may be difficult to perform experiments.

One of the glories of chemical kinetics is that its theories, at least for the present, have to be imperfect. In principle, the equations of quantum mechanics should suffice. In practice, sufficiently accurate solutions remain outside our grasp. Consequently, theoreticians of chemical kinetics and reaction dynamics have to make use of all the basic branches of physical chemistry: quantum mechanics, thermodynamics and statistical mechanics, as well as some branches of physical science, like classical dynamics, which are often thought to have little relevance to chemistry.

In the first part of this book, we describe some of the main recent advances in theory and its applications. Space is too limited to cover all of the theoretical techniques or to provide the kind of technician's manual which will instruct the aspiring specialist in the finer theoretical details. Rather our objective, as elsewhere in the book, is to provide an understanding of a branch of chemical kinetics for those not directly involved as practitioners.

Transition state theory provides the focus for the first chapter in this part of the book. This theory, after a number of years in the wilderness, has been restored to its central place in chemical kinetics and has again become a proper topic for theoretical research. We expect our readers to have a working knowledge of conventional transition state theory based on the supposition of a dynamic equilibrium between activated complexes and reagent species and a treatment of this equilibrium by the methods of statistical mechanics. In Chapter A1, the emphasis is on the dynamical aspects of transition state theory, which provide it with a much more satisfactory foundation than the rather fudged definitions of the transition state which are often presented in undergraduate courses and textbooks.

A second advantage of such a dynamical approach is that it relates transition state theory much more directly to scattering calculations; that is, to calculations which attempt to simulate the complete progress of molecular collisions or re-arrangements. The use of classical equations of motion to compute trajectories has now become fairly routine. The techniques which are used are described in Chapter A2, and the value of such calculations is examined. In this chapter, as in Chapter A1, the main focus is on the simplest $A + BC \rightarrow AB + C$ type of bimolecu-lar reaction occurring over a potential energy barrier separating the reagent and product regions of the system's potential energy hypersurface. More complex systems can be investigated by both methods, but perhaps the major advantage of trajectory calculations is that dynamical aspects of reactive collisions are easily explored. For example, state-to-state rate constants or cross-sections are really no more difficult to calculate than thermal rate constants.

Chapter A3 presents examples of the application of transition state theory and trajectory calculations to data on reaction kinetics and reaction dynamics obtained from experiments. In the case of transition state theory, applications of the theory at two different levels are described. What might be called *ab initio* transition state theory has now been applied to a reaction as complicated (?) as $OH + H_2 \rightarrow H_2O + H$, with considerable success. The calculations rest on an *ab initio* potential energy surface, careful choice of the correct transition state, accurate evaluation of its partition function, and careful allowance for quantum mechanical tunnelling. At the other, more empirical end of the scale, an illustra-tion is given of how Benson's rules of group additivity can be applied to achieve a fairly good estimate of entropies and enthalpies of activation, and hence of the rate constant and its temperature dependence.

In Chapter A4, the last in this part of the book, theories for unimolecular reactions and their applications are considered. As indicated in Professor Simons' introductory chapter, theories of unimolecular reactions now possess a reasonable predictive capability. Therefore, besides explaining the basic principles of these theories in Chapter A4, it is also shown how application of these principles leads to expressions for the rate constants which can be evaluated to a reasonable degree of approximation with a knowledge of some basic information about the energetics and structure of the species involved.

What do we intend a research student to learn by a careful study of Section A—including an attempt at all the problems that are provided? First, a firmer faith in, and a deeper appreciation of, transition state theory. Secondly, a strong indication of how and when various types of theoretical calculation—transition state theory, trajectory calculations, unimolecular rate theory—should be made to support, understand, or extend experimental measurement, and an indication of how these calculations should be performed. Finally, a feel for the predictive power of theory in various different types of reaction.

There is arguably one serious omission from this part of the book dealing with theory: that is a description of the methods used to calculate potential energy surfaces for reactive systems. This is not due to any lack of appreciation of this field: the form of the potential energy surface is the vital factor determining the kinetics and dynamics of any individual reaction. Great strides are being made in calculating such surfaces by *ab initio* methods, and these advances will continue as ingenious quantum chemists are given increasing access to more powerful computers. An especially interesting development is the deployment of gradient techniques, which allows the theoretician, with relatively little effort, to direct his calculations to the crucial parts of the surface along the reaction coordinate and especially at the stationary points (stable configurations and barrier locations). This approach can provide accurate information for full-scale transition state calculations (like that on the $OH + H_2$ reaction referred to above). It can also explain how successive rearrangements of a collision complex can lead elementary reactions to yield products which, at first sight, seem surprising. A prime example is the reaction

$$NH_2 + NO \rightarrow N_2 + H_2O$$

which occurs by the formation and subsequent rearrangements of an H_2NNO complex. Kineticists, with at least a passing interest in theory, should keep an eye open for what should be exciting developments in this area over the next few years.

Chapter A1
Theoretical Foundations

M. S. CHILD

A1.1 Introduction

Modern computational methods do not only provide a powerful tool for the simulation of elementary chemical reactions, as discussed in the following chapter; they have also given deeper insight into the strengths and limitations of existing theories. A particular case in point is transition state theory, which is taken as the focus of this chapter in order to emphasise the dynamical aspects of the theory, as distinct from the statistical emphasis normally encountered in text-books.

The intention is to bring out the fundamental ideas by concentrating on collinear triatomic exchange reactions of the type

$$A + BC \rightarrow AB + C.$$

The following section, A1.2, gives a brief introduction to the factors affecting the forms of potential surfaces governing such reactions, the skewed representation of such surfaces in order best to visualise the classical dynamics, and the classical calculation of reaction cross-sections and rate constants.

The elements of transition state theory are then introduced at their most precise but least physically realisable classical microcanonical level. Here one assumes that the total energy is fixed and that quantum mechanics contributes neither to the reaction dynamics, nor to the quantisation of reactant states. The advantage of these restrictions is that one can obtain some exact results, as discussed in section A1.3.

The next stage is to take the temperature as fixed, rather than the total energy, and to introduce the quantisation of reactant states, leading to a more familiar canonical version of the theory. Unfortunately it is then no longer possible, even in principle, to locate the position of an exact transition state because the location of a reaction bottleneck is generally dependent on energy and angular momentum. The most striking cases occur for 'loose transition states' arising from angular momentum constraints rather than from the presence of an activation barrier. Examples of this type arising in ion–molecule chemistry are discussed in some detail in section A1.4(iii). Loose transition states also arise in an important class of unimolecular reactions, which are considered in Chapter A4.

The final question, addressed in section A1.4, concerns the role of quantum mechanical tunnelling in transition state theory. In particular, what is the best tunnelling path over the potential surface and how may tunnelling transmission factors be estimated? The answers are far from definitive, if only because most available comparisons with exact quantum mechanical theory are restricted to collinear models, but the successful existing results all have an appealing physical characteristic. Just as any rider attempting to negotiate a corner with too high a positive kinetic energy tends to be flung outwards, so a quantum mechanical particle, tunnelling with negative kinetic energy, will tend to cut the corner. This is just one example of the way that classical ideas can provide insight into the quantum mechanics of reactive scattering.

A1.2 Classical reaction dynamics

The classical simulation of a reaction of the $A + BC$ type requires (a) knowledge of a suitable potential energy surface, (b) a convenient representation for the equations of motion, and (c) compaction of the results into a cross-section or rate constant for comparison with experiment. Each of these will be discussed in turn.

(i) Potential energy surface

The key to understanding the form of a potential energy surface is to recognise that a chemical reaction involves a reorganisation of the bonding electrons. Thus one must normally consider the surfaces that correlate with at least two electronic states of the reactants, whose intersection in the interaction region gives rise to the activation barrier.

To see the idea in qualitative terms, consider the reaction

$$H + Cl_2 \rightarrow HCl + Cl$$

as discussed by Herschbach.[1] The electronic configuration of the ground state of Cl_2 may be designated . . . $(3p\sigma_g)^2(3p\pi_u)^4(3p\pi_g^*)^4$ and the electrons have no incentive to combine with the unpaired electron on the H atom; the energy of this state therefore varies very little as the H atom approaches, apart from some exchange repulsion as the electron clouds begin to overlap. Cl_2 also has well known excited states arising from . . . $(3p\sigma_g)^2(3p\pi_u)^4(3p\pi_g^*)^3(3p\sigma_u^*)^1$ for which the potential curves are shown in Fig. A1.1

The excited states include, in particular, the $^1\Pi_u$ state which is dissociative in isolated Cl_2, but in which the unpaired $3p\sigma_u^*$ electron is available to contribute to an HCl bond. Schematically therefore the energy variation of the two states with R_{HCl} (at fixed R_{ClCl}) is shown in Fig. A1.2. The progress of the reaction is determined by the first electronic surface up to the cross-over point at

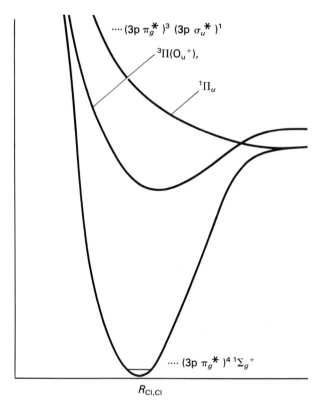

$\cdots$ $(3p\,\pi_g^*\,)^3\,(3p\,\sigma_u^*\,)^1$

$^3\Pi(O_u^+)$,

$^1\Pi_u$

$\cdots$ $(3p\,\pi_g^*\,)^4\,{}^1\Sigma_g^+$

$R_{Cl,Cl}$

Fig. A1.1. Potential curves for the Cl_2 molecule.

$R_{HCl} \simeq 3.5$ Å. The second surface takes over and the dominant process is repulsion between the two Cl atoms, giving rise overall to the collinear potential energy surface depicted in Fig. A1.3.

The two most common approaches to calculating surfaces, the LEPS[2,3] and DIM[3,4] methods, are based on this idea. Both are formulated in valence bond terms in order to ensure the correct dissociation characteristics. The LEPS formulation, stemming from the work of London, Eyring and Polanyi, builds the surfaces from Morse and anti-Morse functions,

$$^1E(R) = D_e\{\exp(-2\beta[R-R_e]) - 2\exp(-\beta[R-R_e])\} \tag{1a}$$

$$^3E(R) = (D_e/2)\{\exp(-2\beta[R-R_e]) + 2\exp(-\beta[R-R_e])\} \tag{1b}$$

chosen to mimic the singlet ground state and the triplet first excited repulsive state of each diatomic. The DIM (diatomics in molecules)[3,4] approach is more sophisticated in employing the known diatomic potential curves and also in a much more careful treatment of angular factors arising from differences between σ and π bonds.

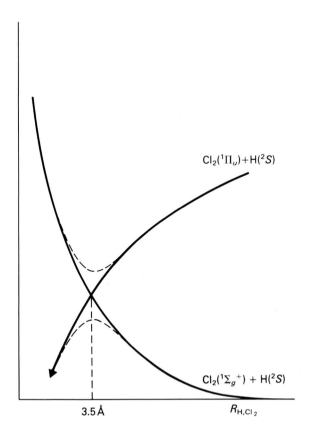

$Cl_2(^1\Pi_u) + H(^2S)$

$Cl_2(^1\Sigma_g^+) + H(^2S)$

3.5 Å

R_{H,Cl_2}

Fig. A1.2. Schematic energy variation of the $Cl_2(^1\Sigma_g^+) + H(^2S)$ and $Cl_2(^1\Pi_u) + H(^2S)$ states as a function of the R_{H,Cl_2} separation.

(ii) Representation of the equations of motion

It is convenient, particularly for collinear simulations, to visualise the dynamics in terms of a particle sliding over the potential energy surface. This requires that the energy is expressed in the form:

$$E = \frac{1}{2} m(\dot\zeta^2 + \dot\eta^2) + V(\zeta, \eta), \tag{2}$$

with a common mass for the two degrees of freedom, and no cross-terms, $\dot\zeta\dot\eta$, in the kinetic energy. The simple minded form in terms of interatomic distances R_{AB} and R_{BC}, is unfortunately

$$E = \frac{1}{2}[\mu_{A,BC}\dot{R}_{AB}^2 + 2(m_A m_C/M)\dot{R}_{AB}\dot{R}_{BC} + \mu_{AB,C}\dot{R}_{BC}^2] + V(R_{AB}, R_{BC}) \tag{3}$$

where $\mu_{A,BC} = m_A(m_B + m_C)/M$ and $M = m_A + m_B + m_C$, but equation 2 can be

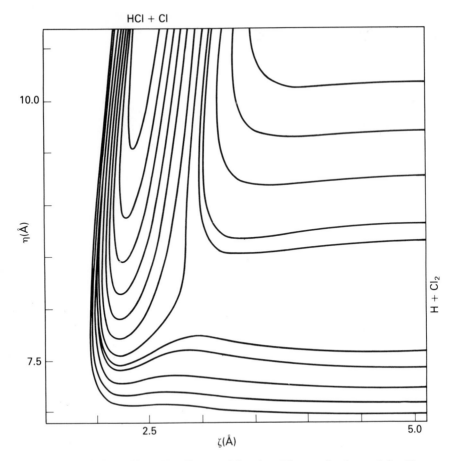

Fig. A1.3. The resulting collinear $H + Cl_2$ potential surface. The axes ζ and η are defined in terms of interatomic distances by equation 4.

recovered by the transformation

$$\zeta = R_{AB} + [m_C/(m_B + m_C)]R_{BC}$$

$$\eta = (\mu_{BC}/\mu_{A,BC})^{1/2}R_{BC}, \tag{4}$$

with

$$\mu_{BC} = m_B m_C/(m_B + m_C). \tag{5}$$

This means that ζ is the separation between A and the centre of mass of BC and η is the bond length of BC, scaled in terms of the ratio of the reactant vibrational to translational reduced masses. It follows that the R_{AB} and R_{BC} axes (i.e. the lines $R_{BC} = 0$, $R_{AB} = 0$, respectively) are inclined in the $\zeta\eta$ plane at an angle

$$\chi = \tan^{-1}[m_B(m_A + m_B + m_C)/m_A m_C]^{1/2}. \tag{6}$$

For $H + Cl_2$ this skewing angle barely differs from the value of $\pi/2$ in the rectilinear surface shown in Fig. A1.3 because $m_A \ll m_B$, m_C, but χ can become very small if $m_B \ll m_A$, m_C.

(iii) Cross-sections and rate constants

The aim of collision dynamics is to predict or interpret the outcome of a fully state-selected chemical reaction, for example

$$A + BC(v, j) \rightarrow AB(v', j') + C$$

at a given relative velocity u. The only generally available approach at present is the classical trajectory method, whereby the classical equations of motion are solved for the trajectory of a representative particle with mass $\mu_{A,BC}$ over an assumed potential surface.

Different samples of trajectories are chosen to simulate different situations. In this chapter motion is artifically restricted to collinear geometries and section A1.3 assumes a *microcanonical* distribution of reactant trajectories which are assumed to be uniformly distributed over the energetically accessible phase space regardless of how the energy is distributed between translational and vibrational motion. The more realistic three-dimensional state-selected reaction is simulated by the quasi-classical trajectory method, discussed at length in Chapter A2. It is however convenient to summarise the main features at this point.

A quasi-classical trajectory is defined as one starting from a defined vibrational–rotational reactant state. It is however necessary to specify as internal variables, not only the initial vibrational and rotational energy of BC, but also two angles (θ, ϕ) to specify the orientation of BC (see Fig. A1.4) and a further angle q to specify the phase of the oscillation. It is also necessary to define an impact parameter b, as shown in Fig. A1.4, which will be taken to have a cut-off value b_{max} above which no reaction will occur, giving an upper bound on the cross-section of πb_{max}^2.

A trajectory run then consists of N trajectories, sampled with equal weighting between 0 and 1 with respect to each of the variables

$$x_b = (b/b_{max})^2, \quad x_\theta = (1 + \cos\theta)/2,$$

$$x_\phi = \phi/2\pi, \quad x_q = q/2\pi, \tag{7}$$

from which the cross-section would be estimated as

$$S_{vj \rightarrow v'j'}(u) = \pi b_{max}^2 N_{vj \rightarrow v'j'}/N \tag{8}$$

where $N_{vj \rightarrow v'j'}$ is the number of trajectories reaching product states between numbers $v' \pm 1/2$ and $j' \pm 1/2$ (see section A2.5).

The next step, given the cross-section $S_{vj \rightarrow v'j'}(u)$ as a function of u, is to obtain

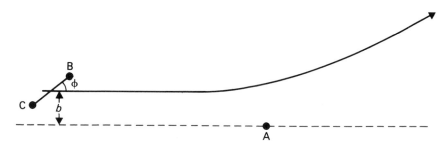

Fig. A1.4. A schematic trajectory; angle ϕ defines the orientation of BC with respect to the plane illustrated, whilst b defines the phase of the vibration of BC.

the rate constant defined as

$$k_{vj \to v'j'}(T) = \langle u\, S_{vj \to v'j'}(u) \rangle_T$$

$$= \int_0^\infty S_{vj \to v'j'}(u)\, u^3 \exp\left(-\frac{mu^2}{2k_B T}\right) du \bigg/ \int_0^\infty u^2 \exp\left(-\frac{mu^2}{2k_B T}\right) du. \quad (9)$$

This means, on substitution from equation 8 for $S_{vj \to v'j'}(u)$, that

$$k_{vj \to v'j'}(T) = \pi b_{\max}^2 [(N_{vj \to v'j'})/N] \int_0^\infty u^3 \exp\left(-\frac{mu^2}{k_B T}\right) du \bigg/ \int_0^\infty u^2 \exp\left(-\frac{mu^2}{k_B T}\right) du$$

$$= \pi b_{\max}^2 [(N_{vj \to v'j'})/N] \bar{u}. \quad (10)$$

A1.3 Classical microcanonical transition state theory

The transition state is properly regarded as a bottleneck on the potential energy surface, the passing of which guarantees reaction. Moreover such a bottleneck must be equally valid for the forward and the back reaction, the forward and backward fluxes through it being necessarily equal. Now suppose one mistakenly assumes the transition state to lie on the reactants' side of the true bottleneck. The forward flux will include not only the reactive flux but also a contribution from trajectories that turn backwards and fail to react. Equally the backward flux will include the true flux for the back reaction and a contribution from the reflected trajectories on their return journey. The forward and backward values will again be equal, but both will exceed the true reactive flux. Hence any transition theory estimate must be an upper bound to the reaction rate in either direction for the assumed potential energy surface.

The main object of the theory is obviously to locate the best transition state. Such bottlenecks commonly lie close to an activation barrier but they can exist without one (as, for example, the loose transition states for unimolecular decomposition which are discussed in section A1.4 and Chapter A4). They can also move as the energy changes, hence the concept of a variational transition state.[5-8] Finally the co-existence of two or more such bottlenecks can have dramatic effects on the reaction rate.

The fundamental ideas are best discussed at a model classical collinear micro-canonical (fixed energy) level for which exact results are available. We shall then examine the complications of the real three-dimensional quantum mechanical world in which temperature is more conveniently fixed than the total energy.

(i) Fundamental ideas

It must be emphasised at the outset that transition state theory is a *dynamical* not a *statistical* theory. To see this, imagine the selection of vibrational coordinates and momenta (r, p_r) for a trajectory study of collinear collisions at given total energy. Figure A1.5 shows that the accessible part of this phase space is bounded by the total energy, and that the area inside can be divided into reactive and non-reactive regions.

The aim of transition state theory is to assess the reactive fraction of the space without troubling to run the trajectories. The task is much easier in a microcanoni-cal theory in which all phase points are assumed equally probable than in a quasiclassical one where the points are restricted to one or other of the (dashed) quantised orbits.

How is this estimate made? The key idea, due to Pechukas,[5] is that if trajectories are to be divided into reactive and inelastic classes the division between them must be in terms of a trajectory which is neither reactive nor inelastic—it must be

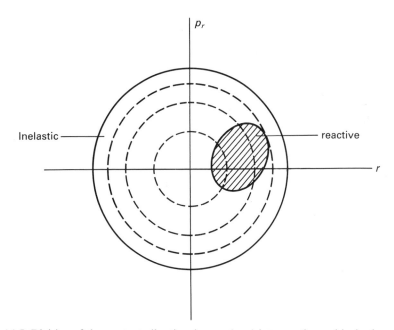

Fig. A1.5. Division of the reactant vibrational space (r, p_r) into reactive and inelastic regions. Dashed curves indicate the classical orbits corresponding to quantised states. The outer boundary is determined by the available energy.

trapped on the potential energy surface as illustrated in Fig. A1.6. This periodic orbit dividing surface (or PODS)[6] constitutes the transition state. Furthermore, if only one such PODS exists, the reactive area in Fig. A1.5 may be shown to be precisely equal to the integral

$$F_X = \oint \boldsymbol{P} \cdot \mathrm{d}\boldsymbol{R} \tag{11}$$

where $\boldsymbol{P}$ is the momentum along the PODS defined by $\boldsymbol{R}$ and the integral is taken around one cycle of the PODS. Thus the reaction probability is exactly given by

$$P = F_X/F_R \tag{12}$$

where F_R is the corresponding integral taken around one cycle of the reactant vibration when all the available energy is in vibrational motion. P is as usual a ratio of quantities applicable to the transition state and to the reactants, F_R and F_X being, in fact, the classical degeneracies of the reactants and transition state respectively; there are no Boltzmann factors, because the energy rather than the temperature is fixed.

Figure A1.7 compares values of P calculated from equation 12 with classical trajectory results for $H + Cl_2$. The agreement is seen to be excellent for $E < 0.2$ eV, but to deteriorate thereafter. The reason for this deterioration is that other PODS spring into life above 0.2 eV as shown in Fig. A1.8. Those labelled 1 and 2 are now flux bottlenecks while the one labelled X is a flux magnet in that all trajectories in the inner region must pass through it; this means that $F_X > F_1, F_2$ where the F_v are defined by the analogue of equation 11. The situation facing a would-be reactive trajectory is that it has a probability P_1 (given by equation 12 with F_1 in place of F_X) of crossing PODS 1 but this does not guarantee reaction because it must also navigate PODS 2. Seen pictorially as in Fig. A1.9, there is a certain area F_X accessible to all penetrating trajectories but only the shaded area which is accessible from reactants via PODS 1 and to products via PODS 2 is occupied by directly

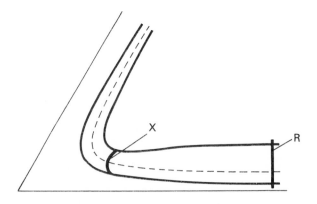

Fig. A1.6. A PODS, X, on the collinear potential energy surface.

reactive trajectories. All others are captured in the interaction region for one or more further crossings through PODS X. Now since F_1 and F_2 are enclosed by F_X there is an inescapable minimum shaded area given by $F_1 + F_2 - F_X$, from which one can define a lower bound to the reaction probability, as

$$P_{LB} = (F_1 + F_2 - F_X)/F_R = P_1 + P_2 - P_X. \tag{13}$$

Notice, however, that P_{LB} becomes relatively useless as a prediction if $F_X > (F_1 + F_2)$. In this case Miller's unified statistical theory (ust)[7] may be applied, by assigning probabilities $P_1 = (F_1/F_R)$ for penetration through PODS 1, $Q_2 = (F_2/F_X)$ for reactive escape, $Q_1 = (F_1/F_X)$ for non-reactive escape and $(1 - Q_1)(1 - Q_2)$ for repeating a previously unsuccessful reactive escape attempt. This situation is illustrated in Fig. A1.10, from which it is seen that the overall reaction probability may be calculated as

$$P_{ust} = P_1 Q_2 \{ 1 + (1 - Q_1)(1 - Q_2) + (1 - Q_1)^2 (1 - Q_2)^2 + \cdots \}$$

$$= \frac{P_1 Q_2}{1 - (1 - Q_1)(1 - Q_2)} = \frac{P_1 Q_2}{Q_1 + Q_2 - Q_1 Q_2}. \tag{14}$$

Note that $P_{ust} \simeq P_1$ if $Q_2 \gg Q_1$ and $P_{ust} \simeq P_1 Q_2 / Q_1 = P_2$ if $Q_1 \gg Q_2$, where $P_2 = F_2/F_R$, by analogy with the definition of P_1.

Returning to Fig. A1.7 one sees that P_1 and P_2 correctly determine upper limits to the reaction probability because both bottlenecks must be passed, and the flux

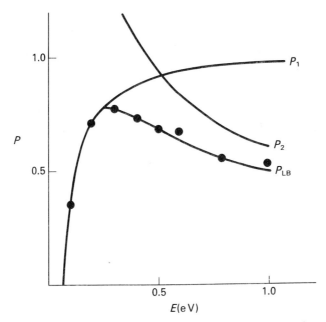

Fig. A1.7. Microcanonical collinear reaction probabilities for the $H + Cl_2$ reaction. P_1 and P_2 are variational estimates derived from the two outer PODS in Fig. A1.8. P_{LB} is given by equation 13 and the points are derived from trajectory runs.

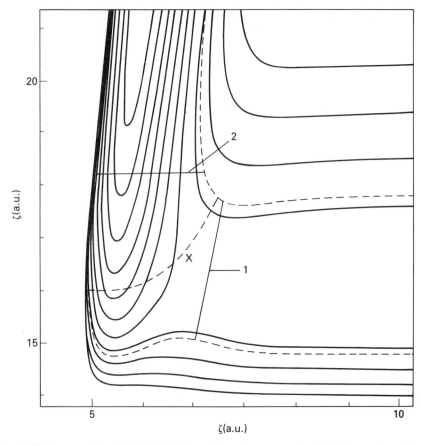

Fig. A1.8. Two repulsive PODS labelled 1 and 2 and an attractive PODS X on the collinear $H + Cl_2$ potential surface.

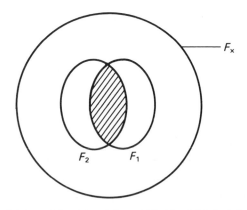

Fig. A1.9. Schematic division of the phase space at X in Fig. A1.8. Points within F_1 and F_2 are directly accessible to reactants and products via PODS 1 and 2 in Fig. A1.8, respectively.

through either singly cannot be less than that through the pair. Two variational aspects of the theory may also be mentioned. In the first place the two outer PODS in Fig. A1.8 move outwards with increasing energy in order to minimise their individual fluxes. Secondly PODS 1 is the best single bottleneck for $E < 0.5$ eV after which the role passes to PODS 2. Of the available estimates, P_{LB} appears to be the most accurate, as well as providing a good lower bound. P_{ust} necessarily lies between P_{LB} and the lesser of P_1 and P_2; it becomes particularly applicable in situations involving a real potential well for which $F_X \gg F_1, F_2$ so that P_{LB} becomes negative.

Details of this PODS approach to a variety of aspects of reaction dynamics are fully described elsewhere by Pollak.[8]

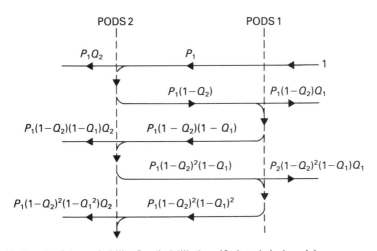

Fig. A1.10. Sketch of the probability flow in Miller's unified statistical model.

A1.4 Canonical variational transition state theory

The essential differences between the true physical situation and the previous classical microcanonical theory are that initial reactant states are quantised, that one commonly works at a fixed temperature rather than at fixed energy, and that quantum mechanical tunnelling may be significant. The last of these is discussed in the final section of this chapter.

Our immediate objectives are to show within a vibrationally adiabatic hypothesis (a) how suitably quantised PODS may be used to define the reaction thresholds for different internal states, and (b) how a tractable variational transition state theory may be developed. Applications to both exchange reactions and unimolecular reactions are outlined with particular reference to what are termed 'loose transition states'.

(i) The vibrationally adiabatic hypothesis

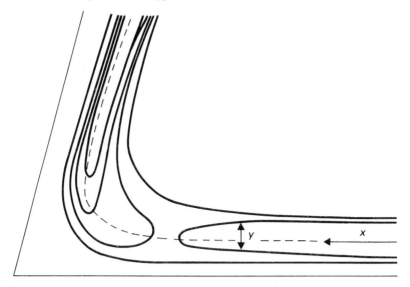

Fig. A1.11. Definition of the reaction path and vibrational coordinates x and y, respectively.

Imagine a potential surface with a defined reaction path along which distance is measured in terms of a reaction coordinate x, as shown in Fig. A1.11. Now cut a section normal to x and quantise the motion in the orthogonal coordinate y. The resulting x dependent energy levels, $W_n(x)$ say, constitute a family of vibrationally adiabatic energy curves as illustrated in Fig. A1.12.

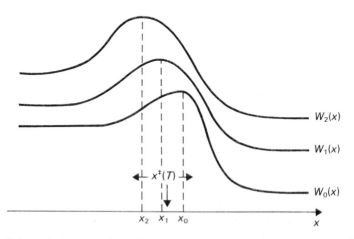

Fig. A1.12. Schematic representation of possible vibrationally adiabatic energy curves $W_n(x)$ showing maxima at different values of the reaction coordinate x_n. The best transition state values $x^*(T)$ vary between the x_n as T changes.

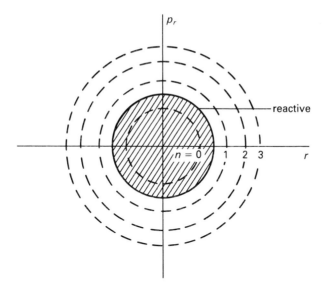

Fig. A1.13. Reactant phase space implication of strict vibrational adiabaticity with the boundary of the reactive zone being concentric with the reactant phase space orbits (compare Fig. A1.5).

This procedure can, in principle, take into account a whole range of orthogonal coordinates, y, to include stretching vibrations, bending vibrations, restricted internal rotations and the overall rotation of the system, but the immediate discussion is limited to a single vibration y.

The relation between this adiabatically quantised picture and the previous microcanonical theory is, in effect, that the reactive boundary in Fig. A1.5 is assumed to be concentric with the system of bound state orbits represented by the dashed lines, i.e., as shown in Fig. A1.13.

In other words each initial state is assumed to be either 100% reactive or 100% non-reactive according to whether its orbit lies inside or outside the shaded area in Fig. A1.13. Consequently an increase in the reactive area as the total energy increases gives rise to a series of thresholds $W_n^{\neq}$ as the reactive area encloses successive orbits in turn. The previous arguments imply that the nth such boundary is associated with a PODS at an energy $W_n^{\neq}$ such that

$$\int_{\text{PODS}} \boldsymbol{P} \cdot \mathrm{d}\boldsymbol{R} = (n + 1/2)h, \tag{15}$$

this also being the quantisation condition for the nth dashed orbit.

This establishes a direct connection between the adiabatic barriers $W_n^{\neq}$ and the PODS of the previous section, A1.3. It also explains why the maxima $W_n^{\neq}$ in Fig. A1.12 are drawn at different values of the reaction coordinate $x_0, x_1, x_2 \ldots$ etc. Evidently the variational aspects of the classical microcanonical theory carry implications for the quantum mechanical canonical theory.

(ii) Canonical variational transition states

The problem posed by the occurrence of different x_n values in Fig. A1.12 is that there is no uniquely defined transition state value $x^{\neq}$. At energies sufficiently low that only $n = 0$ is populated one would naturally associate $x^{\neq}$ with x_0 but thereafter the best one can do is to seek a temperature dependent point $x^{\neq}(T)$ that averages over the x_n according to the populations of the initial states. The solution adopted by variational transition state theory is to define an x dependent partition function[9-11]

$$Q(x,T) = \sum_n \exp\left[-W_n(x)/k_B T\right] \tag{16}$$

which is then minimised with respect to x at a given T, to obtain $x^{\neq}(T)$, so that the partition function of the transition state becomes

$$Q^{\neq}(T) = Q_{\min}(x^{\neq}(T), T). \tag{17}$$

The resulting expression for the rate constant may be written

$$k(T) = (k_B T/h) Q_{\min}^{\neq}(T)/Q_{\text{reac}}(T) \tag{18}$$

with the activation term $\exp\{-W_0[x^{\neq}(T)]/k_B T\}$ concealed within $Q^{\neq}(T)$ if it is assumed that all energies are measured from the reactant $n = 0$ state.

(iii) Implementation

Garrett et al.[9] give details of the implementation of the theory, with tunnelling corrections, for three-atom exchange reactions

$$A + BC \rightarrow AB + C$$

on a known potential energy surface.

Another more empirical approach developed by Quack and Troe[10] for unimolecular reactions involving fission of a single bond is to construct effective vibrationally adiabatic curves by correlation between assumed energy levels of the parent molecule and the known levels of the fragments. The specific algorithm proposed is as follows

$$W_n(x) = W_n^0(x_e) \exp[\alpha(x_e - x)] + W_n^0(\infty)\{1 - \exp[\alpha(x_e - x)]\} + W_{\text{cent}} \tag{19a}$$

$$W_{\text{cent}} = B(x) P(x)[P(x) + 1] \tag{19b}$$

$$P(x) = J \exp[\alpha(x_e - x)] - l\{1 - \exp[\alpha(x_e - x)]\} \tag{19c}$$

where x now measures distance along the reaction coordinate. $W_n^0(x_e)$ and $W_n^0(\infty)$ are energy levels of the parent molecule and fragments, respectively, each sequentially ordered by n. $B(x)$ describes the variation of the rotational constant with x. Finally J and l are the total and orbital angular momenta, respectively. This statistical adiabatic channel model is discussed further in section A4.7.

(iv) Loose transition states

A loose transition state is one whose position is very sensitive to the available energy and angular momentum. One of the simplest examples is seen in the Langevin model for ion–molecule reactions. The assumed long-range potential is attractive.

$$V(R) \sim -q^2\alpha/(4\pi\varepsilon_0)^2 2R^4 \tag{20}$$

where q is the ionic charge, α the molecular polarisability and R the separation between the ion and the molecule. This competes with a repulsive centrifugal term to set up a barrier at an energy dependent radius $R^*(E)$ which can be crossed only for trajectories with impact parameters less than a critical value $b_{max}(E)$, where E is the collision energy of the ion–molecule pair.

The calculation of $R^*(E)$ and $b_{max}(E)$ follows from energy and angular momentum conservation. Thus, in polar coordinates:

$$E = \mu u^2/2 = \mu\dot{R}^2/2 + \mu R^2\dot{\theta}^2/2 - q^2\alpha/(4\pi\varepsilon_0)^2 2R^4 \tag{21}$$

$$L = \mu u b = \mu R^2\dot{\theta} \tag{22}$$

so that on eliminating $\dot{\theta}$

$$E = \mu\dot{R}^2/2 + E(b^2/R^2) - q^2\alpha/(4\pi\varepsilon_0)^2 2R^4. \tag{23}$$

The second two terms constitute an effective potential for radial motion with a maximum value

$$V_{eff}^{max} = (E^2 b^4)(4\pi\varepsilon_0)^2/2q^2\alpha \tag{24}$$

at

$$R^*(E) = \{(q^2\alpha)/(4\pi\varepsilon_0)^2 Eb^2\}^{1/2}. \tag{25}$$

The upper limit for chemical reaction occurs for the b value such that $E = V_{eff}^{max}$; in other words

$$b_{max}(E) = [2q^2\alpha/(4\pi\varepsilon_0)^2 E]^{1/4}. \tag{26}$$

It follows that the reaction cross-section takes the form

$$S(u) = \pi[b_{max}(E)]^2 = k_L/u, \tag{27}$$

where

$$k_L = 2\pi(q^2\alpha/\mu)^{1/2}/(4\pi\varepsilon_0). \tag{28}$$

The inverse velocity dependence in equation 27 has the important consequence that the rate constant $k = \langle uS(u) \rangle$ is independent of temperature, with value k_L.

Similar arguments may be applied to any isotropic attractive potential, $V(R) \sim -C_n/R^n$, with the result that the cross-section varies as $u^{-(4/n)}$ (see problem A1.7). This means that the rate constant, which is proportional to $\langle uS(u) \rangle$, is predicted to increase with increasing temperature for $n > 4$ but to decrease for $n < 4$.

Turning to anisotropic systems, the case of an ion–dipole reaction, as recently analysed by Clary et al.,[12,13] is of particular interest. The potential was assumed to have the long-range form

$$V(R, \theta) = -\alpha q^2/(4\pi\varepsilon_0)^2 2R^4 - q\mu_D \cos\theta/(4\pi\varepsilon_0)R^2 \tag{29}$$

where μ_D is the molecular dipole moment. The interesting feature is that, as R decreases, the ion–dipole term leads to an increasing preference for collinear geometry, $\theta = 0$, while at the same time the bending vibrational frequency about this preferred configuration also increases as R^{-2}. Consequently there are different centrifugally corrected adiabatic barriers, analogous to $V_{\mathrm{eff}}^{\mathrm{max}}$ in equation 24, for different bending vibrational states and these correlate in turn with different rotational states of the reactant molecule. The net result is that the reaction cross-section and hence the rate constant can differ quite markedly from one rotational state to another.

Clary et al.[12,13] performed adiabatic capture calculations of this type within the 'centrifugal sudden' approximation[14] for a variety of ion–dipole reactions of astrophysical interest. In all cases the rate constant showed a negative temperature coefficient consistent with an attractive R^{-n} potential. Figs. A1.14 and A1.15, which are applicable to $H^- + HCN(j)$,[12] show the typical behaviour for reactions of this type. The rate constant decreases both with increasing temperature and with increasing j; moreover the rate of decrease with j is more strongly marked the lower the temperature. The overall rate constant, $k(T) = \sum_j k_j(T)(2J+1)\exp(-E_j/k_B T)$ also increases with decreasing temperature. Notice that the variation becomes especially marked at temperatures below 50 K, a conclusion which has important implications in regard to the chemistry in interstellar clouds.

A1.5 Quantum mechanical tunnelling

The idea that quantum mechanical tunnelling can enhance the rate of light atom transfer reactions has a long history. Here we discuss (i) the available techniques for calculating tunnelling probabilities along a given tunnelling path and (ii) the choice of an appropriate path in the chemical context.

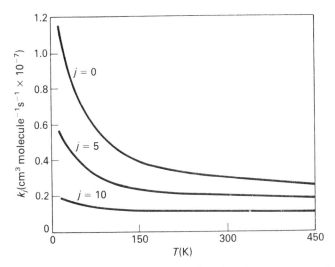

Fig. A1.14. Temperature variation of the rotationally selected rate constants $k_j(T)$ for the $H^- + HCN(j)$ reaction. Taken from reference 12 with permission.

(i) Tunnelling probabilities

The quantum mechanical problem of the motion of a stream of particles of mass μ in one dimension at energy E under a parabolic barrier

$$V(x) = V_{max} - \kappa\, x^2/2 \tag{30}$$

is exactly soluble, with the following results for the reflection and transmission probabilities, R and T respectively[15]

$$R = 1/(1 + e^{2\pi\varepsilon}) \tag{31}$$

$$T = 1/(1 + e^{-2\pi\varepsilon}) \tag{32}$$

where

$$\varepsilon = (E - V_{max})/\hbar\omega^*$$

$$\omega^* = (\kappa/\mu)^{1/2}. \tag{33}$$

The resulting energy dependences of R and T are shown in Fig. A1.17.

How can these results be extended to arbitrary potentials? The simplest approach is to recognise that the tunnelling exponent $\pi\varepsilon$ may be identified with the magnitude of the semi-classical action integral between the classical turning points

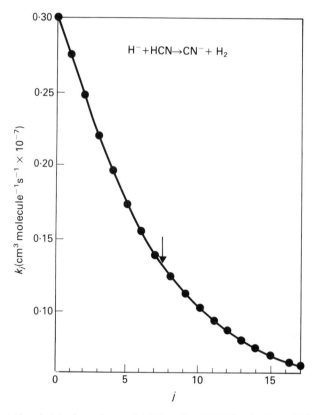

Fig. A1.15. Rotational state dependence of $k_j(T)$ at $T = 300$ K for the $H^- + HCN(j)$ reaction. The arrow marks the Boltzmann averaged rate constant. Taken from reference 12 with permission

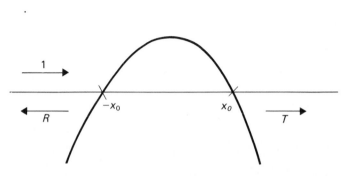

Fig. A1.16. Quadratic barrier problem. T and R are transmitted and reflected amplitudes consistent with a unit incident amplitude from the left.

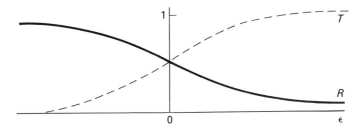

Fig. A1.17. Energy dependence of T and R.

$\pm x_0$ in Fig. A1.16

$$\phi = \frac{1}{\hbar} \int_{-x_0}^{x_0} |p(x)| \, dx = \frac{1}{\hbar} \int_{-x_0}^{x_0} (2\mu[V(x) - E])^{1/2} \, dx$$

$$= \frac{(\mu\kappa)^{1/2}}{\hbar} \int_{-x_0}^{x_0} (x_0^2 - x^2)^{1/2} \, dx = \frac{\pi}{2} \frac{(\mu\kappa)^{1/2}}{\hbar} x_0^2 \tag{34}$$

where

$$x_0^2 = 2(V_{max} - E)/\kappa. \tag{35}$$

Thus

$$\phi = -\pi\varepsilon. \tag{36}$$

The tunnelling exponent for other potentials has been found to be well represented in the general case by changing the form of $V(x)$ in equation 34.

For another illuminating approach[16] consider the time-dependence of the classical motion for the parabolic barrier. For incident motion from $x \to -\infty$

$$x = -x_0 \cosh \omega^* t \tag{37}$$

$$p = -p_0 \sinh \omega^* t \tag{38}$$

where

$$p_0^2 = 2\mu(V_{max} - E). \tag{39}$$

Thus x is negative at all real times, provided $E < V_{max}$, while the momentum p is positive for $t < 0$ and negative for $t > 0$, corresponding to reflection at the barrier. Suppose however that the time is given an imaginary increment $i\tau$ after reaching the barrier at $t = 0$; thereafter

$$x = -x_0 \cosh i\omega^* \tau = -x_0 \cos \omega^* \tau \tag{40}$$

$$p = -p_0 \sinh i\omega^* \tau = +ip_0 \sin \omega^* \tau. \tag{41}$$

x now advances from $-x_0$ at $\tau = 0$ to $+x_0$ at $\tau = \pi/\omega^*$ while p remains imaginary, as expected for penetration into a classically forbidden region. The remarkable thing is that the possibility of quantum mechanical tunnelling is contained within complex time solutions of the classical equations of motion. Moreover the action integral

$$\phi = \frac{1}{\hbar} \int_0^{\pi/\omega^*} |p|\dot{x}\,d\tau \tag{42}$$

again gives the correct tunnelling exponent.

(ii) Choice of a good tunnelling path

Given a potential energy surface, as in Fig. A1.18, one seeks to find a one-dimensional profile to describe the tunnelling. The simplest candidate is the energy variation along the classical path of steepest descent from the saddle point, represented by the dashed line in Fig. A1.18 and commonly called the 'reaction path'. Comparison with accurate quantum mechanical calculations for $H + H_2$ shows however that this classical, one-dimensional prescription, represented by the dashed part of Fig. A1.19 below, underestimates the tunnelling probability by one or two orders of magnitude. One reason for this is that vibrational quantisation has been ignored, and one might prefer instead to choose the relevant vibrationally adiabatic curve as a suitable profile. As seen by comparison between the two parts of Fig. A1.19 the normal reduction in zero-point energy on approaching the saddle point substantially improves the situation. For a given translational energy E_t it is easier to tunnel through the $v = 0$ curve than through the simple profile.

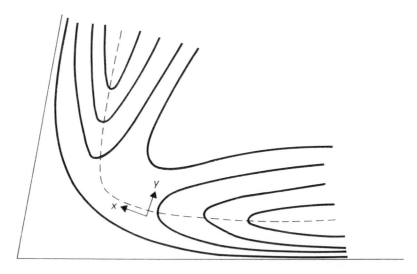

Fig. A1.18. Reaction and vibrational coordinates, x and y, respectively, on a schematic collinear potential surface.

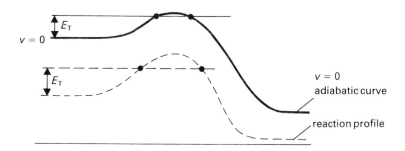

Fig. A1.19. Illustration of enhanced tunnelling possibliities via the $v = 0$ vibrationally adiabatic curve.

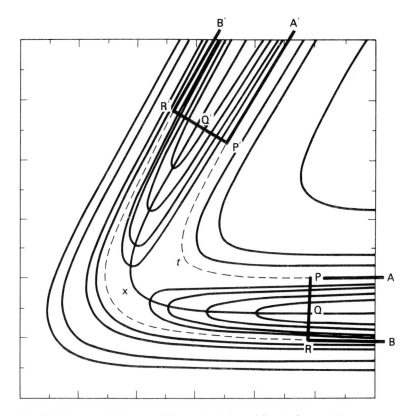

Fig. A1.20. The Marcus–Coltrin tunnelling path t. Adapted from reference 17.

This is not however the end of the story because distance is still measured in terms of the reaction coordinate x taken along the steepest descents path. Marcus and Coltrin[17] argue that there is a shorter and therefore more effective vibrationally adiabatic path. The situation is illustrated in Fig. A1.20. Classical motions relevant to the initial $v = 0$ state are confined to the region bounded by APRB at a given energy E_t, but one can still calculate the dashed extensions PP' and RR' of this region as the loci of quantised $v = 0$ classical motions perpendicular to x. The shortest vibrationally adiabatic path is obviously PP' along which one might define a tunnelling coordinate t. The energy variation will be identical with the upper curve in Fig. A1.19 but the barrier will appear thinner and therefore more penetrable. Fig. A1.21 shows that results obtained in this way are in excellent agreement with the quantum mechanical calculation.

Mention should also be made of a quite different approach due to George and Miller[18] which avoids the vibrationally adiabatic approximation. Instead they

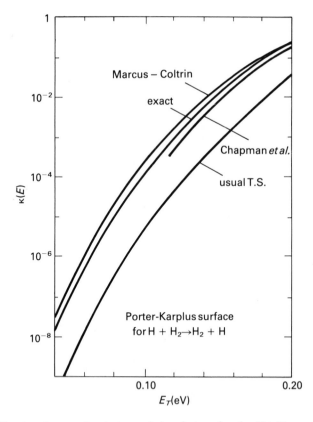

Fig. A1.21. Exact and approximate transmission factors for the $H + H_2$ reaction. Adapted from reference 19. Marcus–Coltrin curve from reference 17 and Chapman *et al.* curve from reference 19.

adopt a complex time approach to allow the trajectory to find its own tunnelling path. The results are shown in Fig. A1.22 for trajectories from $v = 0$ at two different translational energies. Again the trajectories are seen to cut the corner in the potential surface, but the chosen path is quite different to that adopted by Marcus and Coltrin.[17] Nevertheless the results are again in good agreement with the quantum studies. The reason for this good agreement between the Marcus–Coltrin and Miller–George approaches is far from obvious, but the former, being computationally much simpler, is more widely adopted.

A final point concerns an attractive and computationally simple idea due to Chapman et $al.$,[19] who argue that imaginary time mechanics, as required for tunnelling, is equivalent to real time mechanics on the inverted potential. To see this, consider Newton's equations

$$\mu \frac{d^2x}{dt^2} = -\frac{dV}{dx} \tag{43}$$

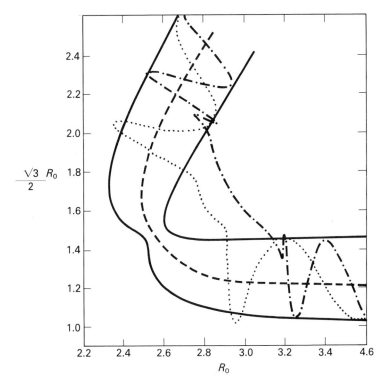

Fig. A1.22. Complex trajectory tunnelling paths for the collinear $H + H_2$ reaction. Taken from reference 18 with permission.

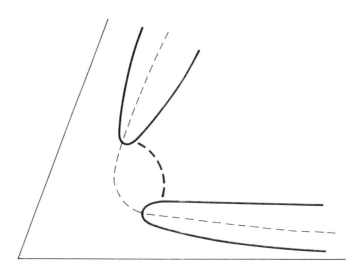

Fig. A1.23. Schematic view of the tunnelling periodic orbit.

and let $t = i\tau$, so that

$$-\mu \frac{d^2x}{d\tau^2} = -\frac{dV}{dx} \tag{44}$$

or

$$\mu \frac{d^2x}{d\tau^2} = -\frac{d\bar{V}}{dx} \tag{45}$$

where

$$\bar{V}(x) = -V(x). \tag{46}$$

Most trajectories generated in this way on a realistic potential surface have no simple physical interpretation, but it is relatively simple at any given energy below the saddle point to find a single periodic tunnelling trajectory as illustrated in Fig. A1.23, which may be thought to determine the authentic tunnelling path. The curve labelled Chapman et al.[19] in Fig. A1.21 shows that this method is also in good agreement with the exact calculations. The conceptual difficulty with this approach is that the tunnelling orbit is fixed solely by the total energy; hence one predicts the same tunnelling probability regardless of the division of this energy between translational and internal reactant motions.

A1.6 Suggestions for further reading

Potential surfaces

Kuntz P. J. (1979). In *Atom–Molecule Collision Theory*, (Ed. by R. B. Bernstein), Plenum Press, New York.

Tully J. C. (1977). In *Modern Theoretical Chemistry*, Vol. VII, (Ed. by G. A. Segal), Plenum Press, New York.

Classical trajectories

Truhlar D. G. and Muckerman J. T. (1979). In *Atomic–Molecule Collision Theory*, (Ed. by R. B. Bernstein), Plenum Press, New York.

Microcanonical transition state theory

Pechukas P. (1976). In *Dynamics of Molecular Collisions, Part B*, (Ed. by W. H. Miller), Plenum Press, New York.

Pollak E. (1985). In *Theory of Chemical Reaction Dynamics III*, Vol. 2, (Ed. by M. Baer), CRC Press, Boca Raton, Florida

Variational transition state theory

Quack M. and Troe J. (1977). Chem. Soc. Specialist Periodical Report, *Gas Kinetics and Energy Transfer*, **2**, 175.

Quack M. and Troe J. (1981). *Theoretical Chemistry: Advances and Perspectives*, **6B**, 199. Academic Press, New York.

Garrett B. C., Truhlar D. G. and Grev R. S. (1981). In *Potential Energy Surfaces and Dynamical Calculations*, (Ed. by D. G. Truhlar), Plenum Press, New York.

A1.7 References

1 Herschbach D. R. (1973) *Faraday Disc. Chem. Soc.*, **55**, 242.

2 Sato S. (1955) *J. Chem. Phys.*, **23**, 592, 2465.

3 Kuntz P. J. (1979) In *Atom–Molecule Collision Theory*, (Ed. by R. B. Bernstein), Plenum Press, New York.

4 Tully J. C. (1977) In *Modern Theoretical Chemistry*, Vol. VII, (Ed. by G. A. Segal), Plenum Press, New York.

5 Pechukas P. (1976) In *Dynamics of Molecular Collisions, Part B*, (Ed. by W. H. Miller), Plenum Press, New York.

6 Pollak E., Child M. S. and Pechukas P. (1980) *J. Chem. Phys.*, **72**, 1669.

7 Miller W. H. (1976) *J. Chem. Phys.*, **65**, 2216.

8 Pollak E. (1985) In *Theory of Chemical Reaction Dynamics III*, Vol. 2, (Ed. by M. Baer), CRC Press, Boca Raton, Florida.

9 Garrett B. C., Truhlar D. G. and Grev. R. S. (1981) In *Potential Energy Surfaces and Dynamical Calculations*, (Ed. by D. G. Truhlar), Plenum Press, New York.

10 Quack M. and Troe J. (1977) Chem. Soc. Specialist Periodical Report, *Gas Kinetics and Energy Transfer*, **2**, 175.

11 Quack M. and Troe J. (1981) *Theoretical Chemistry: Advances and Perspectives*, **6B**, 199. Academic Press, New York.

12 Clary D. C. (1985) *Mol. Phys.*, **54**, 605.

13 Clary D. C., Smith D. and Adams N. G. (1985) *Chem. Phys. Letters*, **119**, 320.
14 Pack R. T. (1974) *J. Chem. Phys.*, **60**, 633.
15 Child M. S. (1974) *Molecular Collision Theory*, Academic Press, London, Appendix C.3.
16 Miller W. H. and George T. F. (1972) *J. Chem. Phys.*, **56**, 5668.
17 Marcus R. A. and Coltrin M. E. (1977) *J. Chem. Phys.*, **67**, 2609.
18 George T. E. and Miller W. H. (1972) *J. Chem. Phys.*, **57**, 2458.
19 Chapman S., Garrett R. C. and Miller W. H. (1975) *J. Chem. Phys.*, **63**, 2710.

A1.8 Problems

Theoretical foundations

A1.1 A combination of three H atoms gives rise to two doublet wavefunctions which may be expressed in the Heitler–London formulation as

$$\psi_1 = 2^{-1/2} A\{a(1)b(2)c(3)[\alpha(1)\alpha(2)\beta(3) - \alpha(1)\beta(2)\alpha(3)]\}$$
$$\psi_2 = 6^{-1/2} A\{a(1)b(2)c(3)[2\beta(1)\alpha(2)\alpha(3) - \alpha(1)\alpha(2)\beta(3) - \alpha(1)\beta(2)\alpha(3)]\}$$

where a, b and c denote $1s$ orbitals on the three atoms and A is the antisymmetrising operator. If overlap is neglected, the 2×2 Hamiltonian matrix has the following elements

$$H_{11} = Q_{AB} + Q_{BC} + Q_{CA} + J_{BC} - 1/2 J_{AB} - 1/2 J_{CA}$$
$$H_{22} = Q_{AB} + Q_{BC} + Q_{CA} - J_{BC} + 1/2 J_{AB} + 1/2 J_{CA}$$
$$H_{12} = (3/4)^{1/2}(J_{AB} - J_{CA})$$

where Q_{AB} and J_{AB} etc. are valence bond coulomb and exchange integrals, which tend to zero in the limit $R_{AB} \to \infty$ etc.

(i) Deduce that if atom A is well separated from molecule BC the system has eigenvalues

$$E_\pm = Q_{BC} \pm J_{BC}$$

and by examination of ψ_1 and ψ_2 that these eigenvalues may be associated with the singlet and triplet states of BC.

(ii) Prove that in general the eigenvalues of H are given by

$$E_\pm = Q_{AB} + Q_{BC} + Q_{CA} \pm \{[(J_{AB} - J_{BC})^2 + (J_{BC} - J_{AB})^2 + (J_{CA} - J_{AB})^2]/2\}^{1/2}.$$

A1.2. (i) Show that the skewing angle for three atoms of equal mass is given by $\chi = \pi/3$ and show that the physical motion following the collinear approach of atom A to stationary BC atoms is correctly described in a hard sphere model (reflection from planes parallel to the R_{AB} and R_{BC} axes) in the skewed axis system.

(ii) Show that a similar hard sphere collision initiated by collinear motion of A alone leads to approach angles of χ, 2χ, 3χ, 4χ . . . at successive reflecting planes and hence that skewing angles $\chi = \pi/2n$ and $\pi/(2n+1)$ lead to non-reaction after $(2n-1)$ internal collisions and $2n$ internal collisions, respectively.

A1.3. The simplest collision model is one in which reaction occurs for all collisions reaching a particle separation d at an energy E_a above that of the reactants. Show that the energy and angular momentum conservation equations

$$E = mu^2/2 = m(\dot{R}^2 + R^2\dot{\theta}^2)/2 + V(R)$$

$$L = mub = mR^2\dot{\theta}$$

imply a closest approach distance R (at which $\dot{R} = 0$) consistent with an impact parameter b_R given by

$$E = Eb_R^2/R^2 + V(R).$$

Hence setting $R = d$, $V(R) = E_a$ show that the cross-section varies as

$$S(u) = \pi b_d^2 = \pi d^2(1 - E_a/E).$$

A1.4. Imagine a symmetrical collinear reaction represented by motion of a particle of mass m over a potential surface having the form

$$V = k(r - r_e)^2/2$$

in the reactants' region and

$$V = E_a + k^\dagger x^2/2$$

along the symmetric stretch coordinate, x, at the saddle point.

Show that the microcanonical reactant and transition state fluxes at total energy E are given by

$$F_R = \oint p(r)\,dr = E/v, \qquad v = \frac{1}{2\pi}\left(\frac{k}{m}\right)^{1/2}$$

$$F_X = \oint p(x)\,dx = (E - E_a)/v^\dagger, \qquad v^\dagger = \frac{1}{2\pi}\left(\frac{k^\dagger}{m}\right)^{1/2}$$

and hence that

$$F_X > F_R \text{ when } E > E_a v^\dagger/(v - v^\dagger).$$

Hints

(i) $\oint p(r)\, dr = 2 \int_a^b [2m(E - V(r))]^{1/2}\, dr$

where $E = V(r)$ at $r = a$ and $r = b$;

(ii) $\int_{-a}^a (a^2 - x^2)^{1/2}\, dx = \pi a^2/2.$

A1.5. Suppose that the potential energy and transition state frequency vary along the reaction coordinate as

$$V(x) = E_a - \alpha x^2$$

$$v(x) = v^\dagger/(1 - \beta x^2).$$

Show by using the results of the previous problem that the flux varies as

$$F(x) = (E - E_a + \alpha x^2)(1 - \beta x^2)/v^\dagger$$

and hence that $F(x)$ is a minimum at the saddle point, $x = 0$, only for $E < E_a + \alpha/\beta$.

A1.6. Show that a Hamiltonian of the form

$$H = mp_r^2/2 + p_\theta^2/2mr^2 - Ar^{-\alpha} + k\theta^2/2r^2$$

gives rise to periodic orbits which follow energy-dependent fixed radii

$$r(E) = [(\alpha - 2)A/2E]^{1/\alpha} \text{ for } \alpha > 2.$$

Deduce that periodic motion is simple harmonic with frequency

$$v = \frac{1}{2\pi[r(E)]^2}\left(\frac{k}{m}\right)^{1/2}.$$

Hints
(i) Set up Hamilton's equations

$$\dot{p}_r = -(\partial H/\partial r), \quad \dot{r} = (\partial H/\partial p_r)$$

and require that $\dot{r} = \dot{p}_r = 0$ when $H = E$. This should yield the formula for $r(E)$.
(ii) Use $H = E$ to show that the residual θ motion satisfies an equation of the harmonic form

$$h(\theta) = \frac{1}{2\mu}p_\theta^2 + \frac{1}{2}\kappa\,\theta^2 = \text{const}.$$

A1.7. Replace the term $q^2\alpha/(4\pi\varepsilon_0)^2 2R^4$ in equation 21 by C/R^n and show that equations 24 and 27 become

$$V_{\text{eff}}^{\max} = f(n)(E^n b^{2n}/C^2)^{1/(n-2)}$$

$$S(E) = \pi[f(n)]^{(2-n)/n} (C/E)^{2n}$$

where

$$f(n) = (2/n)^{2/(n-2)} - (2/n)^{n/(n-2)}$$

A1.8. Consider the problem of reflection at a square barrier

$$V(x) = 0 \text{ for } x < 0 \text{ and } x > a$$

$$= V_0 \text{ for } 0 < x < a.$$

(i) Verify that the Schrödinger equation at $E < V_0$ has a solution

$$\psi = e^{ikx} \qquad\qquad x > a$$

$$= A e^{\kappa x} + B e^{-\kappa x} \qquad 0 < x < a$$

$$= X e^{ikx} + Y e^{-ikx} \qquad x < 0$$

where $k^2 = 2mE/\hbar^2$, $\kappa^2 = 2m(V_0 - E)/\hbar^2$.
(ii) Use the equations of continuity of $\psi(x)$ and $d\psi/dx$ at $x = 0$ and $x = a$ to show that

$$A = \frac{1}{2} P e^{-\kappa a + ika + i\theta}, \qquad\qquad B = \frac{1}{2} P e^{\kappa a + ika - i\theta}$$

$$X = \frac{-i\kappa P}{2k} [A e^{i\theta} - B e^{-i\theta}], Y = \frac{i\kappa P}{2k} [A e^{-i\theta} - B e^{i\theta}].$$

where $P \cos\theta = 1$, $P \sin\theta = (k/\kappa)$.
(iii) The physical interpretation is that X and Y are the incident and reflected amplitudes for $x < 0$ when the solution has unit transmitted amplitude for $x > a$. Deduce that the reflection and transmision coefficients R and T are given by

$$|R|^2 = |Y/X|^2 = (e^{2\kappa a} + e^{-2\kappa a} - 2)/(e^{2\kappa a} + e^{-2\kappa a} - 2\cos 4\theta)$$

$$|T|^2 = 1 - |R|^2 = 4\sin^2 2\theta/(e^{2\kappa a} + e^{-2\kappa a} - 2\cos 4\theta).$$

A1.9. Answers to problems

Theoretical foundations

A1.1. The matrix $\begin{pmatrix} H_{11} & H_{12} \\ H_{21} & H_{22} \end{pmatrix}$ has eigenvalues

$$E_\pm = \frac{1}{2}(H_{11}+H_{22}) \pm \frac{1}{2}\sqrt{(H_{11}-H_{22})^2 - 4H_{12}^2}$$

and eigenfunctions

$$\left.\begin{array}{l} \psi_+ = \cos\theta\,\psi_1 + \sin\theta\,\psi_2 \\ \psi_- = -\sin\theta_1\,\psi_1 + \cos\theta\,\psi_2 \end{array}\right\} \text{ where } \tan2\theta = \frac{2H_{12}}{H_{11}-H_{22}}$$

(i) $J_{AB}=J_{CA}=0$ for R_{AB}, $R_{CA}\rightarrow\infty$
hence $H_{12}=0$ and $\theta=0$.

$$\therefore\ E_+ = H_{11} = Q_{BC}+J_{BC}; \ \psi_+=\psi_1$$

$$E_- = H_{22} = Q_{BC}-J_{BC}; \ \psi_-=\psi_2.$$

It is sufficient for analysis of the wavefunctions to note that

$\psi_1 = A\{a(1)\alpha(1)$ $b(2)c(3)[\alpha(2)\beta(3)-\beta(2)\alpha(3)]\}$
    electron on A   BC electron pair with paired spins

$\psi_2 = A\{2a(1)\beta(1)$ $b(2)c(3)\alpha(2)\alpha(3)$
     $-a(1)\alpha(1)$ $b(2)c(3)[\alpha(2)\beta(3)+\beta(2)\alpha(3)]\}$
    electron on A   BC pair with two types of $\parallel$ spin

The BC terms show that ψ_1 correlates with a singlet and ψ_3 with a triplet.
(ii) Using the above equation for E_+

$$E_\pm = Q_{AB}+Q_{BC}+Q_{CA}\pm\tfrac{1}{2}\sqrt{(2J_{BC}-J_{AB}-J_{CA})^2+2(J_{AB}-J_{CA})^2}$$

$$= Q_{AB}+Q_{BC}+Q_{CA}\pm\tfrac{1}{2}\sqrt{4J_{BC}^2+J_{AB}^2J_{CA}^2-4J_{BC}J_{AB}-4J_{BC}J_{CA}+2J_{AB}J_{CA}}$$

$$+3J_{AB}^2+3J_{CA}^2 \qquad\qquad -6J_{AB}J_{CA}$$

$$= Q_{AB}+Q_{BC}+Q_{CA}\pm\sqrt{J_{BC}^2+J_{AB}^2+J_{CA}^2-J_{BC}J_{AB}-J_{BC}J_{CA}-J_{AB}J_{CA}}$$

$$= Q_{AB}+Q_{BC}+Q_{CA}\pm\sqrt{[(J_{AB}-J_{BC})^2+(J_{BC}-J_{CA})^2+(J_{CA}-J_{AB})^2]/2}.$$

A1.2. (i) Skew angle. $\tan\chi = \sqrt{\dfrac{m_B(m_A+m_B+m_C)}{m_A m_C}} = \sqrt{3}$

$$\therefore\ \chi = \pi/3$$

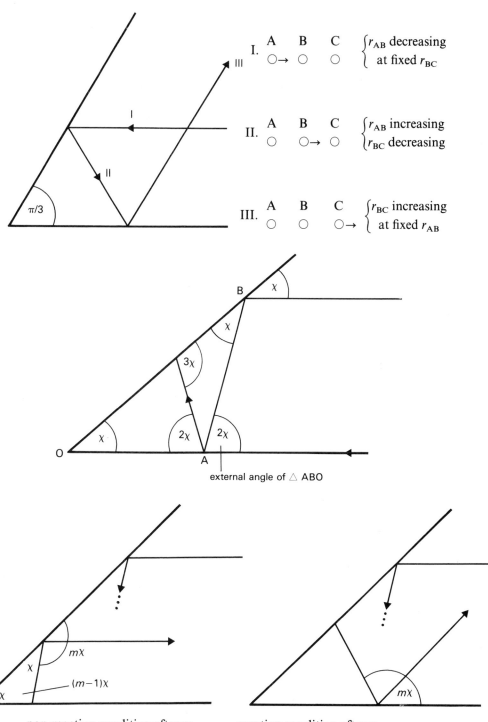

I.
A	B	C
$\bigcirc\to$	$\bigcirc$	$\bigcirc$

$\begin{cases} r_{AB} \text{ decreasing} \\ \text{at fixed } r_{BC} \end{cases}$

II.
A	B	C
$\bigcirc$	$\bigcirc\to$	$\bigcirc$

$\begin{cases} r_{AB} \text{ increasing} \\ r_{BC} \text{ decreasing} \end{cases}$

III.
A	B	C
$\bigcirc$	$\bigcirc$	$\bigcirc\to$

$\begin{cases} r_{BC} \text{ increasing} \\ \text{at fixed } r_{AB} \end{cases}$

external angle of △ ABO

non-reaction condition after m
internal collisions

reaction condition after m
internal collisions

Final path horizontal if $(m+1)\chi = \pi$
$\chi = \pi/(m+1)$
let $m = 2n - 1$ for answer
(non-reacton requires odd number
of collisions with the walls of the
potential)

Final path $\parallel$ product axis
if $(m+1)\chi = \pi$, $\chi = \pi/(m+1)$
let $m = 2n$ for answer
(reaction requires even number of
collisions with the walls of the
potential).

A1.3.

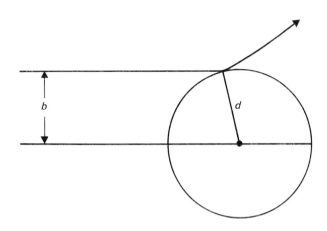

At closest approach $R = d$, $\dot{R} = 0$, $V = E_a$

$$\therefore E = \tfrac{1}{2}mu^2 = \tfrac{1}{2}md^2\dot{\theta}^2 + E_a \tag{1}$$

$$L = mub = md^2\dot{\theta} \tag{2}$$

$$(2) \Rightarrow \dot{\theta} = (ub/d^2)$$

$$(1) \Rightarrow E = \tfrac{1}{2}mu^2(b/d)^2 + E_a = E(b/d)^2 + E_a$$

$$\therefore b^2 = d^2(1 - E_a/E)$$

$$S = \pi b^2 = \pi d^2(1 - E_a/E).$$

A1.4. $F_R = \oint p(r)\, dr = 2 \int_a^b \sqrt{2m(E - \tfrac{1}{2}k(r - r_e)^2)}\; dr$

let $r - r_e = y$, $2E/k = \alpha^2$

$$\therefore F_R = 2\sqrt{mk} \int_{-\alpha}^{\alpha} \sqrt{\alpha^2 - y^2}\; dy = 2\sqrt{mk}\; \pi\alpha^2/2 = E/\nu, \quad \nu = \frac{1}{2\pi}\sqrt{\frac{k}{m}}.$$

Similarly

$$F_X = 2 \int_a^b \sqrt{2m(E - E_a - \tfrac{1}{2}k^\dagger x^2)} \ dx = (E - E_a)/v^\dagger, \ v^\dagger = \frac{1}{2\pi} \sqrt{\frac{k^\dagger}{m}}$$

$$\therefore \ \ F_X > F_R \ \text{if} \ E > vE_a/(v - v^\dagger)$$

$$\frac{E_a}{v^\dagger} < E\left(\frac{1}{v^\dagger} - \frac{1}{v}\right) = \frac{E(v - v^\dagger)}{vv^\dagger}$$

$$\therefore \ \ E > \frac{v^\dagger E_a}{v - v^\dagger}.$$

A1.5. By the argument in question A1.4

$$F(x) = \frac{E - V(x)}{v(x)} = (E - E_a + \alpha x^2)(1 - \beta x^2)/v^\dagger$$

$$\frac{dF}{dx} = \frac{1}{v^\dagger}\{2\alpha x(1 - \beta x^2) - 2\beta x(E - E_a + \alpha x^2)\}$$

$$= \frac{2}{v^\dagger}\{\alpha x - \beta(E - E_a)x - 2\alpha\beta x^3\}$$

$$\left.\frac{d^2 F}{dx^2}\right|_{x=0} = \frac{2}{v^\dagger}[\alpha - \beta(E - E_a)]$$

$$> 0 \ \text{for} \ \alpha > \beta(E - E_a)$$

$$\text{i.e.} \ E < E_a + \alpha/\beta.$$

A1.6. At energy E

(i) $H = p_r^2/2m + p_\theta^2/2mr^2 - Ar^{-\alpha} + k\theta^2/2r^2 = E$ (1)

for fixed radius motion

$$\dot{r} = +(\partial H/\partial p_r) = p_r/m = 0 \qquad \therefore \ p_r = 0 \tag{2}$$

$$\dot{p}_r = -(\partial H/\partial r) = \frac{p_\theta^2}{mr^3} - \alpha Ar^{-\alpha-1} + kr^{-3}\theta^2 = 0. \tag{3}$$

But equation 1 implies $\dfrac{p_\theta^2}{mr^2} = 2E + 2Ar^{-\alpha} - kr^{-2}\theta^2.$

$\therefore \quad \dfrac{1}{r}(2E + 2Ar^{-\alpha} - kr^{-2}\theta^2) - \alpha A r^{-\alpha-1} + kr^{-3}\theta^2 = 0$

$2E = (\alpha - 2)\,Ar^{-\alpha}$

$r = [(\alpha - 2)A/2E]^{1/\alpha}.$

(ii) With this fixed r value and $p_r = 0$, equation 1 becomes

$\dfrac{p_\theta^2}{2mr^2} + \tfrac{1}{2}kr^{-2}\theta^2 = E + Ar^{-\alpha} = E\left(1 + \dfrac{2}{\alpha - 2}\right) = \dfrac{\alpha E}{\alpha - 2}$

cf. $\dfrac{p_\theta^2}{2\mu} + \tfrac{1}{2}\kappa\theta^2 = \text{const.} \qquad \mu = mr^2, \quad \kappa = kr^{-2}.$

A1.7. $V_{\text{eff}}(R) = E(b^2/R^2) - C/R^n.$

at max. $\dfrac{dV_{\text{eff}}}{dR} = 0 = -2Eb^2/R^3 + n\,C/R^{n+1}$

$\Rightarrow \quad \dfrac{1}{R_{\max}} = \left(\dfrac{2Eb^2}{nC}\right)^{1/(n-2)}$

$\therefore \quad V_{\text{eff}}^{\max} = V_{\text{eff}}(R_{\max}) = Eb^2\left(\dfrac{2Eb^2}{nC}\right)^{2/(n-2)} - C\left(\dfrac{2Eb^2}{nC}\right)^{n/(n-2)}$

$= f(n)\,(E^n b^{2n}/C^2)^{1/(n-2)}.$

Orbiting condition at energy E

$E = V_{\text{eff}}^{\max} = f(n)(E^n b^{2n}/C^2)^{1/(n-2)}$

or $b^{2n} = [E/f(n)]^{n-2}(C^2/E^n) = [f(n)]^{2-n}(C/E)^2$

$\therefore \quad S(E) = \pi b^2 = \pi[f(n)]^{(2-n)/n}\,(C/E)^{2/n}$

A1.8.

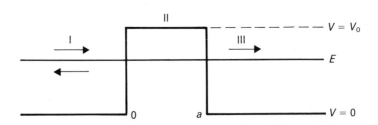

(i) *Schrödinger equation*

Regions I, III $\qquad -\dfrac{\hbar^2}{2m}\dfrac{d^2\psi}{dx^2}=E\psi$ or $\dfrac{d^2\psi}{dx^2}=-k^2\psi,\quad k^2=\dfrac{2mE}{\hbar^2}$

General solution. $\psi = C_1\,e^{ikx}+C_2\,e^{-ikx}$

Chosen solution has $C_1=1$, $C_2=0$ in III to represent an outgoing wave to right of barrier. Problem is then to find $C_1=X$ and $C_2=Y$ to left of barrier, which are incident and reflected amplitudes, respectively.

Region II. $\qquad \left[-\dfrac{\hbar^2}{2m}\dfrac{d^2}{dx^2}+V_0\right]\psi=E\psi$ or $\dfrac{d^2\psi}{dx^2}=\kappa^2\psi,\quad \kappa^2=\dfrac{2m(V_0-E)}{\hbar^2}$

General solution. $\psi=A\,e^{\kappa x}+B\,e^{-\kappa x}$

(ii) *Continuity conditions*

at $x=a$. $\quad \psi(a)=e^{ika}=A\,e^{\kappa a}+B\,e^{-\kappa a}$

$$\psi'(a)=ik\,e^{ika}=\kappa(A\,e^{\kappa a}-B\,e^{-\kappa a}).$$

$\therefore\qquad A=\tfrac12\left(1+\dfrac{ik}{\kappa}\right)e^{-\kappa a+ika}=\tfrac12\,P\,e^{-\kappa a+ika+i\theta}$

$$B=\tfrac12\left(1-\dfrac{ik}{\kappa}\right)e^{\kappa a+ika}=\tfrac12\,P\,e^{\kappa a+ika-i\theta}$$

at $x=0$. $\quad \psi(0)=A+B=X+Y.$

$$\psi'(0)=\kappa(A-B)=ik(X-Y).$$

$\therefore\qquad X=-\dfrac{i\kappa}{2k}\left\{\left(\dfrac{ik}{\kappa}+1\right)A+\left(\dfrac{ik}{\kappa}-1\right)B\right\}=-\dfrac{i\kappa P}{2k}\{A\,e^{i\theta}-B\,e^{-i\theta}\}$

$$Y=-\dfrac{i\kappa}{2k}\left\{\left(\dfrac{ik}{\kappa}-1\right)A+\left(\dfrac{ik}{\kappa}+1\right)B\right\}=\dfrac{i\kappa P}{2k}\{A\,e^{-i\theta}-B\,e^{i\theta}\}$$

(iii) *Reflection and transmission coefficients*

$$X=\dfrac{-i\kappa P^2}{4k}\{e^{-\kappa a+2i\theta}-e^{\kappa a-2i\theta}\}\,e^{ika}$$

$$Y=\dfrac{i\kappa P^2}{4k}\{e^{-\kappa a}-e^{\kappa a}\}\,e^{ika}$$

$$\therefore \ |R|^2 = |Y/X|^2 = \left| \frac{e^{-\kappa a} - e^{\kappa a}}{e^{-\kappa a + 2i\theta} - e^{\kappa a - 2i\theta}} \right|^2$$

$$= \frac{e^{-2\kappa a} + e^{2\kappa a} - 2}{(e^{-\kappa a + 2i\theta} - e^{\kappa a - 2i\theta})(e^{-\kappa a - 2i\theta} - e^{\kappa a + 2i\theta})}$$

$$= \frac{e^{-2\kappa a} - e^{\kappa a} - 2}{e^{-2\kappa a} + e^{2\kappa a} - e^{-4i\theta} - e^{4i\theta}} = \frac{e^{-2\kappa a} + e^{2\kappa a} - 2}{e^{-2\kappa a} + e^{2\kappa a} - 2\cos 4\theta}$$

$$|T|^2 = 1 - |R|^2 = \frac{(e^{-2\kappa a} + e^{-2\kappa a} - 2\cos 4\theta) - (e^{-2\kappa a} + e^{2\kappa a} - 2)}{e^{-2\kappa a} + e^{2\kappa a} - 2\cos 4\theta}$$

$$= \frac{2(1 - \cos 4\theta)}{e^{2\kappa a} + e^{2\kappa a} - 2\cos 4\theta} = \frac{4\sin^2 2\theta}{e^{-2\kappa a} + e^{2\kappa a} - 2\cos 4\theta}.$$

Chapter A2
Quasiclassical Trajectory Methods

I. W. M. SMITH

A2.1 Introduction

As has already been pointed out in section A1.2, any full-scale theory of elementary reactions should incorporate three stages.

(a) The potential energy (hyper)surface (PES), $V(R_1, R_2, R_3, ..)$, for the system should be calculated.

(b) The dynamics of individual molecular events (i.e., collisions between reagents for bimolecular reactions) should be solved.

(c) The results of (b) should be averaged appropriately so that the theoretical results can be compared with those obtained from experiments.

The separation of (a) and (b) implies that the Born–Oppenheimer approximation holds. This is valid at normal collision energies and is usually assumed in chemical problems. In both these stages, one should use the equations and methods of quantum mechanics. In practice, that is beyond the scope of present theoretical methods for all but the simplest systems (which currently means collisions of H atoms with H_2 ($v = 0$) molecules), and approximate methods are valuable. Stage (c) requires calculations based on statistical mechanics and is more tractable.

As explained in Chapter A1, transition state theory concentrates on evaluating the flux of reacting systems through the crucial 'bottleneck' region of the PES. But, despite its successes, transition state theory provides little information about the reaction dynamics; for example, how energy released in a reaction is partitioned among the degrees of freedom of the products. For trajectory calculations, the whole PES is required. It may be based on *ab initio*, quantum mechanical calculations in the few cases where these have been performed with sufficient accuracy (e.g., $H + H_2$, and possibly $F + H_2$). More frequently, semi-empirical methods are used to generate a suitable PES.

Having chosen a PES, individual molecular collisions are simulated by computationally solving the classical equations for the relative motions of the nuclei. Averaging is built into the selection of initial ($t = 0$) values of the position and momentum coordinates of the nuclei using pseudo-random or Monte Carlo techniques. The trajectory is computed until the 'products' of the collision (these may be identical to the reagents if the collision is non-reactive) have retreated to a separation at which they no longer interact significantly. In classical trajectories

(the results of which can be compared with those from classical transition state theory), the selection is made on the basis of the internal energies being allowed a continuous range of values.

This chapter outlines the methods and procedures used in quasiclassical (QCL) trajectory calculations, and examines some of the results of such calculations, as well as the status of the method. Although calculations have been performed on reactions involving more atoms, we shall focus on three-atom bimolecular reactions of the type

$$A + BC \rightarrow AB + C \tag{1}$$

A2.2 Choice of potential energy surface

The choice of a PES for trajectory calculations is crucial, since its form fundamentally controls the molecular dynamics. The choice is especially important when an attempt is being made to simulate the behaviour of a particular reaction. Then the criteria on which the PES has been selected should be critically examined. Even when accurate *ab initio* calculations have been carried out, as for H_3,[1] the calculated values of V have to be fitted to an analytical expression,[2] since to solve the equations of motion (see equation 5 below) gradients of V are required for all accessible geometries. Even for the reaction

$$F + H_2 \rightarrow HF + H \tag{2}$$

whose dynamics have been studied extensively, the PES is not yet satisfactorily defined by *ab initio* methods,[3] despite its relative electronic simplicity and considerable theoretical effort.

When a study has broader objectives—e.g., to examine connections between the form of the PES and the attributes of a reaction[4]—then a relatively simple flexible expression for the PES suffices. One that has often been used for three-atom systems is the LEPS (London–Eyring–Polanyi–Sato) function.[4] It is based on the London equation for H_3 and the Q (Coulomb integrals) and J (exchange integrals) are evaluated as functions of the internuclear separations (R_{AB}, R_{BC}, R_{CA}) by equating valence bond expressions for the pairwise interactions to the Morse and anti-Morse diatomic potential functions given below, the Morse constants being derived from spectroscopic data. Although the S_i are nominally overlap integrals, they are assumed independent of the R_i and used as adjustable parameters to modify the form of the PES

$$V(R_{AB}, R_{BC}, R_{CA}) = \sum_{i=1}^{3} \frac{Q_i}{1+S_i} - \left\{ \sum_{i=1}^{3} \frac{J_i^2}{(1+S_i)^2} - \sum_{i,j} \frac{J_i J_j}{(1+S_i)(1+S_j)} \right\}^{1/2} \tag{3}$$

with

$$\frac{Q_i + J_i}{1 + S_i} = D_{e,i} \{\exp(-2\beta_i \Delta R_i) - 2\exp(-\beta_i \Delta R_i)\} \tag{3a}$$

and

$$\frac{Q_i - J_i}{1 - S_i} = (D_{e,i}/2) \{\exp(-2\beta_i \Delta R_i) + 2\exp(-\beta_i \Delta R_i)\} \tag{3b}$$

where $\Delta R_i = R_i - R_{e,i}$.

Even for a system of only three atoms, it is impossible to represent the full potential energy hypersurface diagrammatically, since V depends on three variables; e.g., R_{AB}, R_{BC} and R_{CA}. Contour line plots of restricted geometries are often presented. LEPS expressions always lead to a minimum barrier to reaction for collinear collisions ($R_{CA} = R_{AB} + R_{BC}$) and the variation of V for such collinear geometries is frequently displayed. Figure A2.1 shows such a diagram for the Porter–Karplus PES for H_3.[5] This PES was based on a sophisticated LEPS

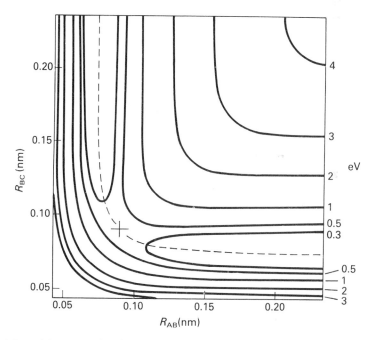

Fig. A2.1. Potential energy surface for the collinear reaction $H_A + H_B H_C \rightarrow H_A H_B + H_C$, as calculated by Porter and Karplus.[5] The contour lines join points of equal potential energy ($1\,eV \equiv 96.5\,kJ\,mol^{-1}$). The dashed line indicates the reaction path of minimum energy and the cross the highest point on this path.

approach and has been used in many dynamical calculations on $H + H_2$ collisions, including the early quasiclassical trajectory study by Karplus, Porter and Sharma.[6] Their paper is strongly recommended as giving a clear and thorough account of the methods used in QCL trajectory calculations.

A2.3 Dynamics of individual collisions

For three particles, A, B and C, equivalent expressions for the classical Hamiltonian (the total energy) can be written in terms of Cartesian coordinates for each particle and the corresponding components of momenta, equation 4a, or more usefully in terms of an alternative coordinate system which separates relative motion and motion of the centre-of-mass of the entire system, equation 4b

$$H' = \frac{1}{2m_A} \sum_{j=1}^{3} p_j^2 + \frac{1}{2m_B} \sum_{j=4}^{6} p_j^2 + \frac{1}{2m_C} \sum_{j=7}^{9} p_j^2 + V(q_1 \ldots q_9) \tag{4a}$$

$$H = \frac{1}{2\mu_{BC}} \sum_{j=1}^{3} P_j^2 + \frac{1}{2\mu_{A,BC}} \sum_{j=4}^{6} P_j^2 + \frac{1}{2M} \sum_{j=7}^{9} P_j^2 + V(Q_1 \ldots Q_6). \tag{4b}$$

Here, m_A, m_B and m_C are the masses of the individual atoms and μ_{BC}, $\mu_{A,BC}$ and M have the meanings introduced in section A1.2(ii). The momenta and position coordinates, p_j and q_j, are for the individual atoms, whilst P_j and Q_j represent momenta and position in the centre-of-mass frame-of-reference.

In the absence of any external influence, the energy associated with motion of the centre-of-mass, i.e.

$$H_{cm} = \frac{1}{2M} \sum_{j=7}^{9} P_j^2 \tag{5}$$

is conserved and can be separated out, leaving

$$H'_{rel} = \frac{1}{2\mu_{BC}} \sum_{j=1}^{3} P_j^2 + \frac{1}{2\mu_{A,BC}} \sum_{j=4}^{6} P_j^2 + V(Q_1 \ldots Q_6). \tag{6}$$

The classical equations of (relative) motion are:

$$(\partial H_{rel}/\partial P_j) = (\partial T_{rel}/\partial P_j) = \dot{Q}_j \tag{6a}$$

$$-(\partial H_{rel}/\partial Q_j) = -(\partial V/\partial Q_j) = \dot{P}_j. \tag{6b}$$

There are six equations of each kind, and it is these twelve coupled differential equations that must be numerically integrated in a computer to simulate a single trajectory starting from some initial set of P_j, Q_j. $(\partial V / \partial Q_j)$ must be calculated from $(\partial V / \partial R_i)$ using the 'chain rule', i.e.,

$$(\partial V / \partial Q_j) = \sum_i (\partial V / \partial R_i)(\partial R_i / \partial Q_j). \tag{7}$$

In conventional QCL trajectory calculations (see section A2.7 for less conventional approaches), each trajectory is started ($t = 0$) with A far enough from BC for there to be negligible interaction between them. The computation is stopped either when A and BC, having collided, have again separated to some pre-defined separation without reaction or when AB and C, the products of a reactive collision, have separated to a distance at which they do not significantly interact. Figure A2.2 shows the variations of R_{AB}, R_{BC}, R_{CA} in a reactive collision between an H atom and H_2 on the Porter–Karpulus PES.[5] This plot represents a direct reactive collision, since there is only one instant at which $R_{AB} = R_{BC}$.

A2.4 Selection of initial dynamical variables

Usually, the initial P_j and Q_j are not selected directly but are calculated from selected values of quantities such as the relative velocity and impact parameter, whose values are distributed according to well-known statistical laws. Which

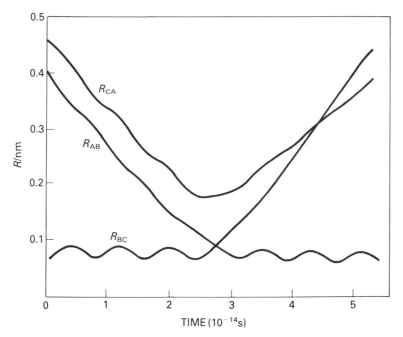

Fig. A2.2. Typical direct reactive classical trajectory for an $H + H_2$ collision.[5] Note that the R_{AB} and R_{BC} curves cross only once.

quantities are chosen pseudo-randomly depends on what one is trying to calculate. For example, to find a reaction cross-section the relative velocity/collision energy would be fixed at a particular value in each of a set of trajectories, but to determine a rate constant the collision energies would be chosen pseudo-randomly from a distribution appropriate to a chosen temperature. The exact method of pseudo-random selection depends on the nature of the distribution function for the variable. Here, four examples are given to illustrate the various methods that can be employed.

(a) The most straightforward case arises when the probability of a variable taking a particular value is the same throughout its entire range. For example, initial values of the angle (ϕ) between the BC axis and a line joining A to the centre-of-mass of BC (see Fig. A2.3) in the plane containing all three atoms are distributed in this way. The angle can be selected by multiplying 2π by a random number (ξ) between 0 and 1 drawn from the computer:

$$\phi = \xi.2\pi \tag{8}$$

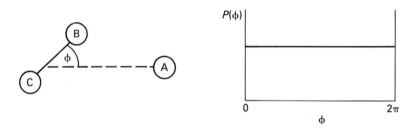

Fig. A2.3. Definition of the angle ϕ and its distribution function.

(b) In certain cases, a parameter itself is not evenly distributed over its full range of values, but some simple function of the quantity is evenly distributed. The impact parameter provides a good example. The probability, $dP(b)$, of the impact parameter having a particular value between b and $b+db$ is proportional to $2\pi b\,db$ (see Fig. A2.4), so that a technique is required that will generate a distribution reflecting the weighting factor, $2\pi b$. To do this, a function $F(b)$ must be found such that b is uniformly distributed in F-space; i.e., such that $dP(b)/dF$ is a constant (say $1/\beta$). Since

$$dP(b)/dF = (dP(b)/db)/(dF/db)$$
$$1/\beta = 2\pi b/(dF/db)$$
$$F = \int dF = \beta \int 2\pi b\,db. \tag{9}$$

That is, F is proportional to b^2 and the magnitude of the impact parameter is chosen using the equation

$$b = (\xi \, b_{max}^2)^{1/2} \tag{10}$$

where ξ again represents a random number and b_{max} is a maximum value of the impact parameter which must be carefully chosen to be large enough not to exclude an appreciable proportion of significant events, but not so large that time is wasted by computing any 'uninteresting' trajectories.

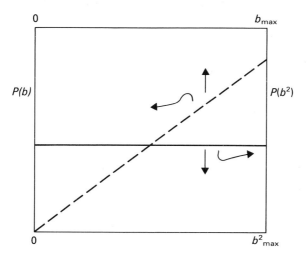

Fig. A2.4. Probability distribution function for impact parameter and (impact parameter)^2.

(c) There is also a Monte Carlo technique for selecting variables for which no closed form to F can be found. A random number is used to provide a value of, for example, the relative velocity, but a decision whether to accept this value is made on the result of a 'game-of-chance' against a second random number. Thus, one chooses

$$u = \xi_1 \, u_{max} \tag{11}$$

where u_{max} is chosen to include all but a very small fraction of the velocity distribution. Then one calculates $P(u)$ relative to $P(u_{max})$, the most probable value of the relative velocity. If

$$P(u)/P(u_{max}) > \xi_2 \tag{12}$$

this value of u is accepted; if not, a new value of u is selected via equation 11 and the cycle continued until a value is accepted via equation 12.

(d) At $t = 0$, R_{BC} is usually set randomly at the maximum or minimum, classically allowed, BC separation for the isolated molecule at its quasiclassically determined vibrational energy. Averaging over the vibrational phase of BC is then achieved by varying the initial distance of A from BC's centre-of-mass by an amount related to the period of BC's vibration.[6] This leads to the correct classical distribution over the phase of vibration.

The application of pseudo-random or Monte Carlo techniques to the selection of initial P_j and Q_j for trajectory calculations means that averaging [stage (c) of the full scale theory delineated earlier] is built into the choice of individual molecular events whose dynamics are solved [stage (b)]. Using Monte Carlo techniques means that results of acceptable accuracy can often be achieved with quite small (10^3–10^4) sets of trajectories.

A2.5 Calculation of final state properties

Once a given trajectory is completed, the P_j and Q_j are transformed to more physically meaningful quantities, such as final translational energy and scattering angle. If A + BC are the 'products' of a non-reactive collision, the procedure used at this point is almost exactly the reverse of that employed to convert the pseudo-randomly selected initial quantities to the P_j, Q_j at selected initial quantities to the P_j, Q_j at $t = 0$. If reaction has occurred, a different but related set of equations must be used.[6]

One difference between the initial and final procedures is associated with the treatment of internal energies. Energies of molecular rotation and vibration are not entirely independent (remember that the moment of inertia of a diatomic molecule depends on the square of the internuclear separation). Moreover, at $t = 0$, internal energies are chosen to correspond to those of quantum states. However, once a trajectory calculation starts, no such quantum mechanical restriction can be included and the internal energy of the product molecule will not correspond to that of a quantum state. The total internal energy is usually partitioned between rotation and vibration using a simple iterative procedure.[7] The results are then 'binned'; that is, a molecule with a given vibrational (rotational) energy is assigned to the nearest quantum state.

Depending on specifications made in the selection of initial quantities, QCL trajectory calculations usually lead to estimates of either reaction cross-sections (S_{reac}) or rate constants (k_{reac}). If N^* trajectories out of N lead to reaction, these quantities are given by

$$S_{reac} = \pi b_{max}^2 (N^*/N) \tag{13}$$

$$k_{reac}(T) = \pi b_{max}^2 \bar{u} (N^*/N) \tag{14}$$

where $\bar{u}$ is the magnitude of the mean relative velocity at temperature T. At the 68% confidence level, the error in (N^*/N) is

$$\Delta(N^*/N) = (N^*/N)[(N - N^*)/N^*N]^{1/2} \tag{15a}$$

or, expressed as a fraction

$$\Delta(N^*/N)/(N^*N) = [(N - N^*)/N^*N]^{1/2}. \tag{15b}$$

Thus, if $N = 10^4$ and $N^* = 100$

$$(N^*/N) = 0.01 \pm 0.001.$$

A2.6 Importance sampling

Trajectory calculations are expensive in computer time and it is therefore important to make them as efficient as possible. Two ways to do this are;
(a) to concentrate the sampling of an initial variable around those values which make the largest contribution to the derived quantity;
(b) to make use of information obtained for one set of conditions to predict results for other conditions.

As an example of (b) we consider the derivation of thermal rate constants. Remember that when a rate constant is to be evaluated, the initial translational and rovibrational energies for trajectories are selected pseudo-randomly from the appropriate Maxwell–Boltzmann distributions for a defined temperature T. The rate constant is then calculated from equation 14. The same results can be used to calculate the rate constant at any other temperature T' by assigning to each trajectory a weight $w_i(T')$, which corresponds to the probability that the selected initial conditions would occur at T' relative to their probability at temperature T.[8] The rate constant at T' is given by:

$$k(T') = \pi b_{\max}^2\, \bar{u}\left[\sum_{N^*} w_i(T') \bigg/ \sum_{N} w_i(T') \right]. \tag{16}$$

'Importance sampling', i.e., (a) above, has been discussed in detail by Muckerman and Faist.[9] It is illustrated here with regard to the choice of relative velocity and impact parameter.

The first goal must be to exclude trajectories that will 'inevitably' lead to uninteresting results. Since one is usually concerned only with reactive trajectories (and with establishing the fraction of trajectories that lead to reaction), one seeks to establish the minimum (or threshold) value of u and the maximum value of b that can lead to reaction. Sampling of relative velocity and impact parameter is subsequently confined within the limits thereby established.

Secondly, one can bias the sampling, even within the 'allowed' range, away from the purely statistical. Two methods are possible and can be exemplified by considering the selection of impact parameter. Remembering that the reaction cross-section is given by

$$S_{\text{reac}} = \int_0^{b_{\max}} P_{\text{reac}}(b)\, 2\pi b\, db \qquad (17)$$

it is clear that one wishes to minimise error especially at those values of b where $P_{\text{reac}}(b)\, 2\pi b$ is relatively large. One way to do this is to select b within successive segments b to $b + \Delta b$ and to calculate a number of trajectories within each segment to give the same absolute error in $\int_b^{b+\Delta b} P_{\text{reac}}(b)\, 2\pi b\, db$. Alternatively one can sample in accordance with some prior expectation, $P^0_{\text{reac}}(b)$, of the form of $P_{\text{reac}}(b)$, the opacity function. An extension of simple collision theory, incorporating an orientation-dependent energy barrier,[10] provides a suitable estimate of $P^0_{\text{reac}}(b)$ for direct $A + BC$ reactions (see Fig. A2.5). One can then sample pseudo-randomly over a distribution $P^0_{\text{reac}}(b)\, 2\pi b$ and calculate the reaction cross-section according to

$$S_{\text{reac}} = \pi b^2_{\max} \left[\sum_{N^*} w_i(b) \middle/ \sum_N w_i(b) \right] \qquad (18)$$

where $w_i(b)$ is the probability that a particular value of b is chosen from $2\pi b$ relative to that from $P^0_{\text{reac}}(b)\, 2\pi b$.

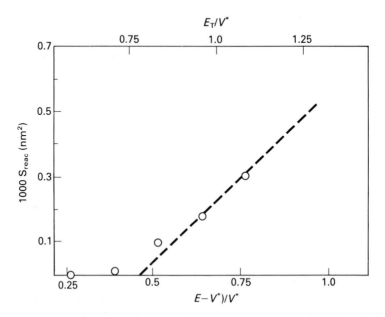

Fig. A2.5. Plots of $2\pi b$, $P^0_{\text{reac}}(b)$, and their product against impact parameter b. The form of $P^0_{\text{reac}}(b)$ is that suggested by an extension to simple collision theory which incorporates an orientation-dependent barrier.[10]

A2.7 Combining trajectories and transition state theory

The most effective way of eliminating the need to compute 'uninteresting' trajectories is to start trajectories in the transition state region. Such methods of combining trajectories and transition state theory were pioneered by Anderson and his coworkers[11] and applied to the reactions between $F + H_2$ and $H_2 + I_2$. Classical phase space was sampled in the strong interaction region (although no attempt was made to find the optimum transition state) and trajectories were run forward and backward in time (i.e., towards products and towards reagents). The results—for example, with regard to distributions of scattering angle and energy among the products—agreed with those from conventional classical trajectories within statistical error, as of course they should given the deterministic nature of classical dynamics.

Anderson's calculations, and others referred to in Chapter A1, emphasise the close connection between 'statistical' transition state theory and 'dynamical' trajectory calculations. The relationship between the estimates of rate constants by the classical 'combined' method (CLT–TST), by classical transition state theory (CLTST), and by classical trajectories (CLT) is summarised in the relationship

$$k_{CLTST} \geqslant k_{CLT-TST} = k_{CLT} \tag{19}$$

that is, the rate constant calculated via classical transition state theory (when the transition state is found by variational means and its phase integral is evaluated without secondary assumptions) exceeds or is equal to the rate constants given by classical trajectories whether these start at the transition state or in the reagents' phase space. The inequality arises because transition state theory makes no allowance for trajectories that having crossed the transition state region in the direction reagents to products, recross and do not lead to reaction (see Chapter A1).

Similar comparisons of conventional quasiclassical trajectories starting at a 'standard' transition state, in which allowance is made for quantisation of motions orthogonal to the reaction coordinate, have been made,[12] but as yet only for collinear collisions. The combined method can include reactions of vibrationally excited species, and the agreement with conventional calculations is generally good, especially when the vibrationally adiabatic barriers occupy an 'early' position along the reaction coordinate.

A2.8 Tests of quasiclassical trajectories

In any trajectory calculations, it is important to check that the numerical integration of the equations of motion is being performed correctly. At a deeper level, it is necessary to examine the general validity of the trajectory methods of simulating reaction dynamics and, in particular, the assumption of classical dynamics.

The correct choice of time step and subsequent computation of trajectories can be checked by ensuring that the total energy and angular momentum is conserved in each trajectory. A further test that is often applied is to reverse the direction of time at the end of a trajectory and check whether it returns to the initial P_j, Q_j point in phase space. However, there are situations in which the failure to satisfy this test does not invalidate the results of the trajectory calculations.

With simple direct collisions, trajectories initiated from neighbouring points in phase space follow closely parallel paths. In such circumstances, reaction occurs from a limited number of large volumes of phase space (see Fig. A1.5), and trajectories back-integrate very close to their original starting points. For indirect collisions, especially those proceeding via a deep minimum on the PES, the situation can be very different. For example, in the limit where long-lived collision complexes form, it is expected (see Chapter A4) that systems lose all memory of their mode of formation except total energy and angular momentum. This seems at odds with the determinism of classical dynamics. What it means is that trajectories beginning from neighbouring phase space points can diverge dramatically. As a result of small rounding errors, the forward and back integration of the equations of motion can lead to very different trajectories. However, this does not invalidate the results derived from an ensemble of calculated trajectories.[13]

Two distinct kinds of evidence can be used to assess the validity of the QCL trajectory method of simulating reaction dynamics. First, the results can be compared with those from accurate quantum scattering calculations on the same system. For collisions in three dimensions, this comparison can only be made for the H_3 system. The results of such a comparison, shown in Fig. A2.6, are encouraging. However, the level of agreement is fortuitous; it appears that the neglect in the trajectory calculations of quantum mechanical tunnelling associated with motion along the reaction coordinate is very nearly balanced by the neglect of quantisation of the motions orthogonal to the reaction coordinate. In most systems, both effects will be smaller than in H_3 so there is reason to hope that QCL trajectories provide a good estimate of dynamic quantities.

A second reason to be encouraged comes from comparisons of experimental results on isotopically related reactions. Infrared chemiluminescence experiments (see Chapter B.2) yield the relative rates at which the $F + H_2$, $F + D_2$ reactions populate excited rovibrational ($v'j'$) levels of the HF, DF products. The energy partitioning can be represented on triangular plots like those in Fig. A2.7. These diagrams are prepared by deriving a supposed continuous distribution of molecules over classical internal energies from the actual distribution over quantised v', j' states. This procedure is the reverse of the 'binning' (see section A2.5) used to find quantum state distributions from the results of QCL trajectory calculations, and the similarity of the two plots in Fig. A2.6 increases one's faith in that method.

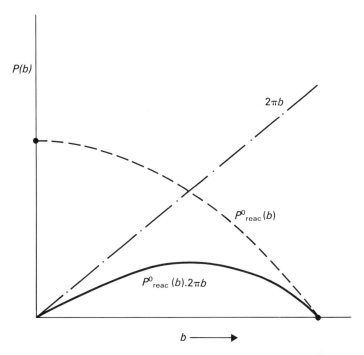

Fig. A2.6. Comparison of the reaction cross-sections for $H + H_2$ $(v = 0, j = 1) \rightarrow H_2 + H$ from three-dimensional quantum mechanical scattering calculations[14] ($\bigcirc$) and three-dimensional quasiclassical trajectories.[6] $(E - V^*)$ is the total energy in excess of the barrier height V^*; E_t is the translational energy.

To summarise: QCL trajectory calculations appear to give a reasonably good description of molecular collision dynamics but one should be cautious about accepting the results in 'threshold' or 'classically forbidden' regions: for example, close to the threshold collision energy for reaction or near the edges of a distribution over quantum states.

A2.9 Results and conclusions from QCL trajectory calculations

Quasiclassical trajectory calculations were first performed about 20 years ago and, with the increasing availability of high speed computers, they can now be carried out quite routinely. The major achievements of the method have arisen through its ability to relate general properties of potential energy surfaces to specific aspects of reaction dynamics. The work of Polanyi and his coworkers has been especially important showing, for example, that the existence of an early barrier to reaction (see Fig. A3.6) leads to attractive energy release and high excitation of the vibrational motion in the newly formed molecule, whereas a late barrier leads to energy being released as the products repel one another and high translational excitation.[4,16]

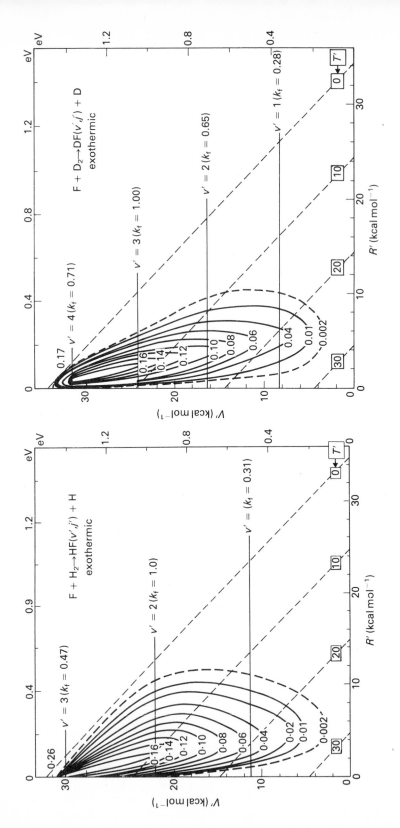

Fig. A2.7. Triangular plots showing product energy distributions. The contour lines join states which have an equal probability of being populated in the thermal reactions. Vibrational and rotational energies (V' and R') are plotted along rectilinear axes ($1\,\mathrm{kcal\,mol^{-1}} \equiv 4.184\,\mathrm{kJ\,mol^{-1}}$). Because the energy released to the products is approximately constant, values of the final relative translational energy can be represented by the dashed lines. [Reproduced, with permission, from Polanyi J. C. and Woodall K. B (1972) *J. Chem. Phys.*, **57**, 1574].

Trajectory calculations on particular chemical systems have been less valuable. This is mainly because of the difficulty of establishing a reliable PES either from *ab initio* calculations or from available experimental data. The difficulty is exemplified by efforts to characterise the dynamics of collision between $F + H_2$ and $H + FH$. It is now clear[3] that many QCL trajectory calculations have been carried out on PES which are known to be incorrect.

However, trajectory calculations do appear to have established that different regions or properties of the PES are responsible for different reaction attributes. Thus, the location of the saddle-point and the form of the minimum energy path from there to products largely determines how energy is partitioned between translational and vibrational motions (the 'specificity of energy release'), whereas the form of the path from reagents to the saddle point decides reagent energy 'usage' (the 'selectivity of energy consumption'), that is whether reaction is preferentially promoted by translational or vibrational excitation of the reagents. Finally, the barrier height and its dependence on the $A\hat{B}C$ angle controls values of the reaction cross-section and rate constants.

A2.10 Suggestions for further reading

Kuntz P. J. (1976) In *Dynamics of Molecular Collisions, Part B* (Ed. by W. H. Miller), Chap. 2. Plenum Press, New York.

Porter R. N. (1976) In *Dynamics of Molecular Collisions, Part B* (Ed. by W. H. Miller), Chap. 1. Plenum Press, New York.

Polanyi J. C. and Schreiber J. L. (1974) In *Physical Chemistry: An Advanced Treatise*, Vol. VIA, Chap. 6 (Ed. by H. Eyring, D. Henderson and W. Jost). Academic Press, New York.

Smith I. W. M. (1980) *Kinetics and Dynamics of Elementary Gas Reactions*. Butterworths, London.

Kuntz P. J. (1979) In *Atom–Molecule Collision Theory* (Ed. by R. B. Bernstein), Chap. 3. Plenum Press, New York.

Truhlar D. G. and Muckerman J. T. (1979) In *Atom–Molecule Collision Theory* (Ed. by R. B. Bernstein), Chap. 16. Plenum Press, New York.

Raff L. M. and Thompson D. L. (1985) In *Theory of Chemical Reaction Dynamics III* (Ed. by M. Baer), CRC Press, Boca Raton, Florida.

A2.11 References

1 Liu B. (1973) *J. Chem. Phys.*, **58**, 1925; Liu B. and Siegbahn P. (1978) *J. Chem. Phys.*, **68**, 2457.

2 Truhlar D. G. and Horowitz C. J. (1978) *J. Chem. Phys.*, **68**, 2466 and (1979) **71**, 1514.

3 Schaeffer III, H. F. (1985) *J. Phys. Chem.*, **89**, 5336.

4 Kuntz P. J., Nemeth E. M., Polanyi J. C., Rosner S. D. and Young C. E. (1966) *J. Chem. Phys.*, **44**, 1168.

5 Porter R. N. and Karplus M. (1964) *J. Chem. Phys.*, **40**, 1105.

6 Karplus M., Porter R. N. and Sharma R. D. (1965) *J. Chem. Phys.*, **43**, 3259.

7 Muckerman J. T. (1971) *J. Chem. Phys.*, **54**, 1155.

8 Smith I. W. M. (1977) *Chem. Phys.*, **20**, 437.

9 Faist M. B., Muckerman J. T. and Schubert F. E. (1978) *J. Chem. Phys.*, **69**, 4087; Muckerman J. T. and Faist M. B. (1979) *J. Phys. Chem.*, **83**, 79.
10 Smith I. W. M. (1982) *J. Chem. Educ.*, **59**, 9.
11 Anderson J. B. (1973) *J. Chem. Phys.*, **58**, 4684 and (1974) **60**, 2566 and (1975) **62**, 2446;
   Henry J. M., Anderson J. B. and Jaffe R. L. (1973) *Chem. Phys. Letters*, **20**, 138; Anderson J.
   B. (1974) *J. Chem. Phys.*, **61**, 3390; Jaffe R. L., Henry J. M. and Anderson J. B. (1976) *J. Amer.
   Chem. Soc.*, **98**, 1140, and (1973) *J. Chem. Phys.*, **59**, 1128.
12 Smith I. W. M. (1981) *J. Chem. Soc. Faraday Trans.* 2, **77**, 747; Frost R. J. and Smith I. W. M.
   (1987) *Chem. Phys.*, in press.
13 Schlier Ch. (1983) In *Energy Storage and Redistribution in Molecules*, (Ed. by J. Hinze),
   Plenum Press, New York.
14 Schatz G. C. and Kuppermann A. (1976) *J. Chem. Phys.*, **65**, 4642 and 4668.
15 Polanyi J. C. and Woodall K. B. (1972) *J. Chem. Phys.*, **57**, 1574.
16 Polanyi J. C. (1972) *Acc. Chem. Res.*, **5**, 161.

A2.12 Problems

Quasiclassical trajectory methods

A2.1. The thermally averaged rate constant for a reaction is related to the reaction
cross-section or excitation function, $S_{reac}(u)$, by the equation

$$k(T) = \int_0^\infty u.S_{reac}(u) . P(u) \, du$$

where $P(u) \propto u^2 \exp(-\mu u^2/2k_B T)$.

Therefore using Monte Carlo methods one draws from a velocity distribution
in which

$$P'(u) \propto u^3 \exp(-\mu u^2/2k_B T).$$

(i) What is the most probable value of u from this distribution?

(ii) What is the statistical probability of having any particular value of u
relative to the probability of having the most probable velocity from this
distribution?

A2.2. An extended version of simple collision theory in which steric effects are
approximately included[10] suggests that the probability of reaction at a particular collision energy E varies with impact parameter b according to:

$$P(b, E) = [E(1 - b^2/D^2) - E^0]/2E'$$

(Here, E^0 and $E^0 + 2E'$ are the minimum energies for collinear and 'sideways'
reactions.)

(i) What is the value of b at which reaction is most likely to happen?

(ii) How could one use the result from extended collision theory to incorporate 'importance sampling' into a quasiclassical trajectory study of a reaction?

A2.13 Answers to problems

Quasiclassical trajectory methods

A2.1. (i) $\dfrac{d}{du}[u^3 \exp(-\mu u^2/2k_B T)]$

$= [3u^2 - u^3 \mu u/k_B T] \exp(-\mu u^2/2k_B T)$

for $u_{m.p.}$ $[3u^2 - u^3 \mu u/k_B T] = 0$

i.e. $u_{m.p.} = (3k_B T/\mu)^{1/2}$.

(ii) Substitution into $u^3 \exp(-\mu u^2/2k_B T)$

yields $(3k_B T/\mu)^{1.5} \exp(-1.5)$

$\therefore$ Probability of any other velocity relative to probability of the most probable velocity is

$$P_{rel}(u) = u^3 (3k_B T/\mu)^{-1.5} \exp\{1.5 - [\mu u^2/2k_B T]\}.$$

A2.2. Probability of reaction at any particular impact parameter b is proportional to the product of $2\pi b$ and $P(b, E)$. Then differentiate with respect to b.

(i) $\partial[2\pi b\, P(b, E)]/\partial b = 2\pi[E(1 - b^2/D^2) - E^0]/2E' - 4\pi b^2 E/D^2 2E'$

Equate the right-hand side to zero, then,

$[E(1 - b^2/D^2) - E^0] = 2b^2 E/D^2$.

$E - E^0 = 3b^2 E/D^2$.

So the value of b at which reaction is most probable is given by

$b = D[(E - E^0)/3E]^{1/2}$.

(ii) One might

(a) use the extended collision theory to estimate a value of b_{max}, the maximum impact parameter to be used in the trajectory calculations.
To do this, put $E(1 - b^2 - D^2) - E^0$ equal to zero

$\rightarrow b_{max} = D[(E - E^0)/E]^{1/2}$.

(b) Secondly, one could sample impact parameters not from a $P(b) \propto b$

distribution but from a

$$2\pi b[E(1-b^2/D^2) - E^0]2E'$$

distribution. This would require a 'game-of-chance' routine. Furthermore, the result of any 'successful' trajectory would not be given the weight 1 in the final result but that of

$$\frac{2\pi b[E(1-b^2/D^2) - E^0]/2E'}{2\pi b}$$

i.e., the ratio of the probability of the choice of b in the 'biased' sample to that in the unbiased statistical case.

Chapter A3
Applications of Theory to Bimolecular Reactions

G. HANCOCK

A3.1 Introduction: Reaction dynamics

The detailed study of a bimolecular reaction—the ways in which energy (translational, internal, chemical) in the reactants is converted into that of the products—is the province of the reaction dynamicist with his armoury of molecular beam and laser techniques. Experiments on state-to-state reaction dynamics are considered in Chapter B2. Here we concentrate on the interpretation of molecular beam experiments in terms of the forces (potential energy surfaces) governing the molecular interactions, and review some key experiments which point towards the type of information on these interactions which can be extracted from observations of energy distributions in the products.

A3.2 Molecular beam experiments on reactive scattering

Molecular beam experiments can be conceived in which both internal and translational energies of reactants and products are selected. Most experiments using beams are limited to selection of initial directions and velocities of reactants, and measurements of velocity-resolved angular distributions of products. In these cases the pertinent information on the dynamics of scattering can be judged by looking at a contour plot of the product velocities, measured relative to the centre-of-mass of the collision pair. Such velocities are not measured directly in the laboratory, and transformations from the LAB to CM frame must first be carried out by methods which are tedious but standard.[1]

Figure A3.1 shows the type of result obtained for the reaction between K atoms and Br_2 molecules, and illustrates the scattering pattern observed for one particular type of reaction, that characterised by forward scattering, where the product containing the new bond (KBr) is scattered in a forward direction with respect to the centre-of-mass velocity of the incoming atom (K). K and Br_2 approach each other head on (this is the picture that an observer would see if he were sitting on the centre-of-mass of the system, taken as the origin in Fig. A3.1; remember that the velocity of the centre-of-mass remains constant during the collision event and this can be subtracted from the laboratory behaviour before and after the collision in order to allow the dynamics to be clearly visualised). The KBr product velocity

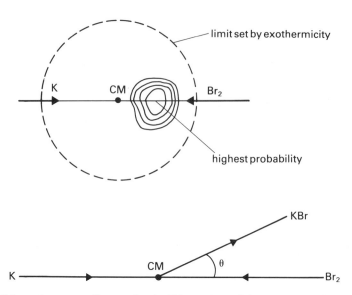

Fig. A3.1. Schematic contour diagram for the KBr product of the $K + Br_2$ reaction. Reactants approach each other head on in the centre-of-mass (CM) frame. With the position of the CM at the origin, the dashed circle shows the maximum value of the KBr CM velocity, determined from the reaction energetics. The solid contour lines show the observed KBr velocities, illustrating that forward scattering occurs (with respect to the K beam) and that the majority of the energy liberated in the reaction appears as internal excitation in the products.

is measured as a function of scattering angle from the initial straight line approach of the centre-of-mass velocities of K and Br_2, and is plotted as a contour diagram, with the height of the contour being proportional to the probability that a reactive event will end up with the KBr product velocity (a vector quantity having both magnitude and direction), with a value given by its coordinates. The dashed circle in Fig. A3.1 represents the maximum possible velocity of the KBr product, determined purely by energy conservation, i.e. the centre of mass velocity of KBr if *all* of the available energy appears as translation in the fragments.

The results for $K + Br_2$ show that the KBr product is scattered forward with respect to the initial direction of the K atom, and that the most probable translational energy is considerably smaller than the maximum possible—implying that most of the available energy appears as internal excitation in the products. These two features, plus the added observation that the reaction cross-section (proportional to the reaction probability) is large ($50–200\ \text{Å}^2$), typify what is known as a stripping reaction, taking place by what has been called the harpoon mechanism. Figure A3.2 illustrates an explanation for two of these features, the magnitude of the reaction cross-section and the forward scattering dynamics. As neutral K and Br_2 approach, the normal potential energy curve showing their interaction would be expected to show a small van der Waals minimum and then rise at short internuclear distance R. However, this curve will be crossed, at an internuclear separation R_c, by an ionic curve showing the Coulombic attraction between K^+ and Br_2^-. Initially, at large R, the ionic curve lies above that for the neutral species (the ionisation energy, I_K, of K is greater than the electric affinity, E_{Br_2}, of Br_2), but naturally this curve falls at smaller but still appreciable values of R due to the Coulombic attraction between the oppositely charged species. At the critical distance R_c, the difference $I_K - E_{Br_2}$ is just balanced by the Coulombic interaction $e^2/4\pi\varepsilon_0 R_c$ (in this simple treatment we neglect dispersion forces). This leads to an estimate for R_c (~ 6 Å in this case) in good accord with the experimental reaction cross-section, if the latter is assumed to be equal to πR_c^2. What is envisaged is that at R_c, the distance at which the electron transfers, Coulombic attraction leads automatically to reaction; the Br^-—Br bond is broken, and K^+ carries off a Br^- ion with no significant repulsion between the bromine atom and the bromide ion.

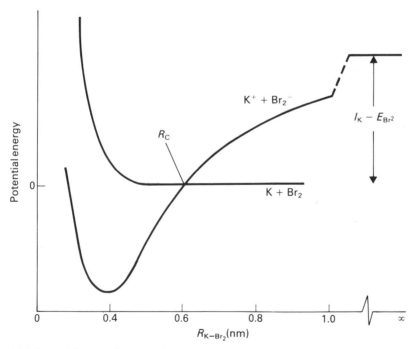

Fig. A3.2. Potential energy diagram to illustrate the harpoon model for the $K + Br_2$ reaction. Two schematic one-dimensional diabatic potential energy curves are shown, for the ionic $(K^+ + Br_2^-)$, and covalent $(K + Br_2)$ interactions, respectively. At large internuclear separation, the ionic curve lies above the covalent: the diabatic curves cross at R_c where the difference between the ionisation energy of K and the electron affinity of Br_2 is matched by the Coulombic attraction between the ions. Switching of potential curves at R_c leads to reaction with cross section $\sim \pi R_c^2$.

Thus the product is scattered in the direction of the incident K beam. This is the so-called harpoon model (the K atom throws the harpoon—the electron), with stripping (i.e. forward scattering) dynamics.

The observation that the majority of the available energy in the $K + Br_2$ reaction goes into internal excitation of the products might suggest an experiment designed to measure directly the distribution of this energy amongst the rotational and vibrational degrees of freedom of KBr. Laser induced fluorescence (see Chapter B1) would seem an obvious choice, but cannot be applied straightforwardly in this case, as KBr does not have a suitable bound upper state to be excited by the technique. Evidence for the extent of internal excitation has however been found by subsequent reactions of the KBr product. The process

$$KBr + Na \rightarrow K^* + NaBr$$

where K^* is the first excited ^2P multiplet is $170 \, kJ \, mol^{-1}$ endothermic, and thus could only occur with considerable excitation in the KBr reactant. The $190 \, kJ \, mol^{-1}$ exothermicity of the $K + Br_2$ reaction can provide this excitation: emission from $K^*(^2P \rightarrow ^2S)$ in a reacting mixture of K, Br_2 and Na has been observed and used to confirm the interpretation of the reactive scattering measurements.

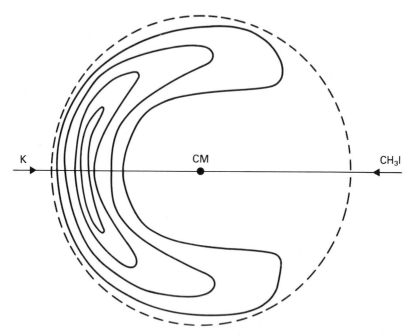

Fig. A3.3. Schematic contour diagram for the KI product of the $K + CH_3I$ reaction. Backward scattering with respect to the K beam occurs, with the majority of the available energy appearing as product translational energy.

A second type of scattering behaviour is typified by the process

$$K + CH_3I \rightarrow KI + CH_3.$$

Figure A3.3 shows a centre-of-mass contour diagram for the KI product. Here backward scattering is apparent, with a large fraction of the available energy appearing as product translation. Furthermore, the reaction cross-section is considerably smaller than in the $K + Br_2$ case (typically $< 50\,\text{Å}^2$ for a backward scattered process), and the dynamics imply that strong interactions only take place when reactants are closer to each other than in the $K + Br_2$ system.

A second example of 'rebound' dynamics, illustrating interesting behaviour, is shown in Fig. A3.4 for the reaction $F + D_2$. The DF product is back scattered, but in this case 'islands' of product intensity at specific values of the translational energy are seen, corresponding to the formation of vibrationally excited DF. There appears to be little rotational excitation in the product: if there was, then the vibrational states would not appear as resolved peaks in the translational energy distributions. The observations can be explained by considering the way in which product rotation arises in such a system. Conservation of angular momentum in the collision tells us that the angular momenta of the reactants and products must be equal. Two contributions towards the reactant angular momentum can be identified, namely that due to relative motion of the two species, l, and that due to rotational motion of the diatomic, j. In magnitude, $l = \mu u b$, where μ is the reduced mass, b the impact parameter and u the relative velocity. The first of these quantities is small for the $F + D_2$ system. The magnitude of j is also small for the widely spaced D_2 rotational levels populated at room temperature. The result is that there is little angular momentum in the reactant system, hence little to

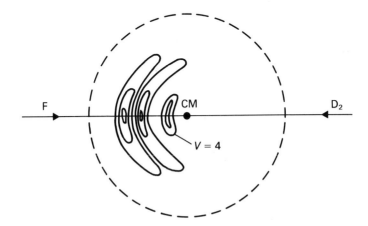

Fig. A3.4. Schematic contour diagram for the DF product of the $F + D_2$ reaction. DF is backscattered, and 'islands' of product intensity are seen corresponding to the formation of DF with up to four quanta of vibrational energy.

partition in the products and thus low rotational energy in the DF fragment. Reactions of this kind, in which internal state resolution is available via translational energy distributions, are rare but important, in that spectroscopic means of quantum state detection (by infrared chemiluminescence from the DF product in this case) can be used to check the results.

A third type of product scattering is observed in the reaction

$$Cs + SF_6 \rightarrow CsF + SF_5,$$

and Fig. A3.5 shows the contour diagram. Scattering here is seen to be symmetric in the forward and backward directions. What occurs is that a $CsSF_6$ collision complex is formed, and this lives for many rotational periods before breaking up into products, which show symmetric scattering about $\theta = 90°$. We shall see shortly that reactions of this kind can be treated theoretically in a relatively straightforward way to predict the distribution of energy in the products: a statistical distribution of some kind should appear if the complex lives long enough to allow equilibration of its internal degrees of freedom.

A3.3 State-to-state kinetics and potential energy hypersurfaces

Much information has now been gathered on the detailed dynamics of reactive scattering from the kind of results outlined above for molecular beam experiments, and from spectroscopic observations of product quantum states described elsewhere in Chapter B1. More details can be found in the general references[1-3] given at the end of this chapter. What controls the outcome of a chemical reaction is clearly the forces between individual atoms, represented by a potential energy

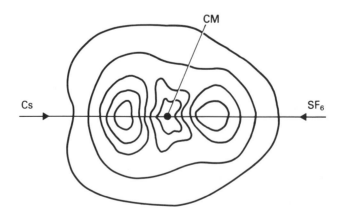

Fig. A3.5. Schematic contour diagram for the CsF product of the $Cs + SF_6$ reaction. Both forward and backward scattering of the CsF product are seen, corresponding to the formation of a long-lived complex.

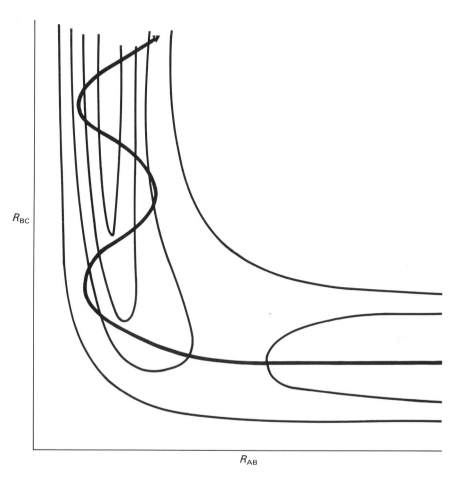

Fig. A3.6. Attractive potential energy surface for the reaction $A + BC \rightarrow AB + C$. The activation energy barrier occurs in the entrance valley, with energy starting to be released at large R_{AB} and before R_{BC} has changed appreciably from its equilibrium value. Energy is channelled efficiently into vibrational excitation of the AB product, as shown by the trajectory in the exit valley.

surface (in an appropriate number of dimensions). Calculations of trajectories on a variety of different empirical surfaces have enabled some general conclusions to be drawn about the relationship between surface features and the scattering dynamics. Figure A3.6 illustrates one such case, for the reaction

$$A + BC \rightarrow AB + C$$

where reactants approach in a collinear fashion. The (small) activation energy barrier for this surface is seen to occur at a quite large value of R_{AB}, the distance between atoms A and B, and at a value of R_{BC} which is little different from its equilibrium value in the isolated molecule. This is a so called 'early barrier'; energy starts to be released before R_{BC} changes appreciably, and trajectory calculations on the surface show that the outcome of the reaction is that energy is efficiently released as vibration of the AB product. This is termed 'attractive' energy release, and Fig. A3.6 shows what is generally described as an attractive potential energy surface. A second type of energy release, repulsive, occurs when the activation barrier is not in the 'entrance valley' as in Fig. A3.6, but in the 'exit valley', i.e. after the trajectory has turned the corner, and the B—C bond is being extended. Energy in this case is transferred more efficiently into translational motion of the products. Mixed energy release takes place by B—C repulsion whilst the nascent A—B bond is still extended from its equilibrium position. The extent of this release depends not only on the surface characteristics but also on kinematic effects due to the relative masses of the atoms.[2]

Intuitively it is satisfying to think of trajectories on a potential energy surface being represented by the motion of a point mass sliding over the topographical features in the way that a small sphere would move on a three-dimensional model of the surface. Figure A3.6 does not quite represent this motion correctly: in order to do this the axes need to be scaled and skewed, and the effect of skewing is shown in Fig. A3.7. The skew angle, χ, depends upon the relative masses of the atoms A, B and C and is given by equation 6 of Chapter A1:

$$\chi = \tan^{-1}[m_B(m_A + m_B + m_C)/m_A m_C]^{1/2}.$$

Figure A3.7 shows a trajectory on a surface with small χ appropriate for a mass combination of heavy (H) and light (L) atoms reacting according to

$$H + LH \rightarrow HL + H.$$

In Fig. A3.7 the surface that is illustrated is largely repulsive ('late barrier' in the exit valley), yet considerable vibrational excitation occurs in the AB product because of the kinematic effect of the point mass rounding the tight turn near the activation barrier. Such dynamical mass effects apparently operate in the reactions:

$$Cl + HI \rightarrow HCl + I,$$

and

$$H + Cl_2 \rightarrow HCl + Cl$$

for which the values of χ are 10.7° and 83°, respectively. Both surfaces are probably 'repulsive', yet in the first of these considerably more energy is converted into HCl vibration than in the second.

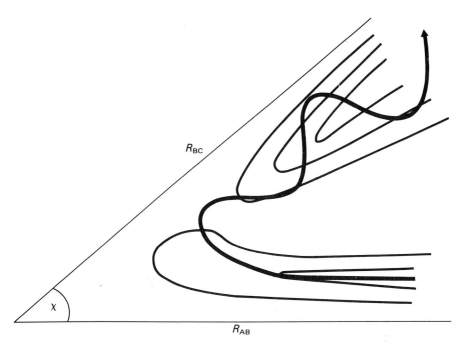

Fig. A3.7. The effect of skewed axes. A predominantly repulsive surface can channel energy efficiently into vibration in the products due to the nature of the trajectory turning the tight corner as energy is released.

Classical mechanics is, for most cases, satisfactory in determining trajectories on potential energy surfaces. Such trajectories can of course run backwards as well as forwards, and thus should be able to give information on the way in which reactant excitation can overcome potential energy barriers from observations of product excitation in the reverse reaction. For example, the reaction

$$Cl + HBr \rightarrow HCl(v') + Br \qquad \Delta H_0^{\ominus} = -69 \, kJ \, mol^{-1}$$

produces vibrationally excited HCl in a population inversion with relative populations for $v = 0 : 1 : 2$ of $0.6 : 1.0 : 0.4$. Vibrational excitation in $HCl(v)$ should thus be efficient in overcoming the activation barrier for the reverse (endothermic) process

$$Br + HCl \overset{k_v}{\rightarrow} HBr + Cl \qquad \Delta H_0^{\ominus} = 69 \, kJ \, mol^{-1}.$$

Specific rate constants k_v for this process have been measured, in experiments in which $HCl(v)$ was produced by prereaction of Cl atoms with HI^4 or by direct optical excitation.[5] What is of dynamical interest is how efficiently a specific form of energy is used to overcome a barrier—vibrational excitation of a reagent increases the total amount of energy available, but would an increase in rate constant also have occurred if the same amount of energy was present in relative translation of the reactants? A case where the different efficiencies of these two forms of energy have been clearly demonstrated is for the reaction

$$K + HCl(v) \rightarrow KCl + H. \qquad \Delta H_0^{\ominus} \simeq 4 \, kJ \, mol^{-1}.$$

Excitation of HCl to $v = 1$ increases the total cross-section for this reaction by about two orders of magnitude over that for the ground state reagents.[6] When the same amount of energy was put into the translational degrees of freedom of the reactants, the cross-section again increased, but by far less than that observed for vibrational excitation.

The effects of reagent rotation on reaction dynamics have been less studied than those for vibration or translation. For the $K + HCl$ ($v = 1, j$) reaction, the cross-section is seen to decrease with increasing j.[7] Calculations on the $Li + HF$ reaction show an initial decrease of S_{reac} with increasing j, attributed to a preferred orientation of the reactions being disrupted as the rotational frequency increases, then followed by an increase in S_{reac}.[8] The latter effect was thought to be due to centrifugal distortion at high j increasing the HF bond distance and thus facilitating H atom abstraction.

Other examples of the effects of reagent excitation appear in Chapter B2, and have been comprehensively reviewed.[1,9] Space precludes their discussion here, except for a final example. This concerns the effect of reagent rotation on the product states of the reaction[10]

$$Ca + HF(v = 1, j) \rightarrow CaF(v'j') + H.$$

In this study it was found that increase in rotational quantum number, j, in HF simply seemed to have the effect of increasing the total energy in the system so that it could be distributed statistically in the internal states of the CaF product; as each CaF(v') state became energetically accessible, it was formed. Energy distributions for reactions of this kind can be calculated with (almost) no knowledge of the potential energy surface controlling the reaction: what is required is that an intermediate 'complex' is formed which allows internal energy to be redistributed statistically amongst degrees of freedom of the products. Calculations can be performed at various levels of sophistication.

(a) Calculation of the prior distribution,[1] in which the probability of forming given product states is proportional to the statistical weight of the channel, i.e. for a rotational level j' the probability is proportional to $(2j' + 1)$, for a given translational energy E the probability is proportional to $E^{1/2}$.

(b) Phase space theory[11] (see section A4.8), in which the prior distribution is modified to allow for angular momentum conservation in the products—this needs to include some information on the long range attractive part of the (one-dimensional) intermolecular potential between the separating fragments.

(c) Statistical adiabatic channel theory[12] (see section A4.7), which is like phase space theory except that the angular part of the intermolecular potential is also taken into account in deciding the way in which bending vibrations in the complex are converted to angular momentum.

Statistical theories such as these have been applied in some detail to the products of photodissociation and these applications have been recently reviewed.[13]

A3.4 Reagent alignment and orientation

Energy in reagents can overcome an activation energy barrier to reaction: intuitively it would seem reasonable to assume that the relative orientation of reagents, or the alignment of vector quantities associated with them (their rotational angular momentum for example) would also influence the dynamics. Partial orientation of the nuclear framework has been possible in molecular beam experiments, for example in the reaction

$$Rb + CH_3I \rightarrow RbI + CH_3.$$

Here the Rb prefers to attack the I end of the molecule[14]: attack within a cone of $45°$ angle at the C end of the molecule produces essentially no reaction. Care should be taken in formulating generalisations about such 'steric factors' however; no such asymmetry effect has been seen in the $K + CF_3I$ reaction[15] where immediate intuition would predict it. More recently orientation of reagents has been achieved by first forming a van der Waals complex between two molecules, then photolysing one of them to propel a fragment in a well-defined direction towards the other. The reaction

$$H + CO_2 \rightarrow OH + CO$$

has been studied in this way,[16] with photolysis of the HBr component of a weakly bound CO_2HBr complex providing the initial reagent orientation.

The absorption of radiation is an anisotropic process and can be used to produce species whose transition moments are aligned in the laboratory frame. For molecules, the transition moments are linked to the symmetry axes, and thus to the molecular framework, and absorption has been used to prepare species with some degree of alignment with respect to incoming reagents. An example is in the reaction

$$Sr + HF(v = 1) \rightarrow SrF + H$$

where absorption of polarised light from a laser was used to prepare $HF(v = 1)$ molecules so that their collisions with a beam of Sr atoms occurred either predominantly 'head on' or 'broadside'.[17] The latter configuration promoted reaction into specific quantum states of SrF more efficiently.

Orbital (as opposed to nuclear framework) alignment has also been used to prepare reagents. An example is in the reaction

$$Ca(^1P_1) + HCl \rightarrow CaCl(B^2\Sigma^+, A^2\Pi) + H.$$

Polarised light was used to excite the $Ca(^1P_1 \leftarrow ^1S_0)$ transition so that the p orbital lobes were aligned either along or perpendicular to the relative velocity vector of the reagents.[18] Some of the resultant emission from the excited CaCl states was found to depend upon this alignment, and a mechanism involving an electron jump model was able to reproduce qualitatively the observed effects.

Finally, we note that studies of alignment in the fragments of a 'half collision', i.e., photodissociation, are starting to provide an almost unprecedented amount of detail on the dynamics of an elementary process. As an example, we consider the photodissociation of HONO.[19] Quantum state-resolved observations of the OH product yields scalar information on the distribution of available energy into the degrees of freedom of the products, and vector information on the fragment recoil with respect to the transition moment, including orbital alignment, as revealed by the relative populations of the OH Λ doublets. Experiments of this kind on bimolecular collisions can be expected.

A3.5 Applications of theory to thermal rate data

In the remainder of this chapter, we consider how measurements of thermal rate data may be compared with the values of rate constants which are predicted theoretically. As Chapter A2 described, if the form of the potential energy surface is known, trajectories on the surface can be calculated; appropriate averaging for the range of collisions that are to be expected can be incorporated to produce a cross section, and thus a thermally averaged rate constant for comparison with measured data can be obtained. Naturally this approach also produces distributions of energy in the fragments (because the trajectories are classical, these have to be converted, as described in section A2.5, into suitable quantised distributions), and these can also be compared with the results of molecular dynamics experiments. For some simple reactions this has been done. For the simplest of all reactions

$$H + H_2 \rightarrow H_2 + H$$

an *ab initio* potential surface exists,[20] trajectory calculations have been run to produce cross-sections,[21] and during the last few years, quantum state-resolved product detection has been possible and experimental results for (v', j') distributions in the HD product of

$$H + D_2 \rightarrow HD + D$$

have been compared with those predicted theoretically.[22]

What happens if we do not have an *ab initio* potential energy surface to hand (which is certainly the case most of the time), and our main aim is understanding the magnitudes of thermal rate constants rather than quantum state distributions? We may apply transition state theory as outlined in Chapter A1. A bimolecular rate constant for the process

$$A + BC \xrightarrow{k} AB + C$$

is given by the expression

$$k = \left(\frac{k_B T}{h}\right) \frac{Q^{\neq}_{ABC}}{Q_A Q_{BC}} \exp\left(-\frac{\Delta E_0^{\neq}}{RT}\right) \tag{1}$$

where the Q's represent partition functions of the transition state and reactants, and $\Delta E_0^{\neq}$ is the appropriate zero-point energy difference between them. If we have a potential energy surface available, then k can be calculated—for example, $Q^{\neq}_{ABC}$ can be found by locating the transition state (as described in Chapter A.1), and thus obtaining its geometrical parameters, and by expanding the potential energy V around the transition state to obtain force constants, $k = -\partial^2 V/\partial y^2$, for motions orthogonal to the reaction coordinate and thus vibrational frequencies $\nu = (1/2\pi)(k/\mu)^{1/2}$. The correct surface will naturally reproduce the correct value of $\Delta E_0^{\neq}$ and thus the observed activation energy E_{act} for the process. Generally we need to proceed with approximate surfaces (or representations of the way the potential energy varies along the reaction coordinate) and use adjustable parameters in them to reproduce the experimentally observed activation energy, E_{act}. One such surface is the LEPS as described earlier (sections A1.2 and A2.2): we briefly mention below a second approach, the Bond-Energy-Bond-Order (BEBO) method which has been used for H atom transfer reactions involving linear transition states. Transition state calculations carried out using these approximate methods and an *ab initio* surface will then be compared with a series of experimental results.

The BEBO method[23] considers the electronic energy $V(R_1, R_2)$ along the path of minimum energy for the reaction

$$A + HB \rightarrow A \ldots \ldots H \ldots \ldots B \rightarrow AH + B$$

$$R_1 \qquad R_2$$

as being given by

$$V(R_1, R_2) = -E_1(AH) - E_2(HB) + E_3(AB)$$

where E_1 and E_2 are bonding interactions, and E_3 is an anti-bonding interaction: in E_1 and E_2 electron spins are anti-parallel, whereas they are parallel in E_3. For $E_1(AH)$, Pauling's empirical relationship between bond order n and bond length r is used:

$$E_1 = E_{1s} n^p$$

$$R_1 = R_{1s} - 0.26 \ln n$$

where E_{1s} is the bond energy for the isolated AH species of bond length R_{1s}, n is the bond order ($n = 0$ for no bond, $n = 1$ is for $R_1 = R_{1s}$, i.e. a complete AH diatomic bond) and p is an empirical constant.

It is assumed that the total bond order is one at all points along the reaction path, so that the HB bond energy is given by

$$E_2 = E_{2s}(1 - n)^{p'}.$$

$E_3(AB)$ is an 'anti-Morse' function (see equation 3b in Chapter A2.2) representing the AB (triplet) repulsion energy. When empirical values of p and p' are assumed, then $V(R_1, R_2)$ can be evaluated for a number of values of n between 0 and 1 (i.e., as the reaction proceeds); the maximum value of $V(R_1, R_2)$ represents the activated complex. In practice, the thermal rate constants thus calculated will not generally have the right value of E_{act}, and the parameters p and p' are adjusted to rectify this.

Calculations of rate constants using the transition state theory approaches outlined above have been carried out for a number of systems, and in reference 24 a comparison has been made of experiment and theory for one of these, the reaction

$$OH + H_2 \rightarrow H_2O + H.$$

In particular, the results of calculations on LEPS, BEBO and an *ab initio* surface were compared. Figure A3.8 shows the geometry of the HOHH transition state, with Table A3.1 indicating the differences between the molecular geometries found in the semi-empirical[25] and *ab initio*[26] approaches. In the former case, α, the H—O—H angle, was assumed to be as in the H_2O molecule, and β, the O—H—H angle, was assumed to be 180°: the *ab initio* results predict slightly different values for these angles, but bond lengths are in good agreement. What is of importance is the transition state predictions for the rate constants. In the semi-empirical case,

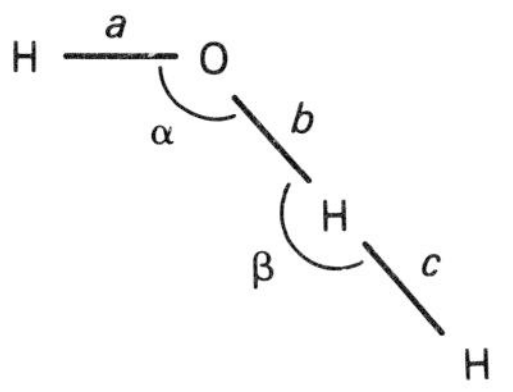

Fig. A3.8. HOHH transition state geometry for calculation on semi-empirical and *ab initio* surfaces. *a*, *b* and *c* are bond lengths, α the HOH bond angle, and β the OHH out-of-plane bond angle

calculations were made to fit the experimentally observed activation energies by varying the adjustable parameters (e.g., p and p' in the BEBO case). For the *ab initio* surface the correct activation energy should emerge, and this was found to be the case provided that $Q_{ABC}^{\neq}$ was evaluated accurately, not just in the rigid rotor harmonic oscillator approximation, and that a tunnelling correction was included.[26] Figure A3.9 shows the final results.[24] The experimentally determined rate constant for the reaction can be represented by the expression

$$k = 1.66 \times 10^{-16} \, (T/\text{K})^{1.6} \exp(-1660\text{K}/T) \text{ cm}^3 \text{ molecule}^{-1} \text{ s}^{-1}$$

over the temperature range 300–2500 K, and both *ab initio* and semi-empirical calculations reproduce this functional form very well. However, caution is required in interpreting the results of the semi-empirical approach:[24] the frequencies and geometries may not truly represent the transition state (other structures may give the same rate constants) and the approximations made in the theory may give large inaccuracies. Methods of locating the transition state in both semi-empirical and *ab initio* surfaces, and of calculating its partition functions (including, for example, the effects of vibrational anharmonicities) have been reviewed.[27]

A3.6 Non-linear Arrhenius behaviour

As Fig. A3.9 shows, the Arrhenius plot for the $OH + H_2$ reaction is distinctly curved, i.e., the activation energy [as defined by $-\partial(\ln k)/\partial(1/RT)$] is temperature dependent. We examine several possible causes of this.

Table A3.1. Calculated geometries for HOHH transition state

	BEBO[i]	LEPS[i]	*ab initio*[ii]
$a/\text{Å}$	0.96	0.96	0.98
$b/\text{Å}$	1.24	1.21	1.33
$c/\text{Å}$	0.85	0.85	0.85
α	104.6	104.6	97.6
β	180	180	165

(i) Reference 24, (ii) reference 26.

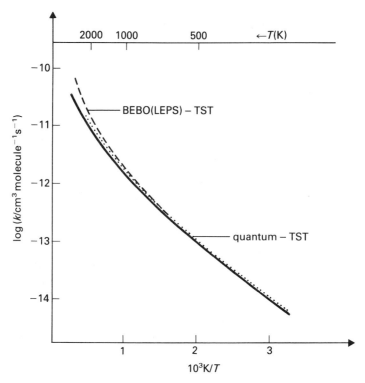

Fig. A3.9. Arrhenius plot for the $OH + H_2 \rightarrow H_2O + H$ reaction. The solid curve shows experimental results; the dashed curve (indistinguishable from the experimental results on this scale below ~ 600 K) is the semi-empirical transition states calculation using BEBO and LEPS surfaces; the dotted line is the *ab initio* calculation with a tunnelling correction. Adapted from reference 24 and reproduced with permission.

(i) State-specific rate constants

Enhancement of the cross-section of a chemical reaction by specific use of internal or translational energy in the reactants has already been mentioned. For the $OH + H_2$ reaction, rate constants have been measured for $H_2(v = 0)$ and $(v = 1)$, both yielding Arrhenius type expressions, but with different A factors and activation energies. As the temperature increases, and thus the contribution from reaction from $H_2(v = 1)$ increases, then the effective activation energy measured would be expected to change. One of the problems at the end of this chapter sets out the data, and the solution shows the curvature introduced. Vibrational excitation of $H_2(v = 1)$ increases the room temperature rate constant by a factor of approximately 120 over that for $H_2(v = 0)$,[28] whereas putting the vibrational energy into $OH(v = 1)$ only increases the room temperature rate constant by about 50%,[29] illustrating the intuitively satisfying idea that energy needs to be supplied to the bond which will break in the chemical reaction. These effects can be understood at a more quantitative level employing the vibrationally adiabatic version of transition state theory that was outlined in section A1.4.

(ii) Temperature-dependent partition functions

The transition state theory formulation, equation 1, incorporates partition functions for reactants and activated complex, all of which will be temperature dependent in ways which are different for different degrees of freedom. For translational partition functions, Q_{tr} is proportional to $T^{3/2}$; for rotational partition functions Q_r is proportional to T for a linear species and to $T^{3/2}$ for a non-linear species, and the vibrational partition function for each harmonic oscillator is $Q_v = [1 - \exp - (h\nu/k_B T)]^{-1}$. The change in degrees of freedom in moving from reactants to activated complex will introduce temperature dependences into the ratio of the partition functions in equation 1 and leads to non-linearity in the Arrhenius plot: for the $OH + H_2$ reaction, the increase in activation energy between 300 and 2000 K due to the temperature dependence of partition functions has been estimated[24] to be 30 kJ mol^{-1}, a factor of two for this particular case.

(iii) Non-equilibrium behaviour

Our normal idea of a rate constant is that during the course of a reaction it remains invariant with time. However if, in the $OH + H_2$ reaction, the more rapidly reacting $H_2(v = 1)$ molecules were being depleted from the reactant mixture by chemical removal at a rate which was larger than the collisional re-establishment of Boltzmann equilibrium between $H_2(v = 0)$ and $(v = 1)$, then we would expect the measured rate constant to be time dependent, and not reflect the true (i.e., equilibrated reactants) thermal rate constant. Vibrational non-equilibrium effects of this kind have been calculated for the $OH + H_2$ reaction[24] and conditions identified where they introduce non-Arrhenius behaviour to the rate constants: they are only important when a substantial fraction of the reaction occurs via vibrationally excited states, and when vibrational relaxation is slow. Rotational and translational equilibria are in general more rapidly established than vibrational, and are assumed to be maintained in Boltzmann equilibrium in the reactants at all times.

(iv) Tunnelling

Non-linear Arrhenius behaviour due to quantum mechanical tunnelling (see section A1.5) will be most apparent for H atom transfer reactions at 'low' temperatures, where the bimolecular rate constant will be higher than that predicted classically. This effect is best seen in calculated rate constant values on *ab initio* potential energy surfaces. For the $OH + H_2$ reaction, without a tunnelling correction, experiment[24] and theory[26] are in good agreement above ~ 1000 K, but diverge at lower temperatures, theory underestimating the measured rate constant by a factor of ~ 3 at 300 K. Inclusion of a tunnelling correction removes the discrepancy.[26] Reference 27 reviews the methods available using both semi-empirical and full quantum mechanical corrections.

A3.7 Entropies of activation and the prediction of A-factors

Equation 1 shows that, if the structure of the activated complex is known, then the bimolecular rate constant may be estimated via transition state theory. Reasonable guesses as to the structure may be made if no more theoretically sound method is available: this is normally done not through evaluation of partition functions but through estimation of entropies of activation. For a bimolecular process, such as

$$A + BC \rightarrow (ABC^{\neq}) \rightarrow products$$

the thermodynamic formulation of transition state theory expresses the rate constant as $k = (k_B T/h)K^{\neq}$, where $K^{\neq}$ is the equilibrium constant for formation of the activated complex. Since E_{act} is defined as $-d \ln k/d(1/RT)$, it is related to the equilibrium constant, $K^{\neq}$, by the equation:

$$E_{act} = RT - d \ln K^{\neq}/d(1/RT).$$

Substituting for E_{act} in the normal Arrhenius expression enables the experimental A factor to be related to the entropy of activation, $\Delta S^{\neq}$, the difference in entropies between $ABC^{\neq}$ and reactants:

$$A = e^2 \left(\frac{k_B T}{h}\right) \exp\left(\frac{\Delta S^{\neq}}{R}\right)$$

where the standard state for $\Delta S^{\neq}$ is defined in concentration units (e.g. 1 molecule cm^{-3} or 1 mol cm^{-3}).[30]

The usual method of estimating $\Delta S^{\neq}$ is to start with an activated complex similar in structure to a stable molecule, whose third law entropy is known or can be calculated, and make appropriate corrections to allow for a 'looser' transition state than that found in the stable molecule, e.g., for changes in vibrational frequencies and symmetry. Benson's book 'Thermochemical Kinetics'[31] describes the approach in detail and we take an example from it, the bimolecular H-abstraction reaction

$$H + C_2H_6 \rightarrow C_2H_7^{\neq} \rightarrow H_2 + C_2H_5$$

to illustrate the entropic constraints placed on the A factor.

Contributions to $\Delta S^{\neq}$ may be assessed as follows, using C_2H_6 as a model for $C_2H_7^{\neq}$.

(a) The addition of a hydrogen atom makes a minimal difference to the mass and moments of inertia of C_2H_6, so that the contributions to the standard entropy of the complex, $S^{\ominus\neq}$, from translation and rotation, may be set equal to the contributions to $S^{\ominus}_{C_2H_6}$, with the addition of a term $R\ln6$, which takes into account the reduction in symmetry on forming the complex.

(b) A contribution of $R\ln2$ has to be added to account for the increase in electronic degeneracy on forming $C_2H_7^{\neq}$ (a doublet) from C_2H_6 (a singlet).

(c) Finally, there are changes in the vibrational and internal rotational partition functions consequent on forming H—H—C—. Benson[31] illustrates how these positive contributions to $S^{\ominus\neq}$ may be estimated, but we may set a lower limit on $S^{\ominus\neq}$ and on $\Delta S^{\neq}$ by neglecting them:

$$S^{\ominus\neq} \geqslant S^{\ominus}(C_2H_6) + R\ln12.$$

Thus

$$\Delta S^{\neq} \geqslant -S^{\ominus}(H) + R\ln12$$

which leads to an estimated A factor of $\geqslant 2 \times 10^{-11}$ cm^3 molecule^{-1} s^{-1}, an order of magnitude smaller than the experimental value of $\sim 2 \times 10^{-10}$ cm^3 molecule^{-1} s^{-1}. This discrepancy could be accounted for by the inclusion of the new bending vibrations and internal rotation in the partition function for the transition state and Benson obtains estimates of 2×10^{-10} cm^3 molecule^{-1} s^{-1} and 10^{-9} cm^3 molecule^{-1} s^{-1} for linear and non-linear H—H—C configurations in $C_2H_7^{\neq}$, respectively. It can be seen that calculations of this kind can be used to predict the pre-exponential factors for a variety of transition states, and hence to distinguish between various postulated microscopic mechanisms for the reaction.

A3.8 Conclusions

In recent years molecular beam and laser techniques have made the ultimate goal of chemical dynamicists just about possible—namely the measurement of cross-sections for fully state-selected reagents evolving into fully state-resolved products. Few reactions have had more than a part of their 'reaction space' probed, i.e., state resolution of products of thermal reactions has been extensively measured, as has velocity (but generally not internal states) of the products of angle-resolved molecular beam studies. These experiments can be expected to increase in sophistication and, will continue to test the theorists' potential energy surfaces for the relatively simple systems for which both *ab initio* calculations and state-resolved experiments have been carried out. For systems with more than 'a few' atoms (where 'few' is now perhaps 4 or 5) complications arise, both for theory and experiment (particularly when the effects of internal state distributions need to be evaluated). These reactions include many of importance in the applications of gas kinetics in, for example, combustion and aeronomy, and for these the main aim will still be the understanding and measurement of the thermally averaged rate constants.

A3.9 Suggestions for further reading

Bernstein R. B. (1982) *Chemical Dynamics Via Molecular Beam and Laser Techniques*, O.U.P.
Smith I. W. M. (1980) *Kinetics and Dynamics of Elementary Gas Reactions*, Butterworths, London.
Levy M. R. (1979) Dynamics of Reactive Collisions. *Prog. React. Kinet.*, **10**, 1.
Child M. S. (1986) Molecular Reaction Dynamics. *Sci. Prog. Oxf.*, **70**, 73.
Zellner R. (1984) Bimolecular Reaction Rate Coefficients. In *Combustion Chemistry*, (Ed. by W. C. Gardiner Jr.), p. 127. Springer Verlag, New York.
Benson S. W. (1976) *Thermochemical Kinetics*, Wiley, New York.

A3.10 References

1 Bernstein R. B. (1982) *Chemical Dynamics Via Molecular Beam and Laser Techniques*, OUP.
2 Levy M. R. (1979) Dynamics of Reactive Collisions. *Prog. React. Kinet.*, **10**, 1.
3 Child M. S. (1986) Molecular Reaction Dynamics. *Sci. Prog. Oxf.*, **70**, 73.
4 Douglas D. J., Polanyi J. C. and Sloan J. J. (1973) *J. Chem. Phys.*, **59**, 6679.
5 Arnoldi D. and Wolfrum J. (1976) *Ber. Bunsenges Phys. Chem.*, **80**, 892.
6 Odiorne T. J., Brooks P. R. and Kaspar J. V. (1971) *J. Chem. Phys.*, **55**, 1980.
7 Dispert H. H., Geis M. W. and Brooks P. R. (1979) *J. Chem. Phys.*, **70**, 5317.
8 Noorbatcha I. and Sathyamurthy N. (1982) *J. Am. Chem. Soc.*, **104**, 1766.
9 Kneba M. and Wolfrum J. (1980) *Ann. Rev. Phys. Chem.*, **31**, 47; Birely J. H. and Lyman J. L. (1975) *J. Photochem.*, **4**, 269.
10 Altkorn R., Bartoszek F. E., DeHaven J., Hancock G., Perry D. S. and Zare R. N. (1983) *Chem. Phys. Letters*, **98**, 212.
11 Pechukas P. and Light J. C. (1965) *J. Chem. Phys.*, **42**, 3281; Pechukas P., Rankin C. and Light J. C. (1966) *J. Chem. Phys.*, **44**, 794; Light J. C. (1967) *Disc. Faraday Soc.*, **44**, 14; Nikitin E. E. (1965) *Theor. Exp. Chem.*, **1**, 144; Klots C. E. (1971) *J. Phys. Chem.*, **75**, 1526; Klots C. E. (1972) *Z. Naturforsch., Teil A*, **27**, 553; Kinsey J. L. (1970) *J. Chem. Phys.*, **54**, 1206.
12 Quack M. and Troe J. (1974) *Ber. Bunsenges. Phys. Chem.*, **78**, 240; Quack M. and Troe J. (1975) *Ber. Bunsenges. Phys. Chem.*, **79**, 170; Quack M. and Troe J. (1981) *Int. Rev. Phys. Chem.*, **1**, 97.
13 Buelow S., Noble M., Radhakrishnan G., Reisler H., Wittig C. and Hancock G. (1986) *J. Phys. Chem.*, **90**, 1015.
14 Beuhler R. J. and Bernstein R. B. (1969) *J. Chem. Phys.*, **51**, 5305.
15 Brooks P. R. (1973) *Faraday Disc. Chem. Soc.*, **55**, 299.
16 Buelow S., Radhakrishnan G., Catanzarite J. and Wittig C. (1985) *J. Chem. Phys.*, **83**, 444.
17 Karny Z., Estler R. C. and Zare R. N. (1978) *J. Chem. Phys.*, **69**, 5199.
18 Rettner C. T. and Zare R. N. (1981) *J. Chem. Phys.*, **75**, 3636; (1982) **77**, 2416.

19 Vasudev R., Zare R. N. and Dixon R. N. (1984) *J. Chem. Phys.*, **80**, 4863.
20 Siegbahn P. and Liu B. (1978) *J. Chem. Phys.*, **68**, 2457.
21 Blais N. C. and Truhlar D. G. (1983) *Chem. Phys. Letters*, **102**, 120.
22 Marinero E. E., Rettner C. T. and Zare R. N. (1984) *J. Chem. Phys.*, **80**, 4142; Gerrity D. P. and Valentini J. J. (1985) *J. Chem. Phys.*, **82**, 1323.
23 Johnson H. S. and Parr C. (1963) *J. Am. Chem. Soc.*, **85**, 2544; Johnson H. S. (1966) *Gas Phase Reaction Rate Theory*, Ronald Press, New York.
24 Zellner R. (1984) Bimolecular Reaction Rate Coefficients. In *Combustion Chemistry*, (Ed. by W. C. Gardiner Jr.), p. 127. Springer Verlag, New York.
25 Smith I. W. M. and Zellner R. (1974) *J. Chem. Soc. Faraday Trans.* 2, **70**, 1045.
26 Schatz G. C. and Walch S. P. (1980) *J. Chem. Phys.*, **72**, 776.
27 See for example, Garrett B. C., Truhlar D. G. and Grev R. S. (1981) In *Potential Energy Surfaces and Dynamics Calculations*, (Ed. by D. G. Truhlar), p. 587 and references therein. Plenum Press, New York.
28 Zellner R. and Steinert W. (1981) *Chem. Phys. Letters*, **81**, 568; Glass G. P. and Chaturvedi B. K. (1981) *J. Chem. Phys.*, **75**, 2749.
29 Spencer J. E., Endo H. and Glass G. P. (1977) 16*th Symp. Combustion*, 829.
30 Golden D. M. (1971) *J. Chem. Ed.*, **48**, 235.
31 Benson S. W. (1976) *Thermochemical Kinetics*, Wiley, New York.

A3.11 Problems

Applications of theory to bimolecular reactions

A3.1. A 'prior' distribution for the internal energies of products of a chemical reaction can be calculated by writing down the *statistical* probability that particular quantum states can be formed. No dynamical effects are assumed: all energetically allowed product states are populated with a probability proportional to the number of states in the group.

For a reaction

$$A + BC \rightarrow AB(v', j') + C$$

write down the statistical probability of finding AB in a state (v', j') in terms of the available energy for partitioning into products (E), the vibrational energy $(E_{v'})$, and the rotational energy $[Bj'(j'+1)]$, (the degeneracy of translational states is proportional to the translational energy to the power $1/2$).

Integrate this expression over j' (j' ranges from zero to some maximum value determined by energy conservation) to give an expression for the statistical probability $P^0(v')$ of finding AB in the *vibrational* level v', in terms of $E_{v'}$ and E. Assume that AB behaves as a rigid rotor. Now rewrite your expression in terms of $f = E_{v'}/E$, the fraction of available energy going into vibration. Normalise your expression by assuming AB is a harmonic oscillator, and that the energy levels are continuous, so that

$$\int_0^1 P^0(f)\, df = 1.$$

You should find

$$P^0(f) = \frac{5}{2}[1 - f(v')]^{3/2}.$$

Given the following data, calculate the experimental $P(v')$ values for the $CO(v')$ product of $O + CS$ and $O + CSe$ reactions and plot them on the same graph as the prior distribution.

v'	$E(CO, v')/cm^{-1}$	Observed $P(v')$	
		$O + CS$	$O + CSe$
0	0	—	—
1	2 143	—	—
2	4 260	—	—
3	6 350	—	—
4	8 414	—	—
5	10 452	—	—
6	12 464	—	—
7	14 449	0.1	—
8	16 408	0.21	—
9	18 342	0.49	—
10	20 249	0.61	0.25
11	22 131	0.73	0.40
12	23 987	0.91	0.45
13	25 817	1.00	0.47
14	27 622	0.92	0.63
15	29 401	0.30	0.74
16	31 154	—	0.94
17	32 882	—	0.94
18	34 585	—	1.00
19	36 263	—	0.66
20	37 916	—	0.20

[Observed $P(v')$ are scaled to unity at maximum values]

For $O + CS$, $E = 376 \text{ kJ mol}^{-1}$ ($31\,430 \text{ cm}^{-1}$)

$O + CSe$, $E = 495 \text{ kJ mol}^{-1}$ ($41\,377 \text{ cm}^{-1}$).

A function often used to describe the distributions is a so-called 'surprisal plot'. The surprisal, $I[f(v')]$ is defined as

$$I[f(v')] = -\ln[P(v')/P^0(v')]$$

where $P(v')$ is the observed distribution and $P^0(v')$ the prior distribution. Calculate surprisals for each reaction and show that they vary linearly with $f(v')$.

A3.2. A calculation on the curvature introduced into Arrhenius type plots from vibrational enhancement of a simple reaction.
The reaction

$$OH + H_2(v) \rightarrow H_2O + H$$

has rate constants for $v = 0$ and 1 given by

$$k_0 = 9.3 \times 10^{-12} \exp(-18\,000/RT)$$

$$k_1 = 6.0 \times 10^{-11} \exp(-11\,000/RT)$$

where the rate constants are expressed in units of $cm^3\,molecule^{-1}\,s^{-1}$ and the activation energies are in $J\,mol^{-1}$. Calculate the rate constants expected for a Boltzmann vibrational distribution in the reactants at temperatures between 300 and 2500 K and plot the results in the usual Arrhenius form.

A3.12 Answers to problems

Applications of theory to bimolecular reactions

A3.1. The probability $P^0(v',j')$ of finding AB in a state (v',j') will be proportional to

$$(2j'+1)[E - E_{v'} - Bj'(j'+1)]^{1/2}.$$

The two terms represent (a) the degeneracy of the AB rotational level, and (b) the density of the translational states (the term in the square brackets being the energy available for partitioning into translation).

For a probability $P^0(v')$ we integrate this expression from $j' = 0$ to j'_{max}, where j'_{max} is given by $(E - E_{v'}) = Bj'_{max}(j'_{max} + 1)$. We assume rotational energy levels are continuous. The result is

$$P^0(v') = \frac{2k}{3B}(E - E_{v'})^{3/2}.$$

where k is a proportionality constant.

Assume AB is harmonic, with $E_{v'} = v'\varepsilon$ and $E = v'_{max}\varepsilon$. Substituting $f = E_{v'}/E$, we find

$$P^0(f) = \frac{2k}{3B}(\varepsilon v'_{max})^{3/2}(1-f)^{3/2}$$

Normalising,

$$P^0(f) = \frac{5}{2}[1 - f(v')]^{3/2}$$

(see reference 1).

Figures A3.10 and 11 show (a) prior distributions plotted as a function of $f(v')$, together with measured $P(v')$ values for both reactions, where the measured $P(v')$ have been scaled so that they sum to the same value as the prior; and (b) the surprisal plots for the two reactions.

The aim here is to illustrate the similarity between the two vibrational distributions, both of which are inverted and produce stimulated emission from the CO product.

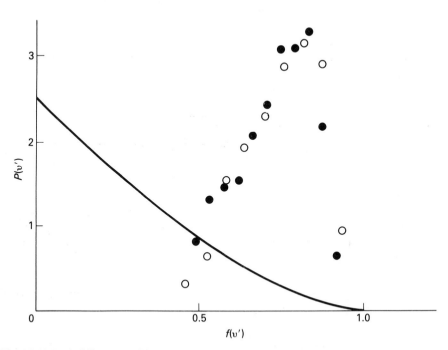

Fig. A3.10. Probability $P(v')$ of finding the product CO in the v'th vibrational level in the O + CS, CSe reactions: $f(v')$ is the fraction of the available energy appearing as vibration. The solid line is the prior distribution $P^0(v')$; the open and filled circles are the experimental values for the O + CS and O + CSe reactions, respectively.

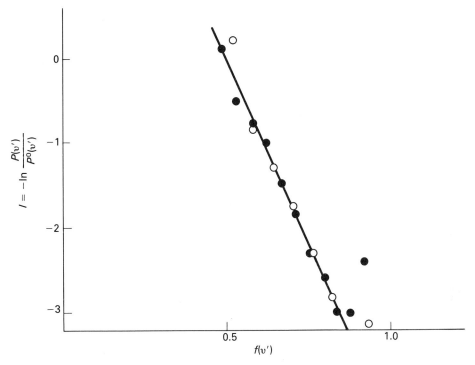

Fig. A3.11. Surprisal plot for the vibrational energy distributions in CO product of the O + CS (open circles) and O + CSe (filled circles) reactions. The surprisal I is defined as $I = -\ln P(v')/P^0(v')$, where $P(v')$ is the measured distribution and $P^0(v')$ the prior distribution. Points for both reactions lie on the same line, consistent with similar dynamics in the two cases.

For the original experimental data see Hancock G., Ridley B. A. and Smith I. W. M. (1972) *J. Chem. Soc., Faraday Trans.* 2, **68**, 2117; Morley C., Ridley B. A. and Smith I. W. M. (1972) *J. Chem. Soc., Faraday Trans.* 2, **68**, 2127.

A3.2. $k = \{k_0 + k_1 \exp[-\Delta E/RT]\}/Q_v$

where $\Delta E = E[H_2(v = 1)] - E[H_2(v = 0)]$

$\qquad = 49.6 \, \text{kJ mol}^{-1}$

Q = vibrational partition function

$\qquad = [1 - \exp(-\Delta E/RT)]^{-1}$.

(For discussion see Zellner R. (1984) Bimolecular Reaction Rate Coefficients. In *Combustion Chemistry*, (Ed. by W. C. Gardiner Jr.), p. 127. Springer Verlag.

The calculated temperature dependence of k is shown in Fig. A3.12 (overleaf).

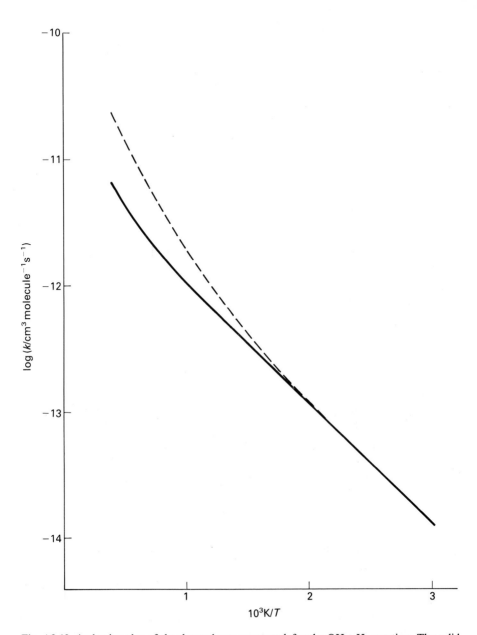

Fig. A3.12. Arrhenius plot of the thermal rate constant k for the $OH + H_2$ reaction. The solid curve is that calculated using measured state-specific rate constants for $H_2(v = 0)$ and $H_2(v = 1)$ as required in the problem. The dashed curve is the experimental variation of the thermal rate constant. Curvature is seen in both plots: the failure of the state specific rate constant at *high T* may be due to the assumption that at *low* temperatures the reaction proceeds without tunnelling: see reference 24 for details.

Chapter A4
Theories of Unimolecular Reactions and their Applications

I. W. M. SMITH

A4.1 Introduction and types of unimolecular reaction

In unimolecular reactions, each microscopic reactive event involves a single isolated molecule in which a rearrangement of chemical bonding leads it to isomerise or dissociate. Theories of unimolecular reaction are based on the notion that reaction, i.e., the rearrangement of atoms, is preceded by redistribution of a molecule's internal energy. The intramolecular energy transfer process transforms molecules with sufficient energy to react (i.e., with $E > E^0$, E representing internal energy and E^0 the minimum internal energy which can bring about reaction) to activated molecules. Activated molecules possess sufficient energy in the correct internal mode, or combination of modes, to carry the system along the reaction path and through the transition state region leading to products.

As well as treating the intramolecular energy transfer process, a full mechanism for thermally induced unimolecular reactions must include bimolecular processes in which intermolecular energy transfer takes place. However, in the limit of high total pressure, collisions become so frequent and the rate of intermolecular redistribution of energy so rapid that the distribution of reactant molecules over internal states is effectively undisturbed by reaction. It will be shown later that the activation energy derived from rate constant measurements under limiting high pressure conditions corresponds quite closely to both the threshold energy E^0 and to the electronic potential energy barrier to reaction.

Tables A4.1 and A4.2 compare the activation energies obtained in the limit of high pressure, E_{act}^{∞}, with the corresponding values of the changes in molar internal energy, $\Delta E_{298}^{\ominus}$, for some prototype unimolecular reactions. Isomerisation and dissociation are the two main types of unimolecular reaction. The examples of dissociation have been subdivided according to whether the species involved are molecules (no unpaired electrons) or radicals (one or more unpaired electrons).

In those cases where only molecular species are involved, that is, isomerisations and dissociations of type (a) (Table A4.2), the overall chemistry is simple and straightforward experiments can be used to investigate the reaction kinetics. The reagent is admitted to a well-regulated, high temperature reactor and the progress of reaction is followed by withdrawing samples for chemical analysis at various

Table A4.1. Energies and entropies of activation for two isomerisation reactions obtained in the limit of high pressure compared with the standard changes in internal energy and entropy for the reactions

	$CH_3NC \rightarrow CH_3CN$	$c\text{-}C_3H_6 \rightarrow CH_3CH{=}CH_2$
T range	470–535 K	690–810 K
$E_{act}^\infty/kJ\,mol^{-1}$	$+160.5$	$+274$
$\Delta E_{298}^\ominus/kJ\,mol^{-1}$	-72	-33
$\Delta S_{act}^\infty/J\,mol^{-1}\,K^{-1}$	$+2.9$	$+35$
$\Delta S^\ominus/J\,mol^{-1}\,K^{-1}$	-1.3	$+29.3$

Table A4.2. Energies and entropies of activation for different types of dissociation reaction obtained in the limit of high pressure compared with the standard changes in internal energy and entropy for the reactions

(a) Molecule $\rightarrow$ molecule + molecule		
e.g.	$C_2H_5Cl \rightarrow C_2H_2 + HCl$	$c\text{-}C_4H_8 \rightarrow 2C_2H_4$
T range	671–766 K	693–741 K
$E_{act}^\infty/kJ\,mol^{-1}$	$+254$	$+262$
$\Delta E_{298}^\ominus/kJ\,mol^{-1}$	$+70$	$+73$
$\Delta S_{act}^\infty/J\,mol^{-1}\,K^{-1}$	$+19$	$+38$
$\Delta S^\ominus/J\,mol^{-1}\,K^{-1}$	$+501$	$+544$

(b) Radical $\rightarrow$ molecule + radical		
e.g.	$CH_3CHCH_3 \rightarrow C_3H_6 + H$	$CH_3CH_2CH_2 \rightarrow CH_3 + C_2H_4$
T range	ca. 300–500 K	ca. 400–500 K
$E_{act}^\infty/kJ\,mol^{-1}$	$+174$	$+142$
$\Delta E_{298}^\ominus/kJ\,mol^{-1}$	$+169$	$+110$
$\Delta S_{act}^\infty/J\,mol^{-1}\,K^{-1}$	$+1$	$+0.5$
$\Delta S^\ominus/J\,mol^{-1}\,K^{-1}$	$+471$	$+497$

(c) Molecule $\rightarrow$ radical + radical		
e.g.	$HNO_3 \rightarrow OH + NO_2$	$C_2H_6 \rightarrow 2CH_3$
T range	220–1200 K	450–1000 K
$E_{act}^\infty/kJ\,mol^{-1}$	ca. $+205$	ca. $+360$
$\Delta E_{298}^\ominus/kJ\,mol^{-1}$	$+205$	$+360$
$\Delta S_{act}^\infty/J\,mol^{-1}\,K^{-1}$	$+30$	$+50$
$\Delta S^\ominus/J\,mol^{-1}\,K^{-1}$	$+528$	$+527$
[standard state: 1 molecule cm^{-3}]		

time intervals. This method cannot be applied so easily to reactions involving free radicals, like dissociations of classes (b) and (c) of Table A4.2, Also it should be noted that usually only a limited range of temperatures can be covered with experiments of this kind. Because the reactions have high activation energies, they proceed at a convenient rate for measurement only through a narrow range of temperature.

The activation energies for the reverse of the reactions given in Tables A4.1 and A4.2 can be calculated and used to establish how the potential energy varies along the path leading from reagent to products. According to the principle of detailed balance, the ratio of rate constants, k_f and k_r, for forward and reverse reactions is equal to the equilibrium constant for a reaction:

$$(k_f/k_r) = K_c. \tag{1}$$

Differentiating both sides of this equation with respect to $(1/RT)$ yields

$$E_{act,f} - E_{act,r} = \Delta E^{\ominus}. \tag{2}$$

It follows that, for an isomerisation or a molecular dissociation, $E_{act,r}$, as well as $E_{act,f}$, is large and there is a well-defined maximum in the profile of potential energy along the reaction coordinate, as depicted in Fig. A4.1. In cases where one bond in a molecule is broken to yield two radical fragments, as in the dissociations of HNO_3 and C_2H_6, the potential energy increases monotonically as the distance between the two fragments increases. There is no maximum in energy along the reaction coordinate, as illustrated in Fig. A4.1c.

A further important feature of reactions of type (c) in Table A4.2, i.e., dissociation to two radicals, is that it is often possible to study the reverse association reaction at low temperature, as well as the dissociation reaction at high temperature. The results must, of course, be related by the principle of detailed balance, so it is possible to obtain kinetic information about some of these systems, including the two examples given in Table A4.2(c), over wide ranges of temperature. The reactions given in part (b) of Table A4.2 are representatives of an intermediate

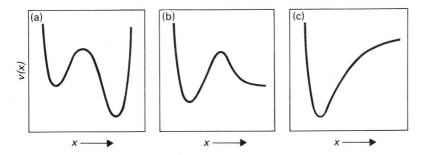

Fig. A4.1. Schematic potential energy profiles for three types of unimolecular reaction: (a) isomerisation, (b) dissociation where there is a high energy barrier and a large activation energy for reaction in both the forward and reverse directions, and (c) dissociation where the potential energy rises monotonically as $x \to \infty$, so that there is no barrier to the reverse association reaction.

category of reaction. The reverse process involves the addition of an atomic or polyatomic free radical to an unsaturated molecule. Such reactions are characterised by small but significant energies of activation and energy barriers.

A4.2 Unimolecular reactions and transition state theory

A fundamental requirement for the application of canonical transition state theory (section A1.4) is that the reagents are distributed over energy states according to the Maxwell–Boltzmann laws. In unimolecular reactions, population is lost from states of high internal energy $(E > E^0)$ as the result of reaction. Transition state theory can only be applied in the high pressure limit. Then collisions are so frequent that a Maxwell–Boltzmann distribution over these high energy states is maintained despite the occurrence of reaction.

For a unimolecular reaction in the limit of high pressure, the standard transition state theory expression for the thermal rate constant is (cf. equation 18 of Chapter A1)

$$k_{uni}^{\infty}(T) = (k_B T/h)(Q^{\neq}/Q_{reag}) \exp(-\Delta E_0^{\neq}/k_B T) \tag{3}$$

where $Q^{\neq}$ and Q_{reag} are partition functions for the transition state (with motion along the reaction coordinate factorised out) and the reagent, and $\Delta E_0^{\neq}$ is the difference in the zero-point energy levels in the transition state and the reagent. The translational partition functions for reagent and transition state will be the same, so differences between $(k_B T/h)$ and the experimentally observed A-factor for a reaction can be approximately equated to the ratio of partition functions associated with internal motions in the transition state and in the reagent. In general, the more loosely bound the transition state compared with the reagent, the greater the increase in rovibrational partition functions and the greater the ratio $(Q^{\neq}/Q_{reag})_{int}$.

A slightly different viewpoint is obtained by using the thermodynamic form of the transition state theory expression (cf. section A4.3):

$$k^{\infty}(T) = (k_B T/h) \exp(\Delta S_c^{\neq}/R) \exp(-\Delta H_c^{\neq}/RT). \tag{4}$$

Now, the deviation of the A-factor from $(k_B T/h)$ is largely associated with $\Delta S_c^{\neq}$, the change in entropy between the transition state species and the reagent. Again, large positive changes in entropy will be associated with 'loose' transition states. These occur especially on potentials, often referred to as Type II potentials,[1] of the kind shown in Fig. A4.1c. By contrast, reactions proceeding via Type I potentials and tight transition states usually exhibit smaller entropies of activation, although other factors such as the 'freeing' of internal rotations (e.g., in the isomerisation of cyclopropane to propene) must be taken into account.

The theories of unimolecular reactions can be applied to all the types of reactions that are illustrated by the examples in Tables A4.1 and A4.2—and to the reverse of all those reactions, including radical–radical association.[2,3] With reactions of type (c) of Table A4.2, occurring over a Type II potential, two particular problems arise. First, it is not obvious where one locates the transition state, since there is no well-defined potential energy maximum on the reaction path. This difficulty is not peculiar to this kind of reaction, and has been considered in section A1.4(iv). However it has to be borne in mind for the remainder of the chapter. The second problem is related to the first and is concerned with the question of whether the concept of a transition state whose 'position' is independent of energy (or temperature) is a useful or valid one when there is no well-defined potential barrier. These questions are answered in slightly different ways by more advanced theories of unimolecular reactions.

A4.3 Theories of unimolecular reaction: beyond Lindemann

Theories of thermal unimolecular reactions are all based on the mechanism involving collisional energy transfer which was originally proposed by Lindemann. This can be represented by the mechanistic scheme:

$$A + M \underset{k_{-1}}{\overset{k_1}{\rightleftharpoons}} A^\dagger + M$$
$$\downarrow k_2$$
$$\text{products} \tag{5}$$

where A represents a reagent molecule, $A^\dagger$ a reagent molecule with $E > E^0$, and M any gas-phase species, i.e., another reagent molecule or a molecule of inert diluent. The shortcomings of the Lindemann model are that it fails to

(a) recognise that the kinetics of each process depends on (at least) the internal energy and angular momentum of the $A^\dagger$ species involved; or

(b) suggest methods whereby the individual rate constants may be estimated.

Theoretical extensions to the Lindemann treatment are based on two fundamental assumptions.

(a) In molecules at high levels of internal excitation, intramolecular vibrational relaxation is rapid and energy is distributed randomly among all the internal modes.

(b) Because of the rapid energy randomisation, the behaviour of highly energised molecules depends only on their total internal energy (E) and total angular momentum (J). Based on these assumptions, the simple Lindemann mechanism can be modified and expressions obtained for the rate constants for reaction of molecules with specified E and J.

$$A + M \underset{k_{-1}(E,J)}{\overset{k_1(E,J)}{\rightleftharpoons}} A^\dagger(E,J) + M$$
$$\downarrow k_2(E,J)$$
$$\text{products} \tag{6}$$

$$\text{Rate}\,(E,J) = \left[\frac{k_2(E,J)k_1(E,J)[M]}{k_{-1}(E,J)[M] + k_2(E,J)} \right][A] \tag{7a}$$

$$k_{uni}(E,J) = \frac{k_2(E,J)[k_1(E,J)/k_{-1}(E,J)]}{1 + k_2(E,J)/k_{-1}(E,J)[M]}. \tag{7b}$$

To obtain an expression for the rate constant which is measured in normal thermal experiments, one must take a properly weighted sum of $k_{uni}(E,J)$ over J (from $0 \to \infty$) and E (from $E^0 \to \infty$). To carry out this calculation, it is necessary to have expressions for $k_2(E,J)$, $[k_1(E,J)/k_{-1}(E,J)]$, and $k_{-1}(E,J)$. Evaluation of $k_2(E,J)$, the specific rate constant for unimolecular reaction is central to theories of unimolecular reaction and is considered below. $[k_1(E,J)/k_{-1}(E,J)]$ defines a function $P(E,J)$ which describes the Boltzmann distribution of molecules over internal energy and angular momentum states. At least in principle, it can be found from the equations of statistical mechanics. $k_{-1}(E,J)$ corresponds to the rate constant for removing energy greater than $E - E^0$ from molecules with sufficient energy to react. This rate constant is usually assumed to be independent of E and J and is written as βZ^0, where Z^0 is the rate constant for all bimolecular collisions between $A^\dagger$ and M and β allows for the fact that not all collisions remove sufficient energy from $A^\dagger$ to reduce the internal energy below E^0. A value of $\beta = 1$ corresponds to the strong collision assumption.

The specific rate constant, $k_{uni}(E,J)$, and the thermal rate constant are pressure-dependent because the second term in the denominator of equation 7b depends on [M]. However, two limiting cases of behaviour can be distinguished and it is useful to consider their treatment first. As $[M] \to \infty$, the term $k_2(E,J)/k_{-1}(E,J)[M]$ can be neglected, the expression for $k_{uni}(E,J)$ becomes $k_2(E,J)P(E,J)$, and the rate constant for reaction under thermal conditions is given by

$$k_{uni}^\infty(T) = \int_{E^0}^{\infty} \sum_{J=0}^{\infty} k_2(E,J)P(E,J)\,dE. \tag{8}$$

Note that, under these conditions, the rate of reaction depends on the rate of the unimolecular chemical step but not on the individual rates of the bimolecular processes (1) and (-1) in which energy is transferred.

In the low pressure limit, i.e., as $[M] \to 0$, the second term in the denominator of the expression for $k_{uni}(E,J)$ becomes much greater than one. Under these conditions, the specific rate constant approaches $k_{-1}(E,J)[M]P(E,J)$ so that

$$k_{uni}^0(T) = \int_{E^0}^{\infty} \sum_{J=0}^{\infty} k_{-1}(E,J)P(E,J)[M]\,dE. \tag{9}$$

This equation demonstrates that the rate constants for unimolecular reactions in the limit of low pressure contain no information about the kinetics of the unimolecular chemical step; the only kinetic information that can be derived is for the energy transfer process (-1).

Before considering the kinetics of unimolecular reactions in the 'fall-off' region between the high and low pressure limits, we discuss how equations 8 and 9 for the rate constants in the high and low pressure limits can be evaluated.

A4.4 Low pressure limit

When one makes the usual assumption that $k_{-1}(E, J) = \beta Z^0$, equation 9 simplifies to

$$k_{uni}^0(T) = \beta Z^0[M] \int_{E^0}^{\infty} \sum_{J=0}^{\infty} P(E, J)\mathrm{d}E \tag{10}$$

where, strictly, E^0 depends on J. This equation can be evaluated by methods described by Troe.[4]

The strategy adopted by Troe is first to evaluate the integral in equation 10 for a non-rotating molecule with harmonic oscillator vibrational modes and a density of internal states which is constant through the energy range making significant contributions to the integral, and then to apply a series of corrections to allow for the approximations involved in the first-order estimate. Making the usual assumption that the vibrational levels comprise a closely packed manifold at internal energies at and above the threshold energy, a first-order approximation for $k_{uni}^0(T)$ is

$$\tilde{k}_{uni}^0(T) = \beta Z^0[M] \int_{E^0}^{\infty} \frac{\rho_{vib,h}(E^0)}{Q_{vib}} \exp\left(-E/k_B T\right)\,\mathrm{d}E \tag{11a}$$

$$= \beta Z^0[M] \frac{\rho_{vib,h}(E^0)k_B T}{Q_{vib}} \exp\left(-E^0/k_B T\right) \tag{11b}$$

where $\rho_{vib,h}(E^0)$ is the density of (harmonic) vibrational energy levels of the reagent molecule at an internal energy corresponding to the threshold energy for reaction, whilst Q_{vib} is the vibrational partition function of the reagent. It has been assumed, in deriving equation 11, that the density of states is constant for $E \geqslant E^0$ and equal to $\rho_{vib,h}(E^0)$.

Q_{vib} can be calculated using the well-known result from statistical mechanics for a system of s independent harmonic oscillators:

$$Q_{vib} = \prod_{i=1}^{s} [1 - \exp(-hv_i/k_B T)]^{-1}. \tag{12}$$

To evaluate $\rho_{vib,h}(E^0)$ it is usual to make use of Whitten and Rabinovitch's[5] empirical modification of Marcus and Rice's[6] semi-classical expression for the density of states associated with a number of harmonic oscillators. Whitten and Rabinovitch's equation is

$$\rho_{vib,h}(E) = \frac{[E + a(E)E_z]^{s-1}}{(s-1)! \prod\limits_{i=1}^{s} h\nu_i} \tag{13}$$

Here, $E_z = \frac{1}{2} \sum\limits_{i=1}^{s} h\nu_i$ is the zero-point energy in a molecule with s harmonic vibrations of frequency ν_i, and $a(E)$ is a constant close to one which can be calculated from the equations

$$a(E) = 1 - bw \tag{14a}$$

$$\log w = 1.0506 \, (E/E_z)^{0.25} \tag{14b}$$

$$b = (s-1) \sum\limits_{i=1}^{s} \nu_i^2 \Bigg/ \left(\sum\limits_{i=1}^{s} \nu_i \right)^2 . \S \tag{14c}$$

For reactions of type (c) (see Table 4.2) occurring on Type II potentials, equation 13 is evaluated with $E = E^0 = \Delta E_0^{\ominus}$.

Other things being equal, $\rho_{vib,h}$ at a given energy increases as s increases. Consequently the A-factors for unimolecular dissociation and the rate constants for radical association, both in the low pressure limit, increase as the total number of atoms in the reacting system increases.

Various computational tests[7] have been performed to examine the error involved in writing $k_{-1}(E, J)$ as βZ^0. It appears that this assumption is not a major source of inaccuracy when interpreting the results of thermal reactions. The parameter β is related to $\langle \Delta E \rangle$, the average energy change in collisions in which energy is transferred (both to and from $A^\dagger$), by

$$\beta/(1 - \beta^{1/2}) = -\langle \Delta E \rangle / F_E k_B T \tag{15}$$

where F_E is a factor which allows for the change in the density of internal energy states with increasing energy (see below). The magnitude of β for a particular M can be estimated in two ways. Either one can compare measured values of $k_{uni}^0(T)$ for different M, assume that $\beta = 1$ for the most effective M and then estimate β for other species accordingly, or one can compare a measured rate constant with the value that is estimated assuming strong collisions, i.e., $\beta = 1$. The values of β derived by these methods correlate roughly with the number of atoms in M but depend rather little on the actual chemical reaction.

The first-order estimate of $k_{uni}^0(T)$, i.e., $\tilde{k}_{uni}^0(T)$, must next be multiplied by factors which correct it for approximations which have been made in deriving

---

§This expression can be modified to take account of active "external" rotations (see p. 130).

equation 11

$$k^0_{uni}(T) = \tilde{k}^0_{uni}(T) F_{anh} F_E F_{rot} F_{rot, int} F_{corr}. \tag{16}$$

The formulae for these factors will be given below, with an indication of their magnitude for various chemical systems. Derivations of the equations have been given by Troe[4,8] who also tabulates numerical results for a large number of reactions of type (c) of Table A4.2.

The correction for neglect of vibrational anharmonicity arises because the vibrational levels of an anharmonic oscillator are not equally spaced but get closer together as the internal energy increases. The correction factor is defined by

$$F_{anh} = \rho_{vib,anh}(E^0)/\rho_{vib,h}(E^0). \tag{17}$$

The largest anharmonic effects are associated with those vibrational modes (of number m) which are 'lost' as the molecule dissociates. Troe[4] has proposed the formula

$$F_{anh} = [(s-1)/(s-\tfrac{3}{2})]^m. \tag{18}$$

The correction is independent of temperature and greatest for small molecules, because they will generally have a higher average number of quanta per mode. For example, for dissociation of $NOCl$ to $NO + Cl$, $s = 3$, $m = 2$ and $F_{anh} = 1.78$, whereas for dissociation of HNO_3 to $OH + NO_2$, $s = 9$, $m = 5$ and $F_{anh} = 1.38$.

The factor F_E allows for the fact that the density of vibrational levels increases through the range of levels which are populated to a significant extent under thermal conditions and is not a constant, $\rho_{vib,h}(E^0)$, as assumed in equation 11. F_E is given by

$$F_E = \int_{E^0}^{\infty} \frac{dE}{k_B T} \frac{\rho_{vib,h}(E)}{\rho_{vib,h}(E^0)} \exp\left[-(E-E^0)/k_B T\right] \tag{19}$$

and Troe derives the expression

$$F_E = \sum_{i=0}^{s-1} \frac{(s-1)!}{(s-i-1)!} \left[\frac{k_B T}{[E^0 + a(E^0)E_z]}\right]^i. \tag{20}$$

The factor F_E increases with temperature because as the temperature is raised the range of important internal energies is increased, and F_E is also greatest for large molecules since their densities of vibrational levels increase most steeply with E. Thus, for $NOCl$ dissociation, $F_E = 1.03$ at 300 K and 1.20 at 2000 K, whilst for dissociation of HNO_3, $F_E = 1.08$ at 300 K and 1.85 at 2000 K.

Equation 11 is derived for reaction of a species with zero angular momentum.

For such a system the requirement to conserve angular momentum, as the system proceeds along the reaction coordinate and its moments of inertia change, introduces no complications. However, for states with $J > 0$, the effective potential energy curves are not parallel either with one another or with that for $J = 0$ which simply describes the variation of electronic potential energy. For a Type II potential, adding the term due to angular momentum conservation leads to curves exhibiting 'centrifugal maxima'. As a result of this effect, both E^0 and $\rho(E^0)$ depend on J, and the factor F_{rot} is greater than one, because as the system moves along the reaction path to products, moments of inertia increase, and energies associated with the overall rotation decrease. Consequently, the difference in energy between $J > 0$ and $J = 0$ effective potential curves is less at the configuration corresponding to the centrifugal maximum on a $J > 0$ curve than it is at the equilibrium geometry of the molecular reagent. This effect is illustrated in Fig. A4.2.

Troe[4] has considered the calculation of F_{rot} for reactions over Type I and Type II potentials. For a molecule with two rotational quantum numbers, J and K, and a rotational degeneracy, $g(J, K) = 2(2J + 1)$, the correction factor is defined by

$$F_{rot} = \frac{1}{Q_{rot}} \int_0^{J_{max}} dJ \int_0^{Min(J, K_{max})} dK\, g(J, K)\, \exp[-E_{rot}(J, K)/k_B T]$$

$$\times \frac{\rho_{vib,h}[E^0(J) - E_{rot}(J, K)]\, \exp[-[E^0(J) - E_{rot}(J, K)]/k_B T]}{\rho_{vib,h}[E^0(J = 0)]\, \exp[-E^0(J = 0)/k_B T]}. \tag{21}$$

For reaction on a Type II potential, the crucial quantity $E^0(J)$ depends rather sensitively on the form assumed for the long-range potential. Using a van der Waals potential, Troe has derived the formula:

$$F_{rot} = [(s - 1)!/(s + \tfrac{1}{2})!][\{E^0 + a(E^0)E_z\}/k_B T]^{3/2}$$

$$\times \left\{ \frac{2.15(E^0/k_B T)^{1/3}}{2.15(E^0/k_B T)^{1/3} - 1 + [E^0 + a(E^0)E_z]/(s + \tfrac{1}{2})k_B T} \right\}. \tag{22}$$

For dissociation of NOCl, application of this formula yields $F_{rot} = 31$ at 300 K and 8.7 at 1000 K: for dissociation of HNO_3, $F_{rot} = 18.6$ at 300 K and 4.6 at 1000 K. The calculation of F_{rot} represents the most uncertain part of the estimate of k^0_{uni}, because it is sensitive to the form of the long-range potential.

Many molecules that undergo unimolecular reaction possess one or more internal modes whose associated quantum states differ radically from those of a harmonic oscillator. These modes include free internal rotations, torsional motions, and torsional modes that become internal rotations at higher energies. These motions should not be included in the evaluation of $\rho_{vib,h}(E^0)$ in equation 11 but their contribution is recognised through the factor $F_{rot\,int}$. Formulae have been

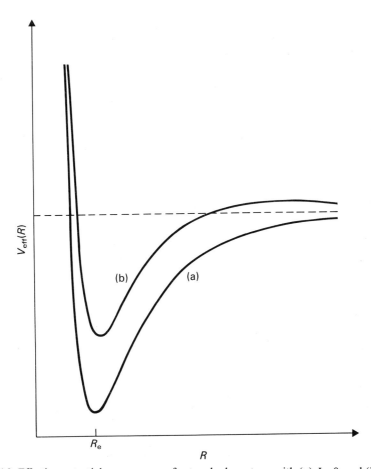

Fig. A4.2. Effective potential energy curves for two-body systems with (a) $J = 0$, and (b) $J > 0$. Note the centrifugal maximum on curve (b) and also the decrease in the difference between the two curves as x increases.

derived by Troe for each of the three types of motion that have just been referred to. For the case of a molecule with s normal harmonic vibrations plus one hindered internal rotation:

$$F_{\text{rot int}} = \frac{(s-1)!}{(s-\frac{1}{2})!} \left[\frac{E^0 + a(E^0)E_z}{k_B T} \right]^{1/2} [1 - \exp(-E^0/sV_0)] \times \left\{ [1 - \exp(-k_B T/V_0)]^{1.2} \right.$$

$$\left. + \frac{\exp(-1.2\,k_B T/V_0)}{(2\pi I_m k_B T/\hbar^2)^{1/2}(1 - \exp[-(n^2\hbar^2 V_0/2I_m(k_B T)^2)^{1/2}])} \right\}^{-1} \tag{23}$$

where V_0 is the barrier to internal rotation, I_m is the reduced moment of inertia associated with the internal rotation, and n is the number of equivalent minima associated with the motion. In the limit where $V_0 \to 0$ and the internal rotation becomes free, equation 23 reduces to

$$F_{\text{rot int}} \to F_{\text{rot int}}^{\text{free}} = \frac{(s-1)!}{(s-\frac{1}{2})!} \cdot \left(\frac{E^0 + a(E^0)E_z}{k_B T}\right)^{1/2}. \tag{24}$$

Finally, if the barrier to internal rotation is high, so that the motion remains a torsional vibration at all energies

$$F_{\text{rot int}}^{\text{tors}} = \frac{1}{s} \cdot \frac{[E^0 + a(E^0)E_z][1 - \exp(-hv_{\text{tors}}/k_B T)]}{hv_{\text{tors}}}. \tag{25}$$

A practical difficulty in applying these formulae may be a lack of information about V_0 or even about v_{tors}, the frequency of the torsional motion. Of our two examples, ClNO has no motions of these types; HNO_3 has one such motion, which treated as a hindered internal rotation with $V_0 = 33 \text{ kJ mol}^{-1}$,[4] has $F_{\text{rot int}} = 6.5$ at 300 K and 2.4 at 1000 K.

F_{corr} is introduced to allow for any error that might arise from the assumption that there is no coupling between the other correction factors. Troe[4] has concluded that this coupling should be slight so that F_{corr} can be put equal to one. In certain reactions of type (c) shown in Table A4.2, two other factors may have to be taken into account.

The first arises because it has been assumed that for reactions proceeding through a loose transition state the threshold energy, E^0, can be equated to $\Delta E_0^{\ominus}$, the difference in zero-point energies between the products and reagents. Although the potential energy V may increase monotonically along the reaction coordinate, it is possible that E_z passes through a maximum. This may be the case in the dissociation of O_3 to $O_2 + O$, but is probably a rarity.

Secondly, one may have to consider the contribution of bound electronically excited states to the overall kinetics.[9] If such states exist, they will contribute to the density of internal states at and near the threshold for reaction. In this situation one should calculate independent values of $k_{\text{uni}}^0(T)$ for each electronic state and add these values to obtain an estimate of the rate constant for the overall thermal reaction. In practice, the density of internal states at E^0 arising from electronically excited states may be quite small. The contribution to the overall rate in the low pressure limit is greatest when:

(a) the excited state is bound almost as strongly as the ground state, and
(b) the molecule is small, since $\rho_{\text{vib,h}}(E)$ increases least steeply with E when s is small.

We are now in a position to estimate $k^0_{uni}(T)$ according to equation 16. Experimental values of $k^0_{uni}(T)$ can be compared with calculated values for about 14 reactions of type (c) in Table A4.2. For about half of these systems, agreement is reached with β for M = Ar between 0.1 and 0.2. For two reactions,

$$SO_2 + M \rightarrow SO + O + M \tag{26a}$$

$$NO_2 + M \rightarrow NO + O + M \tag{26b}$$

particularly large values of β are obtained (1.8 and 0.6 for M = Ar, respectively) if one only includes reaction from the electronic ground state of the triatomic molecules in the calculation. In fact, both SO_2 and NO_2 are known to have electronically excited bound states that must participate in these reactions. Inclusion of the excited states in calculations on reaction 26b lowers the estimate of β_{Ar} from 0.6 to 0.2.[9]

To summarise, the methods described by Troe for estimating the rate constants for molecular dissociation (or radical association) in the limit of low pressure are straightforward to apply and reliable. They should be used to check experimental data. If, after allowing for a reasonable value of β and any possible participation of electronically excited states, the disagreement between experiment and theory is more than a factor of two, possible sources of error in the experiment should be carefully considered.

A4.5 Specific rate constant and microcanonical transition state theory

In order to evaluate equation 8 for $k^\infty_{uni}(T)$, the rate constant in the limit of high pressure, or indeed to use equation 7b to find the rate constant other than at the low pressure limit, an expression is needed for the specific rate constant, $k_2(E, J)$. The general formula for the specific rate constant is derived by microcanonical transition state theory; that is, transition state theory applied to systems of defined energy, rather than at a defined temperature. For the moment, any dependence of the specific rate constant on angular momentum state is ignored and an expression is derived for $k_2(E)$ which is related to the overall thermal rate constant by

$$k_{uni}(T) = \int_{E^0}^{\infty} k_2(E)\, P(E) \left\{ \frac{1}{1 + k_2(E)/\beta Z^0 [M]} \right\} dE \tag{27a}$$

or in the high pressure limit by

$$k^\infty_{uni}(T) = \int_{E^0}^{\infty} k_2(E)\, P(E)\, dE. \tag{27b}$$

Microcanonical transition state theory[10] first considers the probability that a system with energy between E and $E + dE$ is to be found within an element of phase space volume within the transition state (or, in other words, on the critical surface dividing reagents from products). For a system of N atoms, phase space has $6N$ dimensions, and the probability corresponds to the fraction of the total number of systems in the transition state:

$$d^{6N}P^* = n^*/n_{\text{reag}}$$

$$= (d\tau_{6N}/h^{3N})/[g(H^0)dH^0/h^{3N}] \tag{28}$$

where $g(H^0)dH^0$ is the total volume of reagent phase space bounded by the hypersurfaces $H^0 = E$ and $H^0 + dH^0 = E + dE$, and the h^{3N} factors are introduced to reduce phase space volumes to numbers of phase space cells. In the limit of closely spaced energy states the denominator in equation 28 can be written as $\rho(E)dE$.

The rate of reaction, as a result of systems passing through the phase space cells whose number is given by $d\tau_{6N}^*/h^{3N}$ is

$$\text{Rate}\,(E) = d(n^*/V)/dt$$

$$= \left[\frac{n_{\text{reag}}}{V}\right] \frac{(dp_x/h)(dq_x/dt)(d\tau_{6N-2}^*/h^{3N-1})}{\rho(E)\,dE} \tag{29}$$

$$\text{so } k_2(E) = \frac{(dp_x/h)(dq_x/dt)(d\tau_{6N-2}^*/h^{3N-1})}{\rho(E)\,dE} \tag{30}$$

where the derivatives of the phase space coordinates associated with motion along the reaction path x have been factorised out, and V is the total volume.

Now $(dp_x/h)(dq_x/dt)$ corresponds to the flux through a single phase space cell in the direction reagents to products and, since $(dq_x/dt) = p_x/\mu_x$

$$(dp_x/h)/(dq_x/dt) = dE/h \tag{31}$$

so that equation 30 simplifies to

$$k_2(E) = \frac{(d\tau_{6N-2}/h^{3N-1})}{h\rho(E)}. \tag{32}$$

But $(d\tau_{6N-1}/h^{3N-1})$ is simply the number of internal states at the transition state which have energy less than E. This quantity is usually denoted by $N^*(E)$ yielding

$$k_2(E) = N^*(E)/h\rho(E). \tag{33}$$

This simple equation is the fundamental result of microcanonical transition state theory. To apply it to unimolecular reactions, one needs prescriptions to calculate the density of internal states in the molecular reagent and the number of states in the transition state, which are provided by the Whitten–Rabinovitch[5] formulae referred to earlier. Alternatively, and more accurately, they may be calculated using efficient direct-count algorithms.[11] In addition, allowance should be made for angular momentum effects.

Finally, it should be noted that in applying equation 33 one should 'place' the transition state so as to minimise $N^*(E)$, the number of internal states. The importance of this variational principle has already been stressed in Chapter A1 with regard to canonical transition state theory. Obviously, the application of the variational principle to the microcanonical rate constant expression, equation 33, may not lead to a minimum value of $k_2(E)$ at the same position along the reaction path for all internal energies.

A4.6 RRKM theory[6,12]

The Rice–Marcus treatment of unimolecular reactions, usually referred to as RRKM (Ramsperger–Rice–Kassel–Marcus) theory, improved on earlier theory in three ways. In dealing with molecular vibrations, the existence of zero-point energies is allowed for and states are counted with due regard to the actual molecular frequencies. In addition, account is taken of molecular rotation and centrifugal effects.

The need to conserve J is recognised by distinguishing active and adiabatic modes, energy flow to and from the latter being restricted by the conservation of angular momentum. Figure A4.3 shows the active and adiabatic energies in the reagent (E and E_J) and in the transition state (E^+ and E_J^+), the difference being equal to ΔE_0^+, the difference in the zero-point levels

$$E + E_J = E^+ + E_J^+ + \Delta E_0^+. \tag{34}$$

The distribution function $P(E, J)$ in equations 8 and 9 can be written

$$P(E, J) = \frac{\rho(E) \exp(-E/k_B T)}{Q_{vib}} \cdot \frac{(2J+1) \exp(-E_J/k_B T)}{Q_J}. \tag{35}$$

Combining equations 33, 34 and 35 one obtains

$$k_2(E, J)\, P(E, J) = (1/h\, Q_{vib} Q_J) \exp(-\Delta E_0^+/k_B T) \cdot (2J+1)$$

$$\times \exp(-E_J^+/k_B T)\, N^*(E) \exp(-E^+/k_B T) \tag{36}$$

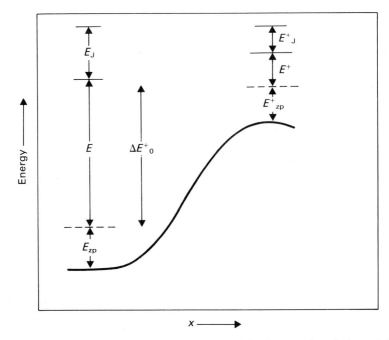

Fig. A4.3. Relationships between the various energy quantities that enter into the formulation of RRKM theory.

and

$$k_2(E, J) = N^*(E^+)/h\rho(E^+ + \Delta E_0^+ + E_J^+ - E_J). \tag{37}$$

To progress further, it is necessary to average over J. This is usually done approximately by taking mean values of E_J^+ and E_J defined as

$$\langle E_J^+ \rangle = l\,k_B T/2 \tag{38a}$$

$$\langle E_J \rangle = (I^+/I)l\,k_B T/2 \tag{38b}$$

where l, the number of adiabatic rotations, is usually taken as 2.

If equation 36 is now substituted into equation 7 and the summation over J performed, one obtains

$$k_{\text{uni}}(T) = (k_B T/h)\frac{Q_J^+}{Q_J Q_{\text{vib}}}\exp(-\Delta E_0^+/k_B T)$$

$$\times \int_0^\infty \frac{N^*(E^+)\exp(-E^+/k_B T)}{1 + k_2(E, J)/\beta Z^0[\text{M}]}\left[\frac{\mathrm{d}E^+}{k_B T}\right]. \tag{39}$$

In the limit $[M] \to \infty$, this equation reduces, as it must, to the transition state theory expression:

$$k_{uni}^{\infty}(T) = (k_B T/h) \frac{Q_J^+ Q_{vib}^+}{Q_J Q_{vib}} \exp(-\Delta E_0^+ / k_B T). \qquad (40)$$

In the fall-off region the integral in equation 39 has to be evaluated numerically. To do this, and indeed to evaluate the transition state theory expression, 'reasonable' choices have to be made for the structure and frequencies of the transition state, so that Q_J^+, Q_{vib}^+, ΔE_0^+ and $N^*(E^+)$ can be calculated and $k_{uni}^{\infty}(T)$ estimated. If the value of $k_{uni}^{\infty}(T)$ is available from experiment, the transition state properties are chosen so that equation 40 is satisfied. Then rate constants at other $[M]$ are computed and calculated 'fall-off curves' are prepared for comparison with experiment.

As was pointed out earlier, kinetic data for reactions occurring on Type I potentials are usually confined to a narrow range of temperatures. Good fits to the fall-off behaviour can be found, but the form of the calculated curves is quite insensitive to the choice of transition state.

For association reactions on Type II potentials, agreement between standard RRKM calculations and experiment is rather less good. Calculations using a fixed transition state predict quite a strong positive temperature dependence of $k_{uni}^{\infty}(T)$ which arises because the low frequency modes in loose transition states cause Q_{vib}^+ to rise steeply with temperature. Such a temperature dependence is not found in practice, and for a number of reactions, data are available over a wide temperature range.

As mentioned earlier, the assumption of a single well-defined transition state is dubious for reactions occurring across a Type II potential. In principle, variational methods can be used to find the minimum value of $k_{uni}^{\infty}(T)$ or of $k_2(E, J)$ and these estimates should serve as upper bounds to the actual rate constants, i.e.

$$k_2(E, J) \leqslant N_{min}^*(E, J)/h\rho(E). \qquad (41)$$

There is no reason to suppose that the minimum estimated values of $k_2(E, J)$ or $k_{uni}^{\infty}(T)$ will be found at the same position for all E, J or all T.

A4.7 Statistical adiabatic channel model

Quack and Troe[13] have proposed a theoretical approach to unimolecular reactions which is quite different from RRKM theory in that the usual notion of a transition state is abandoned. Instead, the variation in adiabatic energy for each channel state along the reaction coordinate (x) is examined. If the total internal energy in the system, i.e., E, exceeds the maximum value of the adiabatic energy in the region between the reagents ($x = x_e$) and products ($x \to \infty$), then that channel is 'open'. For each E, one evaluates the form $V_{ad}(x)$ of all the adiabatic channel states. $N^*(E)$ in equation 33 is then put equal to the total number of the open channels and the specific rate constant for reaction is evaluated.

The adiabatic energy of each channel state is the sum of three contributions:
(a) $V(x)$, the electronic potential energy;
(b) the centrifugal term arising from the need to conserve the total angular momentum;
(c) the energies of vibrational states which change their character as dissociation occurs.

It can be seen that (b) incorporates the same idea as is used to construct effective potential energy curves for two-body systems (see Fig. A4.2), whilst (c) has the same basis as vibrationally adiabatic transition state theory (c.f. section A1.4): namely, that a system stays in the same state along the crucial region of the reaction path but the energy of this state may change. In fact, for the Quack and Troe model to be valid, it is not necessary for individual 'trajectories' to remain adiabatic but only that the assumption of adiabaticity holds on average. Hence, the use of 'statistical' in the title of this theoretical treatment. Examples of adiabatic channel potentials are given in Fig. A4.4.

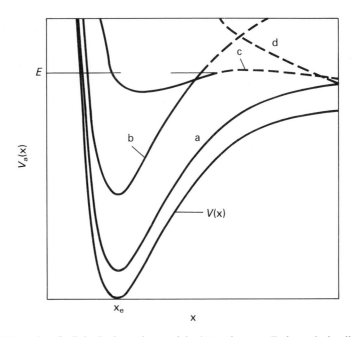

Fig. A4.4. Examples of adiabatic channel potentials. At total energy E, channel a is adiabatically open, but b, c and d are closed.[11]

The magnitudes of the contributions (b) and (c) to the adiabatic energies are difficult to estimate accurately and are undoubtedly sensitive to the form of $V(x)$ at long-range where it is not well-characterised. Quack and Troe[13] use interpolative formulae involving a single length parameter and based on a Morse potential. These formulae are given as equation 19 in Chapter A1 and are similar to those given below in section A4.8, where a method is discussed for estimating the limiting high pressure rate constant $k_{\text{uni}}^{\infty}(T)$.

The statistical adiabatic channel model can be used to predict $k_2(E, J)$, $k_2(E)$ and $k_{\text{uni}}(T)$. If one assumes adiabaticity after systems pass over the maxima on the adiabatic potential curves, then the distributions of reaction products over rovibrational and orbital angular momentum states can also be predicted. In the limit where the length parameter becomes very large, (c) makes a negligible contribution to the channel potentials. In terms of RRKM theory, one has J-dependent, 'loose' transition states located at the centrifugal maxima on the effective potential energy curves. Predictions about the product states become identical to those of 'phase space theory'.

A4.8 Phase space theory[14]

Phase space theory provides a means of predicting the distribution of unimolecular reaction products over rovibrational and orbital states. The basic premise is that the formation of molecules in particular final states is only restricted by the laws conserving total energy (E) and angular momentum (J). The actual state distributions are most likely to match those predicted by phase space theory when reaction proceeds from a true collision complex through very loose transition states and there is only very weak coupling between the separating products.

To compute the distributions predicted by phase space theory, calculations are performed on the dissociation of complexes of defined E, J and then appropriately summed to find distributions that can be compared with experiment. The procedure can be illustrated by considering the simplest example of a triatomic ABC decomposing to AB + C. The quantum numbers for vibration and rotation of AB and for orbital motion of AB + C are represented by v', j' and l'.

For complexes of specified E, J, the quantum numbers l' and j' are limited first by the condition

$$|l' + j'| \geqslant J \geqslant |l' - j'|. \tag{42}$$

In Fig. A4.5, this equation defines a rectangular grid of l', j' energy levels. The curved boundary which completes the definition of allowed states arises from the requirement that total energy is conserved and that the final energy of relative translation, i.e.,

$$E_{l'} = E - E_{v'} - j'(j' + 1)\hbar^2/2I_{AB} \tag{43}$$

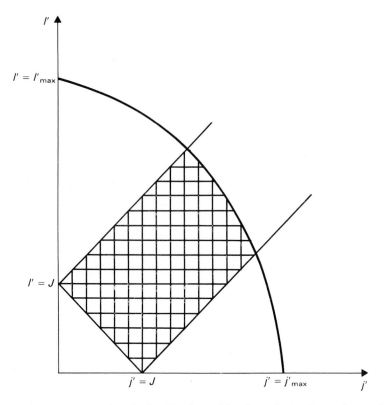

Fig. A4.5. Diagram representing the l', j' levels resulting from the break-up of a species of defined internal energy E and total angular momentum J and allowed by the laws conserving E and J.

exceeds the centrifugal maximum on the effective, two-body, potential energy curve. It is, of course, sensitive to the form assumed for the long-range part of $V(x)$. Each grid-point l', j' on the diagram is associated with $(2l'+1)(2j'+1)$ states and, according to phase space theory, the rates into particular l', j' levels will be proportional to these degeneracies.

The curved boundary in Fig. A4.5 depends, as equation 43 demonstrates, on $E_{v'}$ and hence on the choice of v', the vibrational state of AB. As v' is increased, this boundary encloses a smaller number of l', j' points. Phase space theory always predicts that a product vibrational distribution falls monotonically with v'.

As well as predicting the distribution of product states from thermal and photochemical unimolecular dissociations, phase space theory can be used to calculate state-to-state cross-sections for bimolecular reactions proceeding through long-lived complexes. In the first part of the calculation, one computes cross-sections for the formation of complexes of defined E, J from specified states of the reagents. If the formation of complexes is limited only by centrifugal effects, so that there is unit probability of complex formation for $l < l_{max}$, the total cross-section is

$$S_c(|v,j; E_t) = (\pi/k^2) \sum_{l=0}^{l_{max}} (2l+1) \tag{44}$$

where the wavenumber associated with relative translation is $k = (2\mu E_t)^{1/2}/h$. The cross-section for the formation of complexes of specified J is

$$S_{c,J}(|v,j; E_t) = (\pi/k^2) \sum_{l=0}^{l_{max}} (2l+1) \frac{(2J+1)}{(2l+1)(2j+1)}. \tag{45}$$

The last term represents the fraction of collisions with orbital and rotational quantum numbers l, j that couple to yield a total angular momentum quantum number J.

To determine detailed reaction cross-sections, the cross-sections for formation of complexes with E, J are multiplied by the probabilities that these complexes will fragment to particular final states. The probabilities are calculated by the methods outlined in the first part of this section. Then summations must be performed over J and over final orbital momentum states.

$$S_{reac}(v',j'|v,j; E_t) = \sum_{l'} \sum_{J} S_{c,J}(|v,j; E_t) P_{c,J}(v',j',l'). \tag{46}$$

Equation 46 is most likely to predict the actual state-to-state reaction cross-sections when the potential energy in the entrance channel of the potential energy hypersurface falls monotonically to a deep 'well' and when there is also no maximum in the exit channel leading to products. Of course, if the overall reaction is strongly exothermic, a true collision complex in which energy is randomised may not form. However, the phase space formula provides a prediction against which measured distributions can be compared.

A4.9 High pressure limit: maximum free energy model[15]

Statistical adiabatic channel model calculations are complex and, in view of uncertainties about the long-range potential and its effects on the long-range channel potentials, the computational effort may not be worthwhile. Troe[16] has recently described somewhat simplified methods for calculating $k_2(E, J)$ and $k^\infty_{uni}(T)$ but both procedures remain algebraically complex.

Earlier work by Quack and Troe[15] provides a simple model which seems capable of estimating thermal rate constants in the limit of high pressure. Transition state ideas are fundamental to this method. The statistical mechanical and thermodynamic transition state formula for $k_{uni}^{\infty}(T)$ were given earlier as equations 3 and 4. Applying the variational principle means that one should perform a series of calculations at different points along the reaction path. The minimum calculated value of the rate constant then corresponds to the best transition state theory estimate of the rate constant, and this value will occur when $Q^{\neq} \exp(-\Delta E_0^{\neq}/k_B T)$ has its minimum value or when the local free energy, $\Delta A_c^{\neq}$, has its maximum value. In general, the rate constant calculated by adopting the maximum free energy criterion will exceed that computed from the statistical adiabatic channel model. However, the search for the position of maximum $\Delta A_c^{\neq}$ can be carried out at any temperature, so the method allows for the location of the transition state to vary with temperature.

Quack and Troe[13] devised a number of interpolative formulae which allow one to calculate $Q'(x)$, the partition function of the molecular system at any value of x from x_e (its value in the unimolecular reagent) to ∞ (corresponding to separated products). $Q'(x)$ is written as the product of four separable contributions

$$Q'(x) = Q'_{el}(x)Q'_c(x)Q'_{vr}(x)Q'_{EZ}(x) \tag{47}$$

which are given by the following formulae.

(a) $Q'_{el}(x) = g_{el} \exp[-V(x)/k_B T]$ \hfill (47a)

where g_{el} is the electronic degeneracy of the specified molecular state and it is assumed that $V(x)$ can be represented by the Morse expression:

$$V(x) = D_e\{1 - \exp[-\beta(x - x_e)]\}^2 \tag{47a'}$$

(b) $Q'_c(x)$ is the partition function associated with overall rotations of the system around axes perpendicular to the reaction coordinate. For reactions occurring over a Type II potential.

$$Q'_c(x) = k_B T/B_s \tag{47b}$$

where $B_s = \hbar^2/2\mu x^2$, μ being the reduced mass of the separating fragments.

(c) $Q'_{vr}(x)$ is the partition function associated with the 'active' modes; that is, all the internal motions except those dealt with in (b). To evaluate $Q'_{vr}(x)$, it is necessary to postulate how the energy levels of vibrational modes in the unimolecular reagent change as x increases and some of these motions become less restricted. Based on a formula developed for the statistical adiabatic channel calculations, Quack and Troe propose that the change in partition function can be represented approximately by

$$\ln Q'_{vr}(x) = \{\ln Q'_{vr}(x_e) - \ln Q'_{vr}(\infty)\} \exp[-\gamma(x - x_e)] + \ln Q'_{vr}(\infty) \tag{47c}$$

where γ is 0.75 α (see equation 19 in Chapter A1) and

$$Q'_{vr}(x_e) = Q_{vr}(\text{reagent})\{1 - \exp(-\varepsilon_{RC}/k_B T)\}S_e/Q'_c(x_e) \qquad (47c')$$

with $\varepsilon_{RC} = h\nu_{e,RC}$ the energy associated with one quantum of vibrational excitation in the vibration along the reaction coordinate in the molecular reagent, S_e is a symmetry factor, and $Q'_c(x_e)$ is the partition function for overall adiabatic rotation in the reagent molecule.

(d) Term (a) allows for the variation of electronic potential energy along the reaction coordinate. $Q'_{EZ}(x)$ allows for changes in the zero-point energy level which will be altered as certain motions become less restricted as x increases. The formula proposed is based on a similar assumption about how the energy levels change as equation 47c

$$\ln Q'_{EZ}(x) = -(1/k_B T)\{E_z(x_e) - E_z(\infty)\}\exp[-\gamma(x'-x_e)] + E_z(\infty). \qquad (47d)$$

By comparison of calculated with experimental data, Quack and Troe[13] propose the use of a length parameter, $\gamma = 0.75\,\text{Å}^{-1}$ in both equations 47c and 47d.

Table A4.3 compares the limiting high pressure rate constants for a number of radical–radical association reactions derived from experiment,[2,3] from calculations using the statistical adiabatic channel model (SACM), and from calculations using equation 47. The difference between the SACM and maximum free energy results is quite slight and, given the uncertainties in the long-range part of $V(x)$ and the relative complexity of SACM calculations, it seems reasonable for the experimentalist to rely on maximum free energy estimates of $k_{uni}^\infty(T)$.

Where necessary the contribution of reactions through electronically excited bound states should be included. In recombination at the high pressure limit, such pathways are likely to contribute to the overall rate in rough proportion to their electronic degeneracy.[9] It appears that maximum free energy calculations can provide estimates within a factor of about 2–3 of the experimental rate constants.

A4.10 Fall-off region

The transition or fall-off region between the two limits of low and high pressure behaviour may span several decades of total pressure. It is very difficult to cover such a wide range in experiments, so procedures which make it possible to derive the complete kinetic behaviour of a system from limited experimental information are very useful.

As has already been pointed out, evaluation of $k_{uni}(T)$ in the fall-off region requires that equation 7 or 27a is solved by numerically integrating the right-hand

Table A4.3. Comparison of the rate constants ($k_{rec}^{\infty}/10^{-11}$ cm^3 molecule^{-1}s^{-1}) for radical–radical association in the limit of high pressure with the values estimated using the statistical adiabatic channel SACM and maximum free energy Max.A models. $x^{\neq}$ is the value of the reaction coordinate at which the transition state is located by the maximum free energy criterion.

Reactants	T/K	Expt.	SACM	Max. A	$x^{\neq}/\text{Å}$
Cl + NO	300	5.8	6.2	7.9	6.3
	2100	—	5.2	7.4	4.6
Cl + NO$_2$	300	—	—	4.3	4.8
	2100	—	—	3.9	3.5
OH + NO$_2$	300	$\geqslant 3$	—	2.2	4.0
	1100	0.2	—	2.4	3.3
CH$_3$ + CH$_3$	300	4.8	5.1	4.7	4.7
	1300	3.9	4.8	6.2	3.6

side. The methods that have been discussed provide expressions for $k_2(E, J)$ or $k_2(E)$, $P(E, J)$ or $P(E)$, and $k_{-1}(E, J)$ or $k_{-1}(E)$.

In an RRKM calculation the first stage would be to choose the properties of a (fixed) transition state. If $k_{uni}^{\infty}(T)$ has been measured, this would be done by choosing the structure and frequencies of the transition state so as to satisfy equation 3. Having selected the transition state, $k_2(E)$ and $P(E)$ can be evaluated making use of the Whitten–Rabinovitch formulae, and $k_{-1}(E)$ can be put equal to βZ^0. When the high pressure limit cannot be reached experimentally, successive 'sensible' choices have to be made for the transition state, and fall-off curves calculated until the available experimental data are matched satisfactorily.

An RRKM model can usually be found to reproduce experimental data at any single temperature, but the method is quite laborious and not very revealing about the transition state. In addition, the use of equation 7, with its implicit assumption that weak collision effects can be incorporated by setting $k_{-1} = \beta Z^0$ is not strictly valid. Instead, a master equation should be set up in which the energy coordinate is divided into 'grains'. The master equation consists of a set of coupled differential equations describing reaction from and energy transfer between these grains.[17] Troe has suggested a number of approximate methods which are computationally simpler than using equation 7 and whose validity has been tested against full master equation solutions. The first of these procedures[17] is based on RRK theory which has been shown to match experimental data quite well as long as s, formally the number of oscillators in the molecule, is treated as an adjustable parameter. With $y = (E - E^0)/k_B T$ and $B = E^0/k_B T$, RRK theory leads to the following expression for the 'reduced' rate constant, $k_{uni}(T)/k_{uni}^{\infty}(T)$:

$$\frac{k_{uni}(T)}{k_{uni}^{\infty}(T)} = \int_{E^0}^{\infty} \frac{[y^{s-1}/(s-1)!]\exp(-y)}{1 + k_{uni}(T)\exp(B)[y/(y+B)]^{s-1}/Z^0[M]} \, dy. \tag{48}$$

Troe[17] has suggested how effective values of B and s (B_K and S_K) are estimated and reduced fall-off curves are constructed by numerically evaluating the integral in equation 48. These curves can then be matched to the available experimental data to find absolute values of $k^0_{uni}(T)$ and $k^\infty_{uni}(T)$. S_K is assumed to be a temperature dependent parameter chosen so that the correct high pressure activation energy is obtained. The activation energy is the difference between the average energy of species in the transition state and the average energy of reagent molecules, so

$$E^\infty_{act} = \Delta E^{\neq}_0 + S_K k_B T - U_{int} \tag{49}$$

$$S_K k_B T = k_B T - k_B \partial \ln Q^{\neq} / \partial(1/T) \tag{50}$$

so

$$S_K = 1 + T \partial \ln Q^{\neq} / \partial T. \tag{51}$$

If E^∞_{act} has not been measured (even if it has, some assumption must be made about $\Delta E^{\neq}_0$), S_K has to be estimated from equation 51 and an RRKM model of the transition state. Having found S_K from equation 49 or 51, the corresponding value of B_K must be calculated

$$B_K = \left(\frac{S_K - 1}{s - 1}\right)\left(\frac{E^0 + a(E^0)E_z}{k_B T}\right). \tag{52}$$

The doubly reduced fall-off curves can be plotted by reading values of (k_{uni}/k^∞_{uni}) at different values of $(k^0[M]/k^\infty_{uni})$ from tables published by Troe.[15]

Comparison of the fall-off curves calculated from modified Kassel integrals with those from complete RRKM calculations suggests that the former are reasonably accurate but rather too broad. To correct this defect and provide a simple functional form for the reduced rate constants, Troe has devised a method[16] which echoes the treatment he has developed for the low pressure rate constant. The reduced rate constant is treated as a product of the term given by the Lindemann–Hinshelwood analysis and two correction factors which broaden the curves. Hence

$$(k_{uni}/k^\infty_{uni}) = F_{LH}(k^0_{uni}[M]/k^\infty_{uni})F^{SC}(k^0_{uni}[M]/k^\infty_{uni}; S_K; B_K)$$
$$\times F^{WC}(k^0_{uni}[M]/k^\infty_{uni}; S_K; B_K; \beta). \tag{53}$$

The Lindemann–Hinshelwood term is

$$F_{LH} = \frac{k^0_{uni}[M]/k^\infty_{uni}}{1 + (k^0_{uni}[M]/k^\infty_{uni})}. \tag{54}$$

It is the spread in values of $k_2(E)$ which is chiefly responsible for broadening the fall-off curves beyond the Lindemann–Hinshelwood prediction, especially at high temperatures. This is allowed for by the factor F^{SC}, whilst F^{WC} allows for the smaller effect of weak collisions. The equations for both factors have been derived from extensive comparisons with full RRKM calculations and they are given in the Appendix to this chapter.

The broadening factor, F^{SC}, is a function of $(k^0_{uni}[M]/k^\infty_{uni})$ and also depends on the effective Kassel parameters S_K and B_K. At higher temperatures, a wider range of $k_2(E)$ values contributes to the thermal rate constant. This effect is reflected in increased values of S_K, F^{SC} deviates further from unity, and the range of total pressure between the low and high pressure limit grows. The ability of the method based on equation 53 to match full RRKM calculations in the limit of strong collisions is illustrated in Fig. A4.6.

The main effect of weak collisions ($\beta < 1$) is to move fall-off curves of $k_{uni}(T)$ versus [M] along the [M] axis towards a higher [M]. If reduced variables are used, i.e. (k_{uni}/k^∞_{uni}) versus $(k^0_{uni}[M]/k^\infty_{uni})$, so that the effect of weak collisions is incorporated into $k^0_{uni}[M]$, no change occurs to a first approximation. However, there is some broadening of the curves for small β. As long as $\beta > 0.05$ and S_K and B_K are not too large, this broadening can be allowed for using the equation for F^{WC} given in the Appendix.

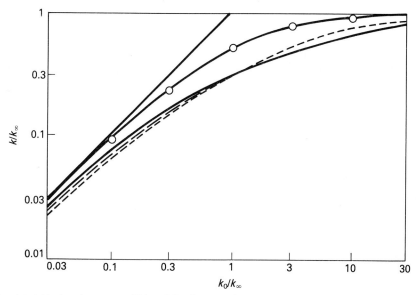

Fig. A4.6. Reduced strong collision fall-off curves for $NO_2Cl \rightarrow NO_2 + Cl$ at 455 K. (—○—) shows the result of a simple Lindemann–Hinshelwood calculation; (-----) is from evaluation of reduced Kassel integrals in equation 48; (– – – –) is based on the simple fall-off expression of equation 55 with 55a; and (———) is the result of both a complete RRKM calculation and the improved fall-off expression of equation A4.

Finally, we note that for some purposes such as atmospheric modelling, simple expressions are required to convey any pressure dependence of the rate constants. Especially for radical association reactions at low temperatures, Troe has suggested[18] that the simple form

$$k_{uni}(T) = k_{uni}^{\infty}(T) F_{LH} F \qquad (55)$$

will be adequate. The factor F is given by

$$\log F = \log F_{cent}/\{1 + \log(k^0[M]/k^{\infty})N\}^2 \qquad (55a)$$

where F_{cent} is a single parameter which is calculated (see Appendix, section A4.11) for $k^0[M] = k^{\infty}$. At least for recombination reactions at room temperature and below, F_{cent} is usually in the range 0.7—0.9.

A4.11 Appendix

The simplest approximation to the broadening factor F^{SC} is

$$\log F^{SC} \simeq \frac{\log F_{cent}^{SC}}{1 + \{\log(k_{uni}^0[M]/k_{uni}^{\infty})/N^{SC}\}^2} \qquad (A.1)$$

with

$$N^{SC} \simeq 0.75 - 1.27 \log F_{cent}^{SC} \qquad (A.2)$$

and F_{cent}^{SC}, the value of F^{SC} where $(k^0[M]/k^{\infty}) = 1$ given by

$$F_{cent}^{SC} = F_1(S_K) + F_2(S_K) \exp(-B_K/19.5)$$
$$+ [1 - F_1(S_K) - F_2(S_K)] \exp\{-2.3[B_K/F_3(S_K)]^{1.5}\} \qquad (A.3)$$

where

$$F_1(S_K) = 1.32 \exp(-S_K/4.2) - 0.32 \exp(-S_K/1.4) \qquad (A.3a)$$

$$F_2(S_K) = 1 - \exp(-S_K/30), \qquad (A.3b)$$

and

$$F_3(S_K) = 7.5 + 0.43 S_K. \qquad (A.3c)$$

For large S_K and B_K, slight asymmetries of the broadening factors have to be taken into account. This can be done by replacing equation A.1 with

$$\log F^{SC}(k^0[M]/k^\infty) \simeq \frac{\log F_{cent}}{1 + \left[\dfrac{\log(k^0[M]/k^\infty) - 0.12}{N^{SC} + \Delta N^{SC}}\right]^2} \tag{A.4}$$

where

$$\Delta N^{SC} \simeq \pm (0.1 + 0.6 \log F^{SC}_{cent}) \tag{A.5}$$

with the $+$ sign for $(k^0[M]/k^\infty) > 1$, the $-$ sign when $(k^0[M]/k^\infty) < 1$.

The broadening factor due to weak collisions can be calculated using the following equations

$$\log F^{WC} = \frac{\log F^{WC}_{cent}}{1 + \left[\dfrac{\log(k^0[M]/k^\infty) + c}{N^{WC} - d\{\log(k^0[M]/k^\infty) + c\}}\right]^2} \tag{A.6}$$

with the weak collision broadening factor at $(k^0[M]/k^\infty) = 1$ being given by

$$\log F^{WC}_{cent} \simeq 0.14 \log \beta \tag{A.7}$$

the width factor,

$$N^{WC} \simeq 0.7 + 0.3 S_K + 0.25 \log \beta \tag{A.8}$$

and the asymmetry terms

$$c \simeq 0.085 S_K = 0.17 \log \beta \tag{A.9}$$

$$d \simeq -0.2 - 0.12 \log \beta.$$

To evaluate equation 55a for F, one should use equation A.2 for N and calculate $F_{cent} = F^{SC}_{cent} F^{WC}_{cent}$ from equation A.3 and equation A.7.

A4.12 Suggestions for further reading

Robinson P. J. and Holbrook K. A. (1972) *Unimolecular Reactions*, Wiley, London.

Forst W. (1973) *Theory of Unimolecular Reactions*, Academic Press, New York.

Quack M. and Troe J. (1977) In Specialist Periodical Report, *Gas Kinetics and Energy Transfer*, (Ed. by P. G. Ashmore and R. J. Donovan), Vol. 2, Chap. 5. Chem. Soc., London.

Troe J. (1978) *Ann. Rev. Phys. Chem.*, **29**, 223.

Troe J. (1979) *J. Phys. Chem.*, **83**, 114.

Smith I. W. M. (1980) *Kinetics and Dynamics of Elementary Gas Reactions*, Chap. 4. Butterworths, London.

Howard M. J. and Smith I. W. M. (1983) *Prog. Reac. Kinet.* **12**, 55.

A4.13 References

1 Waage E. V. and Rabinovitch B. S. (1970) *Chem. Rev.*, **70**, 377.
2 Troe J. (1978) *Ann. Rev. Phys. Chem.*, **29**, 223.
3 Howard M. J. and Smith I. W. M. (1983) *Prog. Reac. Kinet.*, **12**, 55.
4 Troe J. (1977) *J. Chem. Phys.*, **66**, 4745 and 4758.
5 Whitten G. Z. and Rabinovitch B. S. (1963) *J. Chem. Phys.*, **38**, 2466.
6 Marcus R. A. and Rice O. K. (1951) *J. Phys. Colloid. Chem.*, **55**, 894.
7 Troe J. (1973) *Ber. Bunsenges. Phys. Chem.*, **77**, 665.
8 Troe J. (1983) *Ber. Bunsenges. Phys. Chem.*, **87**, 161.
9 Smith I. W. M. (1984) *Int. J. Chem. Kinet.*, **16**, 423.
10 Smith I. W. M. (1980) *Kinetics and Dynamics of Elementary Gas Reactions*, p. 129. Butterworths, London.
11 Stein S. E. and Rabinovitch B. S. (1973) *J. Chem. Phys.*, **58**, 2438.
12 Marcus R. A. (1952) *J. Chem. Phys.*, **20**, 359; Weider G. M. and Marcus R. A. (1962) *J. Chem. Phys.*, **37**, 1835; Marcus R. A. (1965) *J. Chem. Phys.*, **43**, 2658.
13 Quack M. and Troe J. (1974) *Ber. Bunsenges. Phys. Chem.*, **78**, 240 and (1975) **79**, 170.
14 Light J. C. (1967) *Discuss Faraday Soc.*, **44**, 14.
15 Quack M. and Troe J. (1977) *Ber. Bunsenges. Phys. Chem.*, **81**, 329.
16 Troe J. (1981) *J. Chem. Phys.*, **75**, 226 and (1983) **79**, 6017.
17 Troe J. (1974) *Ber. Bunsenges Phys. Chem.*, **78**, 478.
18 Troe J. (1979) *J. Phys. Chem.*, **83**, 114.

A4.14 Problems

Unimolecular reactions

A4.1. From the following data on A-factors in the limit of high pressure, estimate the entropies of activation for these unimolecular reactions. How can the values that you obtain be related to the nature of the transition states?

	A^{∞}/s^{-1} at $T = 500$ K
$C_2F_6 \rightarrow 2CF_3$	4×10^{17}
$CH_3CHO \rightarrow CH_3 + CHO$	1×10^{17}
$CH_2{=}CHF \rightarrow C_2H_2 + HF$	2.5×10^{14}
$CH_3NC \rightarrow CH_3CN$	2.5×10^{13}
$CH_3COOC_2H_5 \rightarrow CH_3COOH + C_2H_4$	2×10^{11}

A4.2. Use Troe's formalism (section A4.4) and the information given below to calculate rate constants for dissociation and association at 298 K in the limits of low pressure and strong collisions for the system:

$$CF_3O_2 + M \underset{k_{ass}}{\overset{k_{diss}}{\rightleftharpoons}} CF_3 + O_2 + M.$$

Given that the experimental value for $k^0_{ass}[N_2] = [N_2] \times 1.8 \times 10^{-29}\,cm^3$ molecule^{-1} s^{-1}, estimate β, the collisional efficiency for energy transfer.

Collision frequency $= Z^0[M]$

$$= [N_2] \times 4.2 \times 10^{-10}\,s^{-1}$$

$E_0 = 141.2\,kJ\,mol^{-1} \equiv 11\,805\,cm^{-1}$

$\tilde{v}(CF_3O_2) = 1303, 1260, 1172, 1092, 870, 692, 597, 580, 448, 286, 190,$ (torsion) $120\,cm^{-1}$.

Following Troe, the torsional motion should be included in calculating E_z, F_{rot} and F_{anh}, but excluded in calculating $\rho_{vib,h}(E^0)$ and Q_{vib}.

$F_{rot\,int}$ can be estimated using figures for CF_3OF: reduced moment of inertia, $I_m = 2.9 \times 10^{-46}\,kg\,m^2$; barrier height to internal rotation, $V_0 = 4.7\,kJ\,mol^{-1}$.

$\Delta H^{\ominus}_{298} = +144.8\,kJ\,mol^{-1}$; $\Delta S^{\ominus}_{298} = +157.7\,J\,mol^{-1}\,K^{-1}$

[standard state $= 1$ atm; 1 atm $= 2.46 \times 10^{19}$ molecule cm^{-3} at 298 K]

A4.3. Use the Maximum Free Energy Model (section A4.9) to estimate the limiting high pressure rate constants k^{∞}_{diss}, k^{∞}_{ass} for the

$$CF_3O_2 + M \rightleftarrows CF_3 + O_2 + M$$

reactions at 298 K.

The rovibrational partition function, Q_{vr}, for O_2 at 298 K is 72.1. For CF_3O_2 and CF_3, Q_{vr} is best estimated using tabulated thermodynamic data and the equation

$$\ln Q_{vr} = 3.665 + \frac{S^{\ominus}(T)}{R} - \frac{[H^{\ominus}(T) - H^{\ominus}(0\,K)]}{RT}$$

$$- 1.5 \ln (m/a.m.u.) - 2.5 \ln (T/K)$$

	$S^{\ominus}(298\,K)$ $J\,mol^{-1}\,K^{-1}$	$[H^{\ominus}(298\,K) - H^{\ominus}(0\,K)]$ $kJ\,mol^{-1}$
CF_3	265.0	2.754
CF_3O_2	322.4	3.100

(actually data for CF_3OF which serves as a reasonable model for CF_3O_2). Take γ as $0.75\,Å^{-1}$.

A4.4. Use the results that you have obtained from questions (2) and (3) and equation 55 to estimate the rate constant for combination of $CF_3 + O_2$ in 1 atmosphere of N_2 at 298 K. Take F_{cent} to be 0.7 and β to be 0.3.

A4.15 Answers to problems

Unimolecular reactions

A4.1. From transition state theory

$$k = (k_B T/h) \exp(-\Delta G_c^{\neq}/RT) \tag{1}$$

also

$$k = A \exp(-E_{act}/RT) \tag{2}$$

$$E_{act} = -d \ln k/d(1/RT).$$

$$= RT^2 \, d \ln k/dT.$$

From (1) $d \ln k/dT = (1/T) + \Delta G_c^{\neq}/RT^2 - (d\Delta G_c^{\neq}/dT)/RT$

$$= (1/T) + 1/RT^2[\Delta G_c^{\neq} + T\Delta S_c^{\neq}]$$

$$= (1/T) + 1/RT^2 . \Delta H_c^{\neq}$$

$$\therefore \; E_{act} = RT + \Delta H_c^{\neq}.$$

Substitution in (2) $\Rightarrow k = A e^{-1} \exp(-\Delta H_c^{\neq}/RT)$

$$\therefore \; A = (k_B Te/h) \exp(\Delta S_c^{\neq}/R)$$

Reagent	$\Delta S_c^{\neq}/\mathrm{J\,mol^{-1}\,K^{-1}}$
C_2F_6	79.4
CH_3CHO	67.9
$CH_2{=}CHF$	18.1
CH_3NC	−1.0
$CH_3COOC_2H_5$	−41.2

The values reflect the structure and 'looseness' of the transition state. Fissions of single bonds (e.g., dissociation of C_2F_6 and CH_3CHO) are characterised by loose transition states and high positive entropies of reaction.

$CH_2{=}CHF \rightarrow C_2H_2 + HF$ passes through a four-centre transition state with a presumably much tighter structure than those for single bond breaking. Likewise the transition state for $CH_3NC \rightarrow CH_3CN$ is 'tight'.

The last reaction is interesting, having a low A factor and a corresponding large decrease in entropy going to the transition state. This can be explained in terms of a cyclic transition state

A4.2. The calculation is carried out for $M = N_2$ in the strong collision limit. Equation numbers correspond to those in Chapter A4.

$$k^0 = Z^0[M][\rho_{vib,h}(E^0)k_BT/Q_{vib}] \exp(-E^0/k_BT)$$

$$\times F_{anh}F_EF_{rot}F_{rot\,int}F_{corr} \qquad (11b \text{ and } 16)$$

Harmonic density of states

The torsional mode is included in calculating b but not in calculating $\rho_{vib,h}(E^0)$. Taking account of active external rotations;[4]

$$b = (s-1)\frac{(s+\tfrac{3}{2})}{s} \sum_{i=1}^{s} v_i^2 \bigg/ \left(\sum_{i=1}^{s} v_i\right)^2 \qquad (14c)$$

$$= 1.354$$

E_z (torsion included) $= 4305\,\text{cm}^{-1}$

$E^0 = 141.2\,\text{kJ mol}^{-1} = 11\,805\,\text{cm}^{-1}$

$\log \omega = 1.0506\,(E/E_z)^{0.25}$

$\quad \omega = 0.0445$ for $E = E^0$

$a(E^0) = 0.940.$

$E^0 + a(E^0)E_z = 15\,851\,\text{cm}^{-1}$

$$\rho_{vib,h}(E^0) = [E^0 + a(E^0)E_z]^{s-1}/(s-1)! \prod_{i=1}^{s} (hv_i)$$

(torsion excluded, $s = 11$)

$$= 2.59 \times 10^4 \text{ states per cm}^{-1} \qquad (13)$$

$Q_{vib} = 2.985$ torsion included $\qquad (12)$

$\therefore$ At 298 K $\quad [\rho_{vib,h}(E^0)k_BT/Q_{vib}] = 1.79 \times 10^6$

$\exp(-E^0/k_BT) = 1.76 \times 10^{-25}$

Correction factors

(i) $F_{anh} = [(s-1)/(s-\frac{3}{2})]^m$ $\hspace{3cm}$ (18)

torsion is included $s = 12; m = 5$

$F_{anh} = 1.26$

(ii) $F_E = \sum_{v=0}^{s-1} \frac{(s-1)!}{(s-1-v)!} \left(\frac{k_B T}{E^0 + a(E^0)E_z}\right)$ $\hspace{2cm}$ (20)

$\hspace{1.2cm} = 1 + 0.143 + 0.0185 + 0.0022 + \cdots\cdots$

$F_E = 1.164.$

(iii) $F_{rot} = [(s-1)!/(s+\frac{1}{2})!][\{E^0 + a(E^0)E_z\}/k_B T]^{3/2}$

$\hspace{1cm} \times \left[\dfrac{2.15\,(E^0/k_B T)^{1/3}}{2.15\,(E^0/k_B T)^{1/3} - 1 + [E^0 + a(E^0)E_z]/(s+\frac{1}{2})k_B T}\right]$ $\hspace{1cm}$ (22)

$\hspace{1cm} = (0.0233)(676.0)(0.616)$

$F_{rot} = 9.72$

(iv) $F_{rot\,int} = \dfrac{(s-1)!}{(s-\frac{1}{2})!}$ $\hspace{4cm} = 0.282$

$\hspace{1.5cm} \times \left[\dfrac{E^0 + a(E^0)E_z}{k_B T}\right]^{1/2}$ $\hspace{2.5cm} = 8.776$

$\hspace{1.5cm} \times [1 - \exp(-E^0/sV_0)]$ $\hspace{2.6cm} = 0.917$

$\hspace{1.5cm} \times [1 - \exp(-k_B T/V_0)]^{1.2}$

$\hspace{1cm} + \dfrac{\exp(-1.2k_B T/V_0)}{\sqrt{2\pi I_m k_B T/\hbar^2}\,[1 - \exp(-\sqrt{n\hbar^2 V_0/2I_m(k_B T)^2})]}$

$\hspace{1cm} \left[0.339 + \dfrac{0.535}{16.14 \times 0.245}\right]$ $\hspace{3cm} = 2.37$

$\hspace{0.5cm} F_{rot\,int} = 0.282 \times 8.776 \times 0.917 \times 2.37$ $\hspace{1.5cm} = 5.56$

$\therefore k_{uni}^0 = [M]\,5.97 \times 10^{-2} \exp(-E^0/k_B T)\ s^{-1}$

$\hspace{1cm} = 1.05 \times 10^{-26}[M]\ s^{-1}$ at 298 K.

Equilibrium constant

At 298 K: $\Delta H_{298}^{\ominus} = 144.8\ kJ\,mol^{-1}$; $\hspace{0.5cm} \Delta S_{298}^{\ominus} = 157.7\ J\,mol^{-1}\,K^{-1}$

$\hspace{2cm} \Delta G_{298}^{\ominus} = 97.8\ kJ\,mol^{-1}$

$$K_p = 7.216 \times 10^{-18} \text{ [atmos]}$$

$$K_c = 1.775 \times 10^2 \text{ molecule cm}^{-3}$$

$$K_c = k_{\text{diss}}/k_{\text{ass}}$$

$$k_{\text{ass}} = k_{\text{diss}}/K_c$$

$$k_{\text{ass}}^0 = [\text{M}] \times 5.9 \times 10^{-29} \text{ cm}^3 \text{ molecule}^{-1} \text{ s}^{-1}.$$

$\therefore$ Third-order rate constant for

$$CF_3 + O_2 + N_2 \rightarrow CF_3O_2 + N_2$$

assuming strong collisions is, at 298 K

$$5.9 \times 10^{-29} \text{ cm}^6 \text{ molecule}^{-2} \text{ s}^{-1}.$$

Comparison with experimental values indicates

$$\beta_{N_2} = 0.30$$

which seems very reasonable.

A4.3. High pressure limit of $CF_3 + O_2(+M) \rightarrow CF_3O_2(+M)$.

$$k_{\text{ass}}^\infty(T) = (k_B T/h)(Q^{\neq}/\Pi Q) \exp(-\Delta E_0^{\neq}/k_B T).$$

Ratio of translational partition functions (per unit volume)

$$(k_B T/h)Q_{\text{tr}}^{\neq}/Q_{\text{tr}, CF_3}Q_{\text{tr}, O_2} = 6.28 \times 10^{-14} \text{ cm}^3 \text{ molecule}^{-1} \text{ s}^{-1} \text{ at 298 K}$$

Ratio of electronic partition functions

$$Q_{\text{el}}^{\neq}/Q_{\text{el}, CF_3}Q_{\text{el}, O_2} = 2/(2 \times 3) = 1/3$$

$$\therefore k = 2.1 \times 10^{-14} \ (Q_{\text{int}}^{\neq}/\Pi Q_{\text{int}}) \text{ cm}^3 \text{ molecule}^{-1} \text{ s}^{-1}$$

where partition functions are compared to the same zero of energy.

To use equation 47, first calculate rovibrational partition functions from the following equation

$$\ln Q_{\text{vr}} = 3.665 + \frac{S^\ominus(T)}{R} - \frac{[H^\ominus(T) - H^\ominus(0\,\text{K})]}{RT}$$

$$-1.5 \ln (m/\text{a.m.u.}) - 2.5 \ln (T/\text{K})$$

and thermodynamic data provided

$$\Rightarrow Q_{\text{vr}}(CF_3O_2) = 8.91 \times 10^6 \ (\text{using data for } CF_3OF)$$

$$Q_{\text{vr}}(CF_3) \quad = 2.95 \times 10^4$$

$$Q_{vr}(O_2) = 72.1$$

(a) electronic contribution: $Q'_{el}(x)$

$$E^0 = 11\,805\,\text{cm}^{-1}; \quad E_z(CF_3O_2) = 4305\,\text{cm}^{-1}$$
$$E_z(CF_3) \quad = 2655\,\text{cm}^{-1}$$
$$E_z(O_2) \quad = 780\,\text{cm}^{-1}$$
$$D_e \equiv 12\,675\,\text{cm}^{-1} \quad r_{C-O} = 1.43\,\text{Å}$$

taking β in the Morse potential as $3\,\text{Å}^{-1}$

$(x/\text{Å}) =$	3.0	3.3	3.5	3.7	4.0
$V(x)\,\text{cm}^{-1}$	227	92.6	50.9	27.9	11.4
$\exp[-V(x)/k_BT]$	2.99	1.57	1.28	1.15	1.06

(b) centrifugal motion: $Q'_c(x) = (2.6685 \times 10^2)(x/\text{Å})^2$

$(x/\text{Å})$	3.0	3.3	3.5	3.7	4.0
$(Q'_c(x))$	2.4×10^3	2.91×10^3	3.27×10^3	3.65×10^3	4.27×10^3

(c)

$$\ln\left\{\frac{Q'_{vr}(x)}{Q_{vr}(\infty)}\right\} = \ln\left\{\frac{Q'_{vr}(x_e)}{Q_{vr}(\infty)}\right\} \exp[\gamma(x_e - x)]$$

$$Q'_{vr}(x_e) = Q_{vr}(x_e)/Q'_c(x_e)$$

$$Q'_c(x_e) = 2.044 \times 10^3 \,[B_s(x_s) = 0.101\,\text{cm}^{-1}]$$

$$\therefore \quad Q'_{vr}(x_e) = 4.36 \times 10^3$$

$$\ln[Q'_{vr}(x_e)/Q'_{vr}(\infty)] = -6.19$$

$(x/\text{Å}) =$	3.0	3.3	3.5	3.7	4.0
$\dfrac{Q'_{vr}(x)}{Q_{vr}(\infty)}$	0.149	0.218	0.270	0.324	0.405

(d) zero-point correction: $E_z(x_e) = 3870\,\text{cm}^{-1}$ (includes reac. coordinate)

$$E_z(\infty) = 3455\,\text{cm}^{-1}$$

At 298 K $[E_z(x_e) - E_z(\infty)]/(-k_BT) = -2.099$

$(x/\text{Å})$	3.0	3.3	3.5	3.7	4.0
$Q'_{EZ}(x)$	0.524	0.597	0.641	0.682	0.737

Product of Factors

$(x/\text{Å})$	3.0	3.3	3.5	3.7	4.0
	560	594	724	928	1351

(Evaluation at $x = 2.9\,\text{Å} \Rightarrow 633.9$)

$$\therefore\ x^{\neq}/\text{Å} = 3.0$$

and $k^{\infty}(T) = 560 \times 2.1 \times 10^{-14}\,\text{cm}^3\,\text{molecule}^{-1}\,\text{s}^{-1}$

$$= 1.2 \times 10^{-11}\,\text{cm}^3\,\text{molecule}^{-1}\,\text{s}^{-1}$$

A4.4. We have

$$\beta = 0.3$$

$$k_{\text{ass}}^{0,\text{SC}} = 5.9 \times 10^{-29}\,\text{cm}^6\,\text{molecule}^{-2}\,\text{s}^{-1}$$

$$k_{\text{ass}}^{\infty} = 1.2 \times 10^{-11}\,\text{cm}^3\,\text{molecule}^{-1}\,\text{s}^{-1}$$

$$F_{\text{cent}} = 0.7$$

$$1\,\text{atm} = 2.46 \times 10^{19}\,\text{molecule}^{-1}\,\text{cm}^{-3}$$

$$F_{\text{LH}} = (43.5/1.2)/[1 + (43.5/1.2)]$$

$$= 0.973$$

From equation 55, $k_{\text{uni}} = k^{\infty}(0.973) \cdot (0.901)$

$$= 1.05 \times 10^{-11}\,\text{cm}^3\,\text{molecule}^{-1}\,\text{s}^{-1}.$$

Section B
Kinetics and dynamics of elementary reactions

Reaction kinetics is an experimentally based science. We have seen in Section A that, although theories of chemical kinetics rest on an increasingly sound foundation and *ab initio* or empirically based models show remarkable success at rationalising experimental data, the predictive capacity of such models is, as yet, limited. Good experimental data take precedence over theoretical estimates and the function of theory is primarily to provide a background of understanding, to provoke experimental study and to enable extrapolation of experimental results to regions where direct measurement is not possible (e.g. at extremes of temperature and pressure).

Progress in experimental kinetics and technical innovation have been closely linked to developments in theory and to the demand for kinetic data in industrial and environmental applications. The 1950's saw the first use of flash photolysis and of the discharge flow technique, both designed to provide experimental data on the elementary reactions of atoms and radicals which are important in various chain reactions. The same decade saw the first experiments on crossed molecular beams and the first direct studies of reaction dynamics. The pulsed and flow techniques also started to make contributions to this field in the 1960's, with the development of infra-red chemiluminescense and of absorption spectroscopy to detect state-resolved products. Monte Carlo simulations of dynamics encouraged further experimentation and demonstrated the importance of the form of the potential energy surface.

There were quite staggering developments in instrumentation during the 1970's. Molecular beams left the 'alkali' age, during which surface ionisation was the major detection technique, and nozzle beams and mass spectrometric detection opened up the method to a wider range of reactive species, although difficulties remained over the production of beams of many important atomic and molecular species. Lasers began to play a major role both as a means of producing radicals and of detecting them. Finally there was the impetus provided by the series of controversies concerning atmospheric chemistry; the introduction of nitrogen oxides into the stratosphere by supersonic aircraft, the impact of halocarbons on the ozone layer and pollution of the troposphere. These problems not only promoted a large injection of funds into kinetics, but also provided a focus for the development of a range of new techniques and for a widening of the temperature range over which laboratory investigations of elementary reactions were commonly conducted.

The developments continue into the 1980's. Variational adiabatic transition state theory, the calculation of *ab initio* potential energy surfaces and the increasing application of the models for unimolecular reactions discussed in Chapter A4 influence the design and interpretation of kinetic experiments, whilst technical advances have been made through the use of new laser methods and combustion provides the new applied focus.

In this central section of the book, we present an overview of the experimental techniques employed by the kineticist. Space precludes a detailed account and we provide instead a critical discussion of the techniques with an ample bibliography and a set of problems. We hope that these, together, will afford the reader an appreciation of the methodology of modern experimental kinetics.

Chapter B1 concentrates on the techniques employed to study thermal reactions. The importance of reaction initiation, via laser photolysis or in a discharge, and reactant detection, in real time, or versus distance in a flow experiment, are self-evident and these topics are discussed at some length. Most of the early experiments concentrated solely on the decay of the reactant, paying little attention to the nature of the products formed. Such an approach is acceptable for many reactions where only one set of products is energetically feasible. For other reactions it is not, and atmospheric chemistry provided a spur to develop methods for identifying product channels and determining their relative yields.

Reaction dynamics is the primary focus in Chapter B2, which examines state selection of reactants and molecular beam techniques. Interaction with theory is at its strongest and most direct in this area—it comes closest to Professor Simons' class of a theory-led field of kinetics; Chapter B2 demonstrates how theoretical understanding and technical advances have combined to generate the illuminating results discussed in Chapter A3.

An unsung revolution in chemical kinetics is that concerned with the recording and analysis of experimental data: digitisation, averaging and on-line computer analysis have all combined to provide the capacity for recording experimental data of high precision. Chapter B3 gives a brief résumé of the statistical aspects of data analysis, including linear and non-linear least squares techniques, the analysis of variance and the quoting of error limits.

Section B concludes with a chapter on case histories where an attempt it made to show how, over a period of years, a variety of techniques—both theoretical and experimental—is applied to a specific kinetic problem and how, over that period, our understanding develops. The examples have been deliberately chosen so that this chapter acts as a bridge between Sections A and B and Section C, which deals with complex kinetic systems.

Chapter B1
Measurements of Thermal Rate Data

G. HANCOCK

B1.1 Introduction

For a complete description of the behaviour of a chemical reaction under thermal conditions (i.e. equilibration of the reactant degrees of freedom at a defined temperature), three sets of experimental data are needed:

(a) A measurement of the concentration of a reactant species (or, in some cases, a measurement of a quantity proportional to reactant concentration) as a function of time following initiation.

(b) Measurements of these rate data as a function of external variables such as total pressure and, of great importance, temperature.

(c) Determination of product identities and their relative yields.

B1.2 Reaction initiation

We first look at the problem in a flow system, where reactant A is introduced into a flow of reactant B (Fig. B1.1). Measurements down the flow tube may be used to convert distance into time: measurements must be started however after mixing is complete. An upper limit for the time taken for this may be estimated from the radial diffusion equation:

$$t = \overline{(r^2)}/4D$$

where $\overline{(r^2)}^{1/2}$ is the root mean square distance that species A will diffuse into B in a time t and D is the diffusion coefficient. Turbulence will lead to shorter mixing times. As an example, if A and B are typical diatomic gases (O_2, N_2) then $D \simeq (0.015/p)$ m^2 s^{-1}, where p is the pressure in Torr. Thus at 1 Torr total pressure, radial diffusion over a distance of 1 cm takes about 2 ms; such calculations for experimental flow tube diameters should be used to ensure complete mixing before concentration measurements commence.

Reaction initiation by photolysis overcomes the mixing problem as generally a homogeneous concentration of the reactant A can be produced, by photolysis of a precursor, in the presence of the (non photolysed) reactant B. Time scales for photolysis are generally limited by the duration of the photolysis source

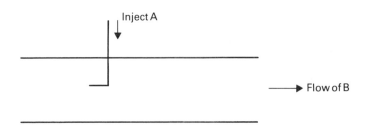

Fig. B1.1. Flow tube mixing reagent A with reagent B.

(although this is not always true: unimolecular decomposition of the reactant precursor may, in the case of very short photolysis pulses, be the rate determining step). Typical time scales are 10 μs for conventional flash photolysis (although this timescale can be drastically shortened at the expense of the total number of photons emitted), 10 ns for UV lasers, and several tens of fs for state-of-the-art pulsed dye laser sources.

Laser sources

UV frequencies are generally needed for photolysis. By far the most common source is now the excimer (or more correctly exciplex) laser, in which a rapid electrical discharge in a high pressure (1–2 atm) gas mixture forms an electronically excited bound state of a species XY^* for which the ground state is dissociative, thus automatically providing a population inversion (see Fig. B1.2). Commonly used fixed frequency sources (with typical output energies from commercial lasers) are listed below:

ArF	193 nm	200 mJ in 10 ns
KrF	248 nm	300 mJ in 10 ns
F_2	157 nm	10 mJ in 10 ns.

Quadrupled Nd:YAG laser systems produce 10–50 mJ at 266 nm at approximately the same pulse width: tunable UV output at energies between 1–5 mJ may be obtained straightforwardly in the wavelength range 260–320 nm by frequency doubling the output of a tunable dye laser. For shorter wavelengths, frequency mixing or Raman shifting techniques are available, generally (but not always) at output energies considerably below 1 mJ. These techniques are generally more useful for detection than photolysis. The advantages of laser photolysis over conventional flash photolysis are threefold: the energy required for photolysis is obtained in a short pulse width, with narrow spectral bandwidth and with high spatial coherence (which, as we shall see, is of importance in reducing scattered light).

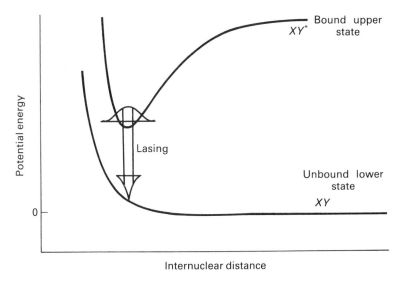

Fig. B1.2. Potential energy diagram for the upper and lower states of an excimer or exciplex laser.

Two potential problems need to be considered in any experiment in which thermal rate constants are measured following photolytic initiation:

(a) Non-radiative energy liberated in a photolysis step will eventually be degraded into heat—this may cause an appreciable (and possibly transient) temperature rise. The addition of a diluent gas can be used to overcome this problem.

(b) The kinetics of an observed quantum level (i) (see Fig. B1.3) can be affected by cascading of population of upper levels (ii) produced in the photolysis step. If the information required is the chemical removal of (i) by reaction, then relaxation of (ii) can reduce the apparent removal rate and can lead to an underestimation of the rate constant.

Problem (b) is of particular importance when single photon absorption produces appreciable population in vibrationally excited levels of the photofragment. A photolysis technique which overcomes this problem is that of infrared multiple

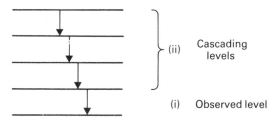

Fig. B1.3. Cascading of population from upper quantum levels. If level (i) is being observed, and levels (ii) collisionally feed this level, then the observed rate of removal of (i) will be affected by the cascade process, and will be an underestimate of the true rate of removal.

photon dissociation (MPD), where an intense pulse of IR radiation (generally from a pulsed CO_2 laser operating at wavelengths between 9–11 μm) causes sequential multiple photon absorption in a polyatomic molecule.[1] Figure B1.4 illustrates the absorption processes driving the molecule from discrete states through a 'quasicontinuum' of high vibrational state density to the dissociation threshold. Normally the lowest thermodynamically allowed dissociation channel predominates, and, because few molecules are excited far above the threshold level, the energy partitioned into fragment excitation is low.

Many examples appear in the literature of the use of photolysis to produce radicals for kinetic studies. Two examples are given below:

(a) $CH_3I + h\nu$ ($\lambda = 248$ nm) $\rightarrow CH_3 + I(^2P_{1/2, 3/2})$.

Photolysis at 248 nm (KrF radiation) dissociates CH_3I into CH_3 radicals and I atoms (in both $^2P_{1/2}$ and $^2P_{3/2}$ states). CH_3 radicals were observed by absorption using an infrared tunable diode laser, and the time-resolved behaviour of the concentration of the radical yielded the $CH_3 + CH_3$ recombination rate constant.[2]

(b) $CH_3OH + nh\nu \rightarrow CH$ (+ other products).

Infrared MPD was used to dissociate CH_3OH and produce the CH radical

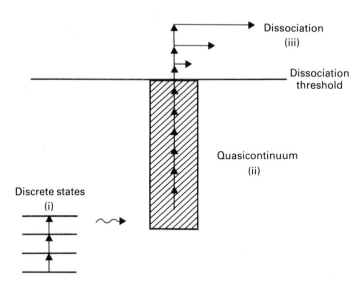

Fig. B1.4. Infrared multiple photon absorption and dissociation. Three regions of absorption are illustrated in this simplified view. In the discrete state region (i), rotational compensation of vibrational anharmonicity and power broadening effects enable absorption of a few monochromatic photons to occur. Region (ii) is of high state density such that absorption is guaranteed in a series of stepwise transitions through the 'quasicontinuum'. In the true continuum, region (iii), further photon absorption competes with unimolecular decomposition.

(amongst other products). The time-resolved behaviour of CH was measured by laser induced fluorescence to generate rate constants, for example, of the O + CH reaction.[3]

B1.3 Time resolved detection

(i) Absorption

Absorption methods usually rely on the Beer–Lambert Law relating the intensity, I, transmitted by a sample of concentration c and pathlength l to the initial intensity, I_0:

$$I = I_0 \exp(-\varepsilon c l).$$

ε is the absorption coefficient (or extinction coefficient or cross section). Authors use a variety of units (and often base 10 instead of base e) and so care is needed in their application. For example, in the detection of CH_3 radicals by absorption near 216 nm, $\varepsilon \simeq 3.5 \times 10^{-17} \, cm^2$ at 300 K (its value is temperature dependent), so 10% absorption in a 50 cm cell occurs for $[CH_3] \sim 2 \times 10^{-3}$ Torr.

Absolute sensitivities of absorption methods depend crucially upon sources of noise and of the time resolution required and are thus difficult to specify. Some of the sensitivity problems are overcome by using multipass optics to increase the pathlength, for which low divergence laser sources have been of particular advantage. Specificity of absorption can be a problem in the UV region of the spectrum, where many species show unstructured continua. Absorption in the IR is naturally far more species selective, and the use of tunable diode lasers in absorption has already been mentioned for kinetics studies of the CH_3 radical.

Molecular modulation spectroscopy is a technique which can achieve high sensitivity by absorption methods: absorptions of 1 part in 10^3 are routinely studied. Figure B1.5 shows the principles of the technique, and an application is outlined in reference 4. Reactants are formed by repetitive pulsed photolysis at a given square wave frequency f. The species monitored by absorption—a free radical, for example—builds up and decays when the photolysis light is on and off, respectively, and absorptions in-phase and in-quadrature with the photolysis frequency show values as a function of f which depend upon the kinetics of radical formation and removal. Experimental results are compared with those from simple kinetic models in order to extract rate constants from the data. For frequencies > 1 Hz a S/N ratio of 1 in a total observation time of 100 s has been measured for an absorption of 3 parts in 10^5.[4] More recent developments have included sampling of the whole waveform (instead of the separate in-phase and in-quadrature components as shown in Fig. B1.5) and the use of tunable IR sources for the absorption measurements.

(a)

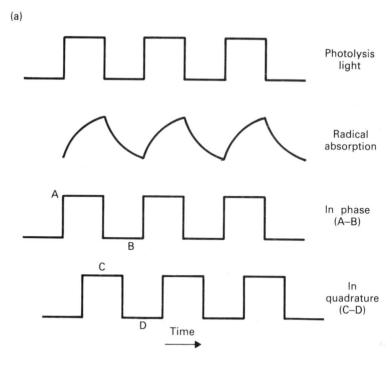

(b)

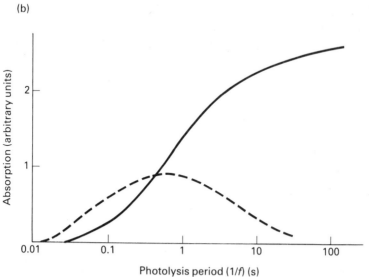

Fig. B1.5. Molecular modulation spectroscopy. (a) shows the square wave modulated photolysis light, with the radical concentration building up during the on period, and decaying when the light is off. Absorption signals are measured both in phase (A–B) and in quadrature (C–D) with the photolysis period, $1/f$. (b) shows typical observed variations of absorption with $1/f$, and the forms of these curves can be related to the rates of the processes occurring (———: in-phase signal; ----: in-quadrature signal).

(ii) Fluorescence

In principle, detection of species by fluorescence presents an advantage over absorption techniques. In the latter, high sensitivity is gained by accurate measurement of I/I_0—two very similar quantities when the absorption is low. In fluorescence, the signal is, in principle, measured against a zero background, and unlike absorption it increases with the intensity of the irradiating light, at least until saturation sets in. Background noise arising from scattered light from the photolysis source, or from the fluoresence excitation source scattered from windows or surfaces within the reaction vessel, is the major limitation in fluorescence measurements. Fluorescence also suffers from two disadvantages: firstly, it is difficult to extract absolute concentrations of species from these measurements, and secondly many species, particularly polyatomic molecules and radicals, have vanishingly small fluorescence quantum yields.

Resonance lamps

A suitable spectroscopic source for a species absorbing on a transition $A \leftarrow X$ is clearly the same species emitting on the transition $A \rightarrow X$, providing that both emission and absorption lineshapes are not affected by dissimilar broadening processes and that the emission source is not subject to severe self-absorption effects. Figure B1.6 shows the experimental arrangement using a so-called resonance lamp operating on the $A \leftrightarrow X$ transition. The lamp consists of a microwave discharge in a flow of appropriate precursor/rare gas mixture directed through baffles at the sample; fluorescence (again through baffles) is detected at right angles. Resonance lamps are normally used for transitions terminating on the ground state X of the species, and are most often used for the detection of atoms where transitions lie in the UV or far UV (and where laser excitation sources—see below—are weak). An example is the O atom resonance lamp operating at 130.2 nm. Sensitivity is high here and 10^9 atoms cm^{-3} can be detected. Molecular resonance lamps are used less often, but an important exception is the OH lamp operating on the $A\,^2\Sigma^+ - X\,^2\Pi$ transition.

Laser induced fluorescence (LIF)

The advantages of laser radiation outlined earlier for photolysis sources apply equally to its use for fluorescence detection. Tunability of laser sources enables a wide range of species to be detected. Commercial tunable dye lasers are available operating at wavelengths between 200–1000 nm with resolution less than or of the order of Doppler widths if required, i.e. with the capability of exciting individual vibration–rotation transitions of most electronic bands. Energies in the visible and near UV ($\gtrsim 350$ nm) can reach the tens of mJ range for pulsed lasers, hundreds of

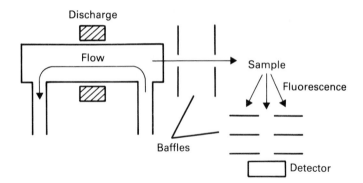

Fig. B1.6. Schematic arrangement for detection of species following excitation by a resonance lamp. The lamp consists of a discharge (often by microwave radiation) in a low pressure of gas either in a flow or a static system. Care needs to be taken in the lamp design to avoid excessive self absorption by ground state species present. The emitted light is incoherent, and substantial baffling of the input and output radiation are needed.

mW for cw lasers, and frequency doubling techniques can produce mJ UV output in the 260–350 nm range (cw operation in the UV via frequency doubling is possible at the mW level, but difficult to achieve routinely). Extension to the far UV by frequency mixing or Raman shifting gives useful output down to ∼ 200 nm with commercial pulsed laser systems. These figures are achieved with dye lasers with a pulsed excimer or Nd : YAG or cw Ar⁺ laser as the pump source. A 'simple' (i.e. less expensive) dye laser system, using a N_2 laser pump, produces output energies of hundreds of μJ in the visible region and 1–$10\,\mu J$ in the UV when frequency doubled. Energies of this magnitude are quite sufficient for LIF. A 10 ns pulse of $10\,\mu J$ focussed to ∼ 1 mm² is often enough to saturate a line in an allowed electronic transition.

What are the problems with such splendid light sources?

(a) *Scattered light.* Sensible design of the laser path in the reaction vessel can reduce scattered light to almost negligible proportions. The main problem is with resonance fluorescence (i.e. when the scattered radiation is at the same frequency as the excitation source). Baffles on the laser beam inlet and outlet systems, and on the collection optics, can reduce the resonance scatter from, for example, a $100\,\mu J$ visible laser pulse (2.5×10^{14} photons) to ∼ 1 scattered photon per pulse. Figure B1.7 shows a carefully designed system, in which scattering from the vessel windows (and secondary scattering from the baffles themselves) is almost entirely removed. If scattered fluorescence can be observed off resonance (Fig. B1.8 illustrates this for a diatomic species), an appropriate blocking filter can further reduce scattered light.

(b) *Saturation.* Generally we expect a LIF signal to increase linearly with input laser intensity. This is true if the lower state population is not appreciably depleted by radiation: if that happens, saturation occurs (Fig. B1.9). The intensity of

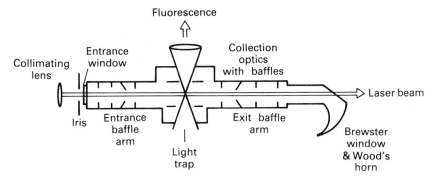

Fig. B1.7. Schematic view of laser detection system, showing baffle arrangements on the incoming and outgoing beams, and on the light collection system. Details of the purpose of each baffle and of their design are given in reference 5, from which the figure is adapted.

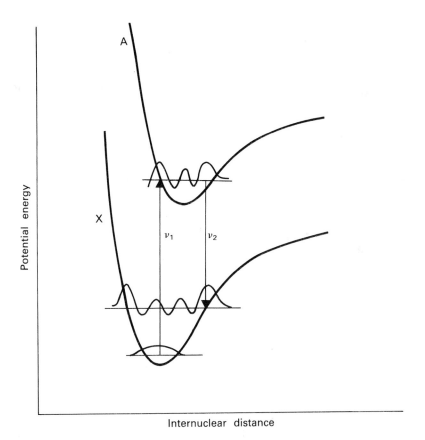

Fig. B1.8. Detection of non resonant fluorescence at frequency v_2 following exitation at v_1 for a diatomic species.

the scattered light is still proportional to the lower state population, but the proportionality 'constant' no longer truly reflects the absorption coefficient of the transition. This can cause problems if

(I) relative populations of specific quantum states are being estimated (line strengths and Franck–Condon factors can no longer be applied straightforwardly in order to relate observed intensities to populations);

(II) fluorescence signals are being normalised to incident laser intensity (for example, where intensity varies with wavelength, or when shot-to-shot intensity variation is large in a pulsed experiment).

Table B1.1 lists the majority of species for which kinetic information has been obtained by using LIF as the time-resolved detection method. The species listed range in their wavelengths of absorption from 226 nm (NO) to ~ 800 nm (C_2O), and unless indicated are in their ground (X) electronic states.

Table B1.1. Species for which LIF is used to obtain kinetic data

C_2 (X and a)	CH_2 ($\tilde{a}^1A_1$)	CH_3O
CH	CHF	C_2H_5O
OH	CF_2	
NH	CFCl	
CN	NCO	
NO	C_3	
	NH_2	
	HNO	
	C_2O	

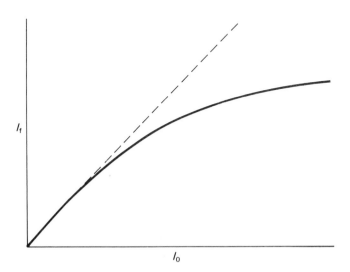

Fig. B1.9. Behaviour of fluorescence intensity I_f, as a function of incident light intensity, I_0, showing how non-linearity sets in, due to saturation of the pumped transition.

The ability of lasers to detect 'a few atoms' is often quoted. In kinetic studies such limits will not be reached due to scattered light problems and to the necessity of making time-resolved observations. 'A few atoms' refers to detectivity of a flux of a few atoms through an irradiated volume (~ 1 cm^3) per second. For practical kinetics where observations are done on a time scale of μs–ms, laser methods can conveniently be applied to concentrations of species $\sim 10^{10}$ cm^{-3} and, with some effort, can be extended to concentrations considerably below this: for example, detectivity of OH radicals (but not for kinetic studies) at concentrations of 10^6 cm^{-3} at atmospheric pressure has been reported.

(iii) Multiple photon excitation and ionisation

When the quantum yield of fluorescence is low (as in the case of many polyatomic species), the use of LIF becomes more difficult. A laser technique which can overcome this problem is that of multiple photon ionisation (MPI) which, as yet, has found its major use in detecting and studying the spectroscopy of free radicals rather than in kinetic applications. Figure B1.10 illustrates its application[6] to the detection of the CH$_3$ radical. Multiple photon excitation to an intermediate state is the first step in the MPI process—illustrated here for the two photon excitation of the $^2A_2'' \leftarrow {}^2A_2''$ transition using 333.5 nm radiation. Two photon excitation processes of this kind require high laser intensities (i.e. short pulses of high energy focussed to small beam sizes), as the rate of the initial process depends on the laser intensity squared. This high intensity often means that absorption by the intermediate state can be saturated, i.e. photon absorption competes effectively with other loss processes such as predissociation, in contrast to the simple fluorescence technique. As shown in Fig. B1.10, absorption by the intermediate state of a further photon at 333.5 nm leads to CH$_3^+$ formation—a so called 2 + 1 Resonantly Enhanced Multiple Photon Ionisation or REMPI process—and detection of the charged species is then a straightforward task. In this example CH$_3$ radicals were detected at ppm sensitivity in an atmospheric pressure flame. Two more examples illustrate the potential application of the REMPI technique.

The ClO radical, of considerable atmospheric importance, has not been detected by LIF (excitation of the A–X transition leads to predissociation), but 3 + 1 REMPI has been reported at wavelengths between 417–474 nm (via various $^2\Sigma$ states as the intermediates).[7] Br atoms have been detected via 2 + 1 REMPI at several discrete wavelengths between 240–285 nm with a sensitivity of $\sim 10^{10}$ cm^{-3} but in a very small volume, such that $\sim 10^{5-6}$ atoms were detected.[8] In this latter case, single photon LIF is possible, but the wavelengths required fall into the vacuum UV and are less readily produced by conventional lasers. Ionisation need not be the route for detection following multiple photon absorption: oxygen atoms have been detected by 2-photon excitation of the $3p\ ^3P$–$2p^4\ ^3P$ transition at

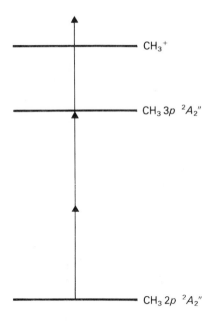

Fig. B1.10. Multiple photon ionisation applied to the detection of methyl radicals. Two photon excitation at 333.5 nm (CH_3) or 333.9 nm (CD_3) excites the ($3p\,^2A_2''-2p\,^2A_2''O\,^0_0$) transition and an additional photon of the same wavelength ionises the radical (a $2+1$ REMPI process).

226 nm followed by 845 nm fluorescence of the $3p\,^3P \to 3s\,^3S$ transition.[9] Note that 226 nm is a wavelength available at the mJ level from Nd : YAG pumped dye laser sources and is produced by mixing the tunable frequency-doubled output of the dye laser with the Nd : YAG 1.06μm fundamental in a non-linear crystal.

(iv) Mass spectrometric detection

Mass spectrometric methods are usually applied in flow systems or in studies employing end product analysis, but important results have been obtained in 'static' experiments where reaction is initiated by photolysis and time resolved mass spectrometric detection of reactants and products is carried out. Figure B1.11 shows schematically the experimental arrangement and the form of the results for the $CH_3 + NO_2$ system.[10] Here infrared MPD was used to produce methyl radicals in the step

$$C_6F_5OCH_3 + nh\nu \to C_6F_5O + CH_3$$

and time resolved photoionisation mass spectrometry used to detect species in the reaction

$$CH_3 + NO_2 \xrightarrow{k} CH_3O + NO.$$

(v) Flow tube methods

Flow tube methods are particularly useful for the generation of atomic species (O, N, H) and for the determination of their concentrations by gas phase titration methods.

N atoms

A discharge (usually by microwave radiation at 2450 MHz) in a flow of N_2 (or N_2/rare gas mixtures) at pressures between 0.1–5 Torr produces N atoms: concentrations of 10 mTorr are straightforwardly available downsteam of the discharge. Vibrationally excited $N_2^{\neq}$ is also formed: a plug of glass wool in the discharge tube is the method enshrined in the discharge flow folklore for decreasing the $[N_2^{\neq}]/[N]$ ratio. Estimation of the concentration, e.g. gas phase titration, relies on observation of emission bands from electronically excited species formed when a flow of NO is added to the gas flow containing N atoms:

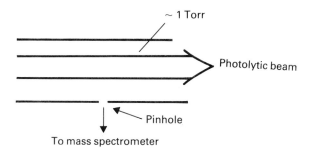

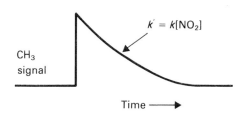

Fig. B1.11. Mass spectrometric detection of species in kinetics. In this example, precursor $(C_6F_5OCH_3)$ plus reactant (NO_2) are subject to a photolysis flash. CH_3 radicals are formed and their relative concentrations monitored, as a function of time, by measurement of the mass spectrometric signal at $m/e = 15$ from radicals passing through a small pinhole in the reaction vessel. The pseudo-first-order rate constant for the decay of the CH_3 signal is given by $k[NO_2]$, where k is the bimolecular rate constant for the $CH_3 + NO_2$ reaction.

$$N+N+M \rightarrow N_2^* + M \quad \text{YELLOW (Rayleigh glow)} \tag{1}$$

$$N+NO \rightarrow N_2 + O \quad (\text{rapid: } k \sim 10^{-10} \text{cm}^3 \text{molecule}^{-1} \text{s}^{-1}) \tag{2}$$

$$O+N+M \rightarrow NO^* + M \quad \text{VIOLET and ULTRAVIOLET}$$
$$(NO \, \beta \text{ and } \gamma \text{ bands}) \tag{3}$$

$$O+NO+M \rightarrow NO_2^* + M \quad \text{GREEN (air afterglow).} \tag{4}$$

Addition of NO causes the yellow Rayleigh afterglow (reaction 1) to be replaced by β and γ band emission from NO, produced by reaction 3, with O atoms formed by the rapid gas phase reaction 2. Addition of NO past the titration end point (i.e. when all N atoms are converted to O) results in the green air afterglow, reaction 4. Figure B1.12 illustrates how the intensity I of the β and γ bands changes with NO flow in the system. I is proportional to $[O][N]$; with the addition of NO the N atom concentration drops, the O atom concentration rises and the intensity thus shows a parabolic dependence on the flow of NO. At the end point, the flow of NO is equal to the initial N atom flow and thus a knowedge of the total pressure and the flows of the species present enables the concentration of N atoms to be found.

O atoms

O atoms are formed quantitatively by addition of NO to N atoms. Higher concentrations (up to 50 mTorr with a straightforward flow system) can be formed

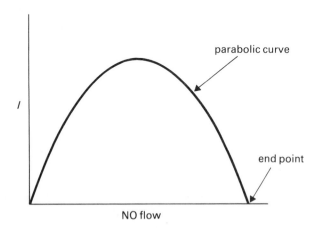

Fig. B1.12. Intensity, I, of the NO γ band emission produced by adding a flow of NO to a flow of N atoms. At the end point the blue γ bands (intensity $\alpha[N][O]$) are extinguished ($[N]=0$) and replaced by the green air afterglow (intensity $\alpha[O][NO]$): the added flow of NO is then equal to the original flow of N atoms.

by a microwave discharge through O_2/Ar mixtures. Metastable O_2 (particularly $O_2{}^1\Delta_g$) is also formed in this process. Estimation of concentrations can be made by adding NO_2:

$$O + NO_2 \rightarrow NO + O_2.$$

Here the green air afterglow has an intensity proportional to $[O][NO]$ and thus when NO_2 has been added in sufficient quantity to convert all the O atoms to NO, an end-point is reached at which the air afterglow is extinguished. Here the flow of NO_2 is equal to the initial O atom flow. Estimating flows of NO_2 is difficult and can be avoided by a method in which both NO_2 and NO are added, but only the flow rate of the latter is required.[12]

Flow tubes are used for two main purposes in kinetics; firstly to convert distance from the point at which one reactant is introduced into a flow of another into time via the flow velocity (typically $1-10\,\mathrm{m\,s^{-1}}$), and secondly to keep radical and atomic species 'alive' at known concentrations in a diluent gas so that kinetic observations can be made. Mass spectrometric detection in flow tube systems (converting distance into time) have been enormously successful in kinetic measurements, and have the advantage that the universal detection system employed allows product formation as well as reactant depletion to be measured both kinetically and in many cases quantitatively. An example is given in reference 12.

An example in which flow tubes are simply used to keep atoms present in a reaction system at a known concentration will be used to illustrate the way in which rate constants are derived from kinetic observations. The example is in the measurement of the $O + OH$ reaction,[13] and Fig. B1.13 illustrates the principles involved. Oxygen atoms are formed in a microwave discharge, a precursor molecule (HNO_3 in this case) is added and photolysed by a UV pulse. This forms OH radicals, which are detected in a time resolved fashion following the UV pulse by an OH resonance lamp. In all cases oxygen atoms are in excess over OH radicals, so that pseudo first-order removal of the latter species occurs:

$$O + OH \overset{k}{\rightarrow} products$$

$$-d[OH]/dt = k[O][OH]$$

and thus $[OH]_t = [OH]_0 \exp(-k[O]t)$.

It is not necessary to know absolute concentrations of OH : $[OH]_t$ and $[OH]_0$ can be replaced by the fluorescence intensities observed at times t and zero, respectively.

What happens if, instead of observing the removal of reactants, we observe the build up of products? We take a hypothetical example

$$A + BC \overset{k_1}{\rightarrow} AB + C$$

$$\overset{k_2}{\rightarrow} AC + B.$$

If we measure BC loss in excess A, then solution of the rate equations yields

$[BC]_t = [BC]_0 \exp\{-(k_1+k_2)[A]t\}.$

If we measure AB formation then (if nothing else happens to it) its time dependence is given by

$$[AB]_t = \frac{k_1[BC]_0}{(k_1+k_2)}\{1-\exp(-(k_1+k_2)[A]t)\}$$

i.e. from the kinetic measurement we still observe (k_1+k_2), which can only be split up into the two components k_1 and k_2 by relative yield measurements of, say, AB and AC. This task is often more difficult than the kinetic measurements. Mass spectrometric sampling can be successful: end product analyses are often used.

(vi) Laser magnetic resonance

Most polyatomic free radical species absorb in the infra red and far IR/microwave regions of the spectrum via vibration–rotation or pure rotational transitions, and absorption at these wavelengths can be highly species selective. If a tunable laser is not available then the transition can be brought into resonance with a fixed frequency source by Zeeman shifting. When this is done inside the laser cavity then the multipass advantage enables low concentrations of radical species to be detected. For example, this detection method has been combined with a flow tube

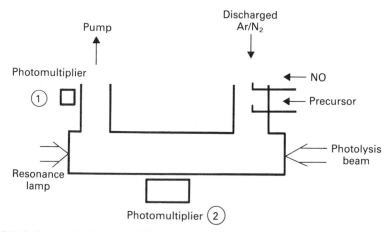

Fig. B1.13. Schematic diagram of the apparatus used for measurement of the O+OH rate constant. Oxygen atoms are formed by titration of a discharged Ar/N_2 mixture with NO, the end point being detected by photomultiplier 1. An OH precursor molecule (HNO_3 in this case) is added to the flow, photolysed to produce OH, and the relative OH concentration measured by resonance fluorescence using photomultiplier 2. Adapted from reference 13.

for kinetic studies of the HO_2 radical,[14] a species of importance in combustion and in aeronomy, which has not been detected by conventional LIF.

B1.4 End product analysis

Relative rate constants for many processes may be derived from product ratios obtained by end product analyses, with gas chromatographic, mass spectrometric or spectroscopic (e.g. Fourier transform infrared) methods used to determine product concentrations. Often such methods (see Section C) rely on selection of a set of chemical processes, construction of a predicted set of results and comparison with experimental results; mechanistic rather than kinetic information is often derived. Many such studies have been reported, particularly with reference to combustion research: we describe briefly two such sets of measurements. In the first, the mechanism was investigated for the decomposition of vibrationally excited ethylene formed by recombination of CH_2 radicals produced by photolysis of ketene.[15] Equimolar mixtures of CH_2CO and CD_2CO were photolysed: of the two pathways for the decomposition of the $CH_2{=}CD_2^*$ species that was formed, namely

route (b) was shown to predominate, proceeding via the vinylidene intermediate, a species whose vacuum UV spectrum could then be assigned.

The second is an example of a pyrolysis reaction carried out at high temperatures (673–813 K) in which ethane decomposition (yielding C_2H_5 radicals) was studied in the presence of O_2.[16] The pressure dependent product yields were used to test the mechanism of decomposition/stabilisation of the nascent $C_2H_5O_2^*$ adduct:

$$C_2H_5 + O_2 \rightarrow C_2H_5O_2^* \overset{M}{\rightarrow} C_2H_5O_2$$
$$\downarrow$$
$$C_2H_4 + HO_2.$$

Products were identified by gas chromatography, and rate constants at one particular temperature were put on an absolute basis by comparison of the rate of formation of C_2H_4 with that of C_4H_{10}, the latter being formed from ethyl radical dimerisation with a known rate constant.

B1.5 Measurement of temperature dependences

A single temperature measurement of a rate constant may be of some practical significance, but gives little information on the reaction mechanism. Theoretically, determination of the activation energy, the pre-exponential factor and any curvature in the Arrhenius plot provides far more understanding, and often measurements over a range of temperatures are of importance in the applications of kinetic information, for example to processes of combustion and aeronomy. We examine briefly three methods of studying the temperature dependences of rate constants.

(i) Heated reaction vessels

Conventional heated vessels operate up to temperatures of ~ 800 K—flow systems with resonance fluorescence detection have been reported as operating at ~ 1900 K.

(ii) Shock tubes

The temperature range for shock tubes is of the order 800–2500 K. Most studies with shock tubes involve unimolecular decomposition measurements (but this can lead to bimolecular recombination rate constants if the equilibrium constant is known). Species can be monitored by absorption or fluorescence methods. Bimolecular reactions have been studied in shock tubes, and two examples are given below. In the first,[17] the recombination of N atoms to form electronically excited N_2^*

$$N + N \xrightarrow{k(T)} N_2^*$$
$$\downarrow hv$$
$$N_2$$

was studied as illustrated in Fig. B1.14. Emission from N_2^* rose to a new value when the temperature dependent recombination rate constant $k(T)$ was increased following shock heating. The ratio of the emission intensities at the two temperatures is related to the rate constant ratios:

$$\frac{I_2}{I_1} = \frac{k(T_2)}{k(T_1)} \frac{[N]_2^2}{[N]_1^2}.$$

In the second example,[18] a flowing mixture containing metastable O_2 ($b^1\Sigma_g^+$ and $a^1\Delta_g$) formed from a discharge in O_2 was shock heated, and, via spectroscopic observations of these states, rate constants for the processes

$$2O_2(a^1\Delta_g) \rightarrow O_2(b^1\Sigma_g^+) + O_2(X^3\Sigma_g^-)$$

$$O_2(b^1\Sigma_g^+) + O_2; \ N_2 \rightarrow \text{products}$$

were measured as a function of temperature.

(iii) Laser heating

Pulsed IR lasers can be used to heat a reaction mixture rapidly to temperatures which are difficult to attain with conventional heated reaction vessels. A strong IR absorber is added to the reactant mixture, pulsed irradiation raises the temperature, and the reaction course is followed by the conventional means already described. Pulsed CO_2 lasers have been used for these experiments; SF_6 and SiF_4, both strong multiple photon IR absorbers, have been used as the sensitisers, and conditions have been chosen such that the absorbed vibrational energy in these species is rapidly converted to equilibrium temperatures by collisional relaxation. Two types of experiment have been performed. In the first, the temperatures achieved have been monitored by adding a 'chemical thermometer' to the system, a molecule which on heating decomposes in more than one channel to form products which can be measured by analysis, the channels having markedly different Arrhenius parameters so that product ratios are characteristic of reactant temperatures.

A second more direct technique has been to monitor the temperature by LIF measurements of the quantum state distributions of a species present in the reactant mixture. The reaction

$$OH + CH_4 \rightarrow products$$

has been studied as a function of temperature between 800–1400 K by this method.[19] Pulsed irradiation of a mixture of IR sensitiser (SF_6), diluent gas (Ar, He), reactant hydrocarbon (CH_4) and a precursor molecule (H_2O_2) was carried out. The precursor is rapidly decomposed at the temperatures reached to form OH radical reactants, whereas the sensitiser and hydrocarbon remain intact. Time resolved measurements of the OH concentration and temperature were carried out by pulsed LIF of the $A\,^2\Sigma^+ - X\,^2\Pi$ transition, and this enabled the complex effects

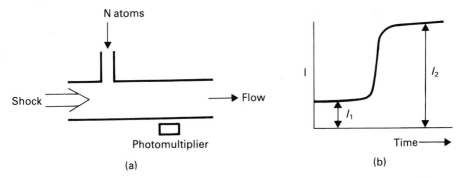

Fig. B1.14. Shock tube study of the N atom recombination rate constant. Nitrogen atoms from a discharge are shock heated in a flow tube [part (a)]. The emission from the Lewis Rayleigh afterglow, I_1, is seen to increase to a new value, I_2, as the shock wave heats the mixture. Relative recombination rate constants as a function of temperature can be obtained from the ratio I_2/I_1.

of shock waves, density changes and temperature variations to be unravelled. Further applications of this potentially extremely useful technique can be expected.

B1.6 Unimolecular reactions

Many of the experimental techniques outlined above can be applied to studies of unimolecular as well as bimolecular kinetics. In this section we consider possible methods of measuring directly the unimolecular decomposition rates of molecules excited above their dissociation thresholds. In normal unimolecular kinetic measurements, collisional activation is used to produce this energised molecule, and the assumption of free energy flow amongst the internal degrees of freedom is used to calculate the decomposition rate. Full details are given elsewhere in this book. Direct measurements of the decomposition rate requires reactant preparation in the absence of collisions, i.e. optically, and several schemes have been devised to achieve this under conditions where the molecule dissociates on the ground electronic state surface (clearly a direct photodissociation step, populating a repulsive excited state of the parent molecule, and leading in the timescale of a molecular vibration to dissociation, is inappropriate). Three methods have been attempted:

(a) Electronic excitation followed by rapid internal conversion to form the energised molecule. Observations have been made by UV absorption spectroscopy,[20] and direct observations of reaction products by ps spectroscopy are now becoming possible.[21]

(b) Excitation of high vibrational overtones of C—H or O—H stretching modes, with time resolved fragment detection, e.g. tBu—O—OH excited at 531.9 nm, $v = 6$ in O—H stretch, decomposing to give OH radicals.[22] The distribution of internal energy in the products of dissociative events of this kind can be of help in deciding if energy randomisation within the parent molecule has occurred.

(c) Infrared multiple photon dissociation of a large molecule with LIF detection of products. Reference 23 reports the infrared MPD of benzylamine:

$$C_6H_5CH_2NH_2 + nh\nu \rightarrow C_6H_5CH_2 + NH_2$$

with the NH_2 fragment detected as a function of time after the IR laser pulse.

Such techniques are still in their early stages of development and have only been applied in rather special cases [although for one example, the decomposition of tetramethyldioxetan, methods (b) and (c) have been used].[24] The study of photodissociation dynamics (of which such measurements are a small subset) is currently an extremely active one: reference 25 lists some recent reviews.

B1.7 Suggestions for further reading

Baggot J. E. and Pilling M. J. (1982) *Ann. Rep. Roy. Soc. Chem.*, 199.

B1.8 References

1   Ashfold M. N. R. and Hancock G. (1982) Roy. Soc. Chem. Specialist Periodical Report, *Gas Kinetics & Energy Transfer*, **4**, 73.
2   Laguna G. A. and Baughcum S. L. (1982) *Chem. Phys. Letters*, **88**, 568.
3   Messing I., Filseth S. V., Sadowski C. M. and Carrington T. (1981) *J. Chem. Phys.*, **74**, 3874.
4   Parkes D. A., Paul D. M. and Quinn C. P. (1976) *J. Chem. Soc., Faraday Trans 1*, **72**, 1935.
5   Pruett J. G. and Zare R. N. (1976) *J. Chem. Phys.*, **64**, 1774.
6   Chou M. S. (1985) *Chem. Phys. Letters*, **114**, 279.
7   Duiganan M. T. and Hudgens J. W. (1985) *J. Chem. Phys.*, **82**, 4426.
8   Arepalli S., Presser N., Robie D. and Gordon R. J. (1985) *Chem. Phys. Letters*, **117**, 64.
9   Bischel W. K., Perry B. E. and Crosley D. R. (1981) *Chem. Phys. Letters*, **82**, 85.
10 Yamada F., Slagle I. R. and Gutman D. (1981) *Chem. Phys. Letters*, **83**, 409.
11 Reeves R. R., Mannella G. and Harteck P. J. (1960) *J. Chem. Phys.*, **32**, 632.
12 Gehring M., Hoyermann K., Schacke H. and Wolfrum J. (1973) 14*th Int. Symp. on Combustion*, 99.
13 Howard M. J. and Smith I. W. M. (1981) *J. Chem. Soc. Faraday Trans. 2*, **77**, 997.
14 Thrush B. A. and Wilkinson J. P. T. (1979) *Chem. Phys. Letters*, **66**, 441.
15 Laufer A. H. (1982) *J. Chem. Phys.*, **76**, 945.
16 Baldwin R. R., Pickering I. A. and Walker R. W. (1980) *J. Chem. Soc. Faraday Trans. 1*, **76**, 2374.
17 Gross R. W. F. (1968) *J. Chem. Phys.*, **48**, 1302.
18 Borrell P. M., Pedley M. D., Borrell P. and Grant K. R. (1979) *Proc. R. Soc. London, Ser. A*, **367**, 395.
19 Smith G. P., Fairchild P. W., Jeffries J. B. and Crosley D. R. (1985) *J. Phys. Chem.*, **89**, 1269.
20 Hippler H., Luther K., Troe J. and Wendelken H. J. (1983) *J. Chem. Phys.*, **79**, 239.
21 Knee J. L., Khundkar L. R. and Zewail A. H. (1985) *J. Chem. Phys.*, **83**, 1996.
22 Rizzo T. R. and Crim F. F. (1982) *J. Chem. Phys.*, **76**, 2754.
23 Reisler H., Pessine F. B. T. and Wittig C. (1983) *Faraday Discuss. Chem. Soc.*, **75**, 284.
24 Cannon B. D. and Crim F. F. (1981) *J. Chem. Phys.*, **75**, 1752; Ruhman S., Anner O. and Haas Y. (1983) *Faraday Discuss. Chem. Soc.*, **75**, 239.
25 (1985) Photodissociation and Photoionisation. In *Advances in Chemical Physics*, Vol. 60 (Ed. by K. P. Lawley, Wiley Interscience, New York); Buelow S., Noble M., Radhakrishnan G., Reisler H., Wittig C. and Hancock G. (1986) *J. Phys. Chem.*, **90**, 1015; Simons J. P. (1984) *J. Phys. Chem.*, **88**, 1287.

B1.9 Problems

Measurements of thermal rate data

B1.1. This problem is designed to allow an estimate to be made of the feasibility of carrying out LIF detection of free radicals following UV dissociation.

You wish to measure the rate constant at 300 K for the process

$$N + CHF \rightarrow products$$

using the apparatus illustrated in Fig. B1.15.

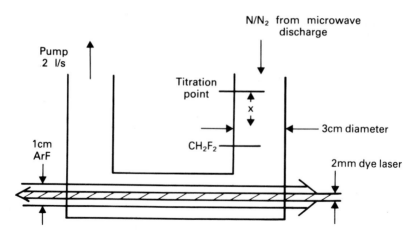

Fig. B1.15.

N atoms are made from a microwave discharge in N_2 at a pressure of 1 Torr: this makes $\sim 10\,mTorr$ N. The concentration is measured by adding NO until the $N + NO$ end point is reached, observed a distance x downstream. Given that the $N + NO$ rate constant is $1 \times 10^{-10}\,cm^3\,molecule^{-1}\,s^{-1}$ calculate x to ensure:

(a) the titration reaction is $> 99\%$ complete *and*
(b) the NO has diffused into the N/N_2 flow.
(Diffusion coefficient, D, for NO in $N_2 = 147\,cm^2\,s^{-1}$.)

Downstream, $10\,mTorr\ CH_2F_2$ is introduced (it does not react with N) and is photolysed at 193 nm by ArF radiation ($10\,mJ$ in a 1 cm diameter beam). For the purpose of this calculation we shall assume that the absorption coefficient of CH_2F_2 at this wavelength is $3 \times 10^{-18}\,cm^2$, and that the quantum yield for CHF production is unity. What concentration of CHF is produced? What would be the situation if a 1J pulse at 193 nm was used?

Given that the process

$$CH_2F_2 \rightarrow CHF + HF$$

has

$$\Delta H^{\ominus}_0 = 360\,kJ\,mol^{-1}$$

calculate the maximum temperature rise caused by photolysis with the 10 mJ beam.

LIF is excited by a counter-propagating pulsed (5 ns) dye laser, 2 mm diameter, $1\,\mu J$ at 470 nm. At this wavelength the CHF absorption cross section is $10^{-17}\,cm^2$; the CHF* radiative lifetime is $2\,\mu s$. Calculate the volume emission rate (photon emitted/$cm^3\,s$) immediately after irradiation with the dye laser.

A photomultiplier views a 1 cm long region of the excited CHF radicals; the 50 mm diameter photocathode is positioned 10 cm from the beam and views scattered photons through a filter (transmission 5%). Its quantum efficiency is 20%, its gain 10^6. The output is fed across a $50\,\Omega$ resistor. Calculate the voltage expected from the initially produced CHF* emission.

B1.2. The reaction between atoms and radicals often produces primary products which themselves are removed by reaction with the atomic species. The following reaction scheme applies.

$$A + BCD \xrightarrow{k_1} AB + CD$$

$$\xrightarrow{k_2} \text{other products}$$

$$A + AB \xrightarrow{k_3} A_2 + B$$

$$\xrightarrow{k_4} \text{other products.}$$

A is an atomic species and is in excess over BCD.

Calculate
(i) the time dependence of the concentration of AB;
(ii) the dependence of the maximum AB concentration on the concentration of A.

B1.3. This problem illustrates the use of very low pressure pyrolysis to determine the equilibrium constant, K, for the process

$$Cl + CH_4 \underset{k_{-1}}{\overset{k_1}{\rightleftharpoons}} CH_3 + HCl$$

k_1 is determined from flash photolysis/resonance fluorescence experiments, and hence k_{-1} obtained.

The equilibrium constant, K, for the reaction

$$Cl + CH_4 \underset{k_{-1}}{\overset{k_1}{\rightleftharpoons}} CH_3 + HCl$$

is determined from measurements of the Cl, HCl and CH_4 steady state concentrations in a very low pressure reactor under the following conditions. Chlorine atoms are admitted at a constant flux, F, (atom s^{-1}) through a pinhole into a low pressure reactor of volume V, and leave the reactor by effusion through a small hole: no other loss processes take place. Steady state conditions apply in the reactor, and thus the input flow rate, F/V, can be related to $[Cl]_0$, the steady state Cl atom concentration with no added HCl or CH_4. On adding these molecules, $[Cl]_0$ changes to a new value, $[Cl]$, due to reactions 1 and -1 taking place: the input flux F remains unaltered.

By considering steady state conditions for Cl and CH_3 show that the ratio $[Cl]/([Cl]_0 - [Cl])$ can be expressed in terms of the steady state values of $[HCl]$

and $[CH_4]$, the rate constants k_1 and k_{-1} and the rate constants for the effusive loss of Cl and CH_3.

The following data were derived from mass spectrometric measurements of $[Cl]$, $[HCl]$ and $[CH_4]$ at 298 K: $[CH_4]$ was kept constant

$$\frac{[Cl]}{[Cl]_0 - [Cl]} \quad 2.85 \quad 3.36 \quad 3.86$$

$$\frac{[HCl]}{[CH_4]} \quad 1.0 \quad 2.0 \quad 3.0$$

Derive a value for K from these data and by making reasonable assumptions concerning the effusive loss rate constants.

In a separate experiment, the $Cl + CH_4$ forward rate constant was measured at 298 K by photolytic production of Cl followed by its observation by resonance fluorescence in excess CH_4.

Cl atom decays were found to be of the form

$$-\frac{d[Cl]}{dt} = k'[Cl]$$

with k' dependent on $[CH_4]$ as given below:

k'/s^{-1}	$[CH_4]/molecule\ cm^{-3}$
1.15×10^4	10^{17}
8.0×10^3	6.5×10^{16}
3.5×10^3	2.0×10^{16}

Determine k_1 and k_{-1} at 298 K.

B1.10 Answers to problems

Measurements of thermal rate data

B1.1. (i) End point is reached when $[NO]_0 = [N]_0 \equiv 10\ mTorr$

$$-\frac{d[N]}{dt} = k[N][NO] = k[N]^2$$

$$\left(\frac{1}{[N]} - \frac{1}{[N]_0}\right) = kt$$

for $[N] < 0.01\ [N]_0$, $t > 3.05 \times 10^{-3}\ s$

$$x > 0.86\ cm.$$

(ii) The diffusion coefficient for NO in $N_2 = 147\ cm^2\ s^{-1}$, we need t such that $(4Dt)^{1/2} \sim$ tube radius:

$$t > 3.8 \times 10^{-3}\ s$$

$$x > 1\ cm.$$

(iii) 10 mJ ArF in 1 cm diameter $= 1.24 \times 10^{16}$ photons cm^{-2}

For a 1 cm path length, $\varepsilon cl = 9.72 \times 10^{-4}$

$\therefore$ $I_{ABS} \simeq I_0 \, \varepsilon cl = 1.2 \times 10^{13}$ photon cm^{-3} = CHF concentration produced.

We started with 3.24×10^{14} molecule cm^{-3} of CH_2F_2, so we have 3.7% dissociation. For 1J the same calculation produces a nonsense (more photons are absorbed than there are molecules present), i.e., c is not constant, and the system is close to saturation. Dissociation should be complete.

(iv) Take the heat capacity as C_v (for the maximum effect) and neglect the contribution from 1% CH_2F_2 and N atoms

i.e. $C_v = \dfrac{5}{2} R$ for N_2 at 300 K.

We photolyse $\dfrac{3.7}{100} \times \dfrac{1}{100}$ mol CH_2F_2 per mol of N_2.

The excess energy $= N_A h\nu - 360 = 260$ kJ mol^{-1}.

i.e. we deposit $3.7 \times 10^{-4} \times 260 \times 10^3 = 96.2$ J mol^{-1} excess energy and the temperature rise is, therefore,

$$\Delta T = \frac{96.2}{20.8} = 4.6 \text{ K}.$$

(v) The initial $[CHF] = 1.2 \times 10^{13}$ cm^{-3}

$$\varepsilon cl = 1.2 \times 10^{-4}$$

A 1 µJ, 470 nm beam of 0.2 cm diameter is equivalent to 7.55×10^{13} photons cm^{-2}

$\therefore$ $[CHF^*] = 9.07 \times 10^9$ radicals cm^{-3}

Volume emission rate $= A[CHF^*] = 4.5 \times 10^{15}$ photons cm^{-3} s^{-1}

Photomultiplier views 3.14×10^{-2} cm^3

$\therefore$ Photons collected $= 4.5 \times 10^{15} \times 3.14 \times 10^{-2} \times \dfrac{2.5^2}{400}$

$$= 2.2 \times 10^{12} \text{ photons s}^{-1}.$$

Allowing for a 5% filter transmission, a 20% quantum yield and a 10^6 gain we produce a current of 2.2×10^{16} electrons s^{-1}

i.e. 3.52×10^{-3} A and 0.176 V across a 50 Ω load.

(This signal would probably saturate the PM tube.)

B1.2. (i) $[AB] = \dfrac{(k_1 + k_2)[BCD]_0}{(k_3 + k_4) - (k_1 + k_2)} \{ \exp[-(k_1 + k_2)[A]t] - \exp[-(k_3 + k_4)[A]t] \}$

where $[A]$ is assumed constant.

(ii) Differentiate and set the resulting expression equal to zero:

$$[AB]_{max} = m \left[\left(\frac{n}{p} \right)^{-n/(p-n)} - \left(\frac{n}{p} \right)^{-p/(p-n)} \right]$$

where $m = \dfrac{(k_1 + k_2)[BCD]_0}{(k_3 + k_4) - (k_1 + k_2)}$

$n = k_1 + k_2$

$p = k_3 + k_4$

i.e. $[AB]_{max}$ independent of $[A]$.

B1.3. Let k_{Cl}, k_{CH_3} be the first-order rate constants for effusive loss of Cl and CH_3. For Cl atoms we have

$$F/V = k_{Cl}[Cl]_0$$

$$F/V = k_{Cl}[Cl] + k_1[Cl][CH_4] - k_{-1}[CH_3][HCl].$$

We are told that F/V is constant—so the RHS of these are equal.

For CH_3 we have

$$k_{CH_3}[CH_3] + k_{-1}[CH_3][HCl] = k_1[Cl][CH_3]$$

Eliminating F/V and $[CH_3]$, and rearranging:

$$\frac{[Cl]}{[Cl]_0 - [Cl]} = \frac{k_{Cl}}{k_1[CH_4]} + \frac{k_{Cl}}{k_{CH_3}} \cdot \frac{k_{-1}}{k_1} \frac{[HCl]}{[CH_4]}$$

as required.

$$\text{Slope} = 0.505 = \frac{k_{Cl}}{k_{CH_3}} \cdot \frac{k_{-1}}{k_1}.$$

Now the ratio k_{Cl}/k_{CH_3} can be estimated from the relative effusive rates of Cl and CH_3. As rate of effusion is first order of the form

$$-\frac{d[N]}{dt} = k[N]$$

where $k = 1/4\,\bar{c}$ per unit area, $\bar{c} = $ mean speed $= \left(\dfrac{8kT}{\pi M}\right)^{1/2}$, i.e. rate constant $\propto M^{-(1/2)}$

$$\frac{k_{Cl}}{k_{CH_3}} \simeq \left(\frac{15}{35.5}\right)^{1/2} = 0.65$$

$$\therefore \frac{k_{-1}}{k_1} = \frac{0.505}{0.65} = \frac{1}{1.29}$$

i.e. $K = 1.29$.

From the data $k_1 = 10^{-13}\,cm^3\,molecule^{-1}\,s^{-1}$

$$\therefore k_{-1} = 7.8 \times 10^{-14}\,cm^3\,molecule^{-1}\,s^{-1}.$$

Chapter B2
State Selection of Reagents and Molecular Beam Techniques

J. P. SIMONS

B2.1 Kinetics of state selected reagents

(i) Introduction

Conventional measurements of gas phase reaction kinetics are conducted on systems at thermal equilibrium; they incorporate the tacit assumption that collisional relaxation processes in the reacting system are sufficiently fast to secure the retention of isothermal conditions, so that the equilibrium population distributions remain effectively undisturbed. When it is possible to measure directly, or deduce, the rates of the contributing elementary reaction(s), determinations of the rate constants $k(T)$ can lead to an estimate of the activation energy E_{act}

$$\frac{\mathrm{d}\ln k(T)}{\mathrm{d}T} = \frac{E_{act}}{RT^2}.$$ (1)

Microscopic information is concealed under the extensive thermal averaging. Although curvature in the Arrhenius plot may be interpreted, in some cases, in terms of microscopic rate constants (see section A3.4), a clear resolution of the individual contributions of separate degrees of freedom in promoting the elementary reaction(s)—for example by reagent vibrational or translational energy—necessarily requires the maintenance of non-thermal equilibrium conditions, at least for the degree(s) of freedom under investigation.

State selected measurements of gas phase reaction rates involve fast pumping of particular internal quantum state(s) or translational excitation of the colliding reagent(s) at a rate sufficient to preclude their subsequent collisional relaxation prior to reaction. By this means the Boltzmann averaging can be broken down to expose the dynamics of individual elementary reactions. Molecular beam experiments provide an ideal 'relaxation free' method in which the colliding reagents are directed into the collision zone as an ordered 'army of molecules' advancing at a pre-selected velocity through a (near) vacuum. Under such conditions the scattered products of single reactive collisions can be monitored, quite free of the blurring due to relaxation processes. The advent of high intensity tunable or line-selectable lasers operating in the infra-red (often pumped by exothermic chemical reaction—see Chapter C6) or in the visible/ultra-violet regions, provides a means for the rapid and selective population of individual internal quantum states in the reagents and, latterly, of controlling their spatial alignment. The new research field of 'dynamical stereochemistry' will become increasingly important in the next few years.

The advent of tunable laser sources also provides a method for studying the kinetics of state selected reagents in bulk systems, with a variety of alternative strategies being employed to avoid collisional relaxation prior to (or after) reaction. These commonly include vibrational or electronic state selection in the reagents (rotational states are usually relaxed far more rapidly), the use of low pressures and cryo-pumping to minimise relaxation in the primary reaction products, or the use of time-resolved laser 'pump and probe' techniques operating on time scales that effectively preclude observation of anything beyond the results of single collision events. This section gives a survey of some of the techniques that have been developed (and are still developing) for studying energy utilisation in reactive collisions—focussing principally on bulk systems—and the kind of information that such experiments can provide. Molecular beam systems are outlined in the second half of the chapter and energy disposal, as opposed to utilisation, is a principal theme of the final Chapter C6, which deals with Chemistry *in* Lasers. When both energy utilisation and disposal are monitored together, the measurements approach the ideal of full 'state-to-state' kinetics, an ideal which draws ever closer for simple bimolecular reactions and has actually arrived for one or two unimolecular (photodissociation) systems.

(ii) Translational state selection: molecular beams

The natural means of controlling reagent translational energies, and also momenta, is based on velocity selected molecular beam sources[1] (see section B2.2). The most common involves a nozzle expansion of the reagent, seeded in a diluent gas from a high 'stagnation' pressure P_0, into a vacuum. A conically shaped skimmer, set beyond the expansion region, selects molecules travelling along, or close to, the beam axis. The primary velocity distributions from which the selection is made can be adjusted by varying the composition and temperature of the gas mix prior to expansion or by changing the mass of the diluent gas. The limiting average terminal velocity of the gas stream achievable through the expansion is

$$u_t = \left[\frac{2k_B T_0}{\bar{m}} \cdot \frac{\bar{\gamma}}{\bar{\gamma}-1} \right]^{1/2} \tag{2}$$

where T_0 and $\bar{m}$ are the temperature and average relative molecular mass of the mixture, and $\bar{\gamma}$ is the average specific heat ratio (C_p/C_v). In practice, the reagent gas may be accelerated to a streaming velocity, u_s, somewhat below u_t but approached more and more closely with increasing stagnation pressure, P_0, or nozzle diameter,

d. If the diluent gas, M, is in very great excess, typically 99% of the initial mix, and the product $P_0 d$ is sufficiently large,

$$u_s \simeq u_t \simeq \alpha_M \left[\frac{\gamma_M}{\gamma_M - 1} \right]^{1/2} \tag{3}$$

where $\alpha_M = (2k_B T_0 / m_M)^{1/2}$. For illustration the nozzle expansion of ethene, seeded in high dilution in H_2 at 300 K, would generate a superthermal beam travelling with an average velocity $u_s \simeq 3 \, \mathrm{km \, s^{-1}}$. The width of the velocity distribution is governed by the gas stream temperature T_s, which, because of the very efficient relaxation of relative translational motion during the supersonic expansion, is typically reduced to temperatures of a few degrees Kelvin or less. A useful indicator of the velocity distribution is the speed ratio $s = u_s / \alpha_s$, where $\alpha_s = (2k_B T_s / m_M)^{1/2}$; for the example of ethene in H_2, $s \simeq 30$ would be typical. Further resolution of the reagent velocity within the overall distribution (as well as the experimental determination of the distribution itself) is achieved by insertion of rotating slotted discs or choppers to effect a time of flight analysis. Finally, in case it be thought that seeded beams are used only to accelerate a reagent species, the technique can lead to deceleration when the reagent is lighter than the diluent gas. For example, oxygen atoms generated in a microwave discharge in an oxygen/rare gas mix would be accelerated to superthermal energies in helium but they would be slowed a little in neon and slowed a great deal further in argon.

As an alternative to nozzle beam expansions, high speed rotors, spinning in a low pressure of the reagent gas, can be used to generate pulsed superthermal molecular beams, rather like a continuously swinging golf club.[2]

Varying the rotor speed provides continuously variable translational velocities up to the point at which the rotor material (or the operator's nerve) fails; for rotors constructed from carbon fibre composites the upper limit corresponds to molecular velocities $< 2.5 \, \mathrm{km \, s^{-1}}$, comparable to those achieved in most seeded nozzle beam expansions. The velocity distribution, $I(u)$, like that of a nozzle source, is approximately Maxwellian but shifted in velocity space by the rotor tip speed, u_0,

$$I(u) \propto u^3 \exp \left\{ -\frac{(u - u_0)^2}{\alpha^2} \right\}. \tag{4}$$

Unlike the nozzle source, however, there is no expansion cooling and the most probable escape velocity normal to the rotor surface is $\alpha = (2k_B T / m)^{1/2}$, where T is the surface temperature. Thus the speed ratio, s, $\simeq (u_0 + \alpha) / \alpha$; for the example of ethene accelerated at $u_0 = 2 \, \mathrm{km \, s^{-1}}$ with $T = 330 \, \mathrm{K}$, $s \simeq 5.5$.

Either technique, used in a crossed-beam or, alternatively, a beam-bulk gas configuration can be used to determine relative excitation functions, $S(E_{CM})$, for reaction, usually averaged over both the reagent and product internal quantum state distributions—but still no mean feat. Absolute determinations of excitation functions are very difficult to achieve with any precision. In one approach, they can be estimated through measurement of the depletion in the elastically scattered incident beam flux at high scattering angles—equivalent to the consumption of the reagent beam colliding at low impact parameters (see section B2.2). Finally, note that the centre-of-mass collision energy, E_{CM}, is determined, of course, by the relative velocity of the collision partners and their reduced mass (see section A2.3): high translational energy in one of the partners does not necessarily imply a high collision energy.

(iii) Translational state selection: molecular photodissociation

Photolytic sources provide an alternative means of generating translationally energetic atomic or simple free radical fragments that can be employed to probe the energy dependence of reaction, or energy transfer, in bulk systems. The technique has been restricted largely to the special (but important) case of H(D) reactions, but could certainly be extended. Tunable monochromatic photolysis of H(D)I, H(D)Br or $H_2(D_2)S$ generates tunably monoenergetic H(D) atoms and momentum conservation ensures their production with very high initial velocities, corresponding to virtually all the available energy remaining after bond fission. The method has been available for at least two decades,[3] but its routine implementation in conventional gas kinetic experiments has been hampered by the need (a) to correct the kinetic data for the effects of moderating non-reactive collisions and (b) by uncertainties in the division of energy in the initial photolysis, e.g. the branching into spin–orbit excited atoms in the dissociation

$$HI + h\nu \rightarrow H + I(^2P_{3/2}, {}^2P_{1/2}). \tag{5}$$

However, the latter difficulty has been resolved by the new generation of photofragment spectroscopy experiments which characterise the distribution of competing dissociative channels, while the problem of translational moderation can be avoided by using pulsed laser 'pump-probe' techniques[4] (see Fig. B2.1). The first UV laser pulse is used to generate the translationally excited atoms or free radicals (e.g. H_t from HBr, OH_t from H_2O_2); the second (briefly) delayed probe laser monitors the secondary reaction products and, with the partial pressures appropriately adjusted, the short pump-probe delay ensures that the reaction is monitored under effectively 'single collision' conditions. The endothermic reaction

$$H_t + O_2 \rightarrow HO(v, j, f) + O, \ \Delta H^{\circ} = 69\,\text{kJ}\,\text{mol}^{-1} \tag{6}$$

studied by Wolfrum,[4] provides an instructive example of the method. Translationally excited H_t atoms travelling at the breakneck speed of $32\,\text{km}\,\text{s}^{-1}$ are generated by ArF laser photodissociation of HBr at 193 nm. The OH radicals generated via reaction 6 are monitored after < 100 ns delay by a tunable dye laser which excites their fluorescence. The experimental parameters ensure the maintenance of 'single collision' conditions and the avoidance of significant relaxation, either of the

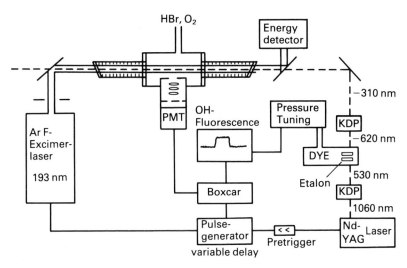

Fig. B2.1. Laser 'pump-probe' apparatus for studying the kinetics of translationally excited atomic hydrogen reactions. (Reproduced, with permission, from Kleinermann K. and Wolfrum J. (1983) *Laser Chem.*, **2**, 339.)

initial translational energy in the reagent H atoms or of the internal energy distributions in vibration, v, rotation, j, or fine structure states, f, in the product OH.

The very high velocity of the H atoms relative to that of the reagent O_2 molecules under the bulk conditions at 300 K, ensures that the O_2 target is a 'sitting duck'; the collision energy $E_{CM} \simeq \frac{1}{2}\mu_{H-O_2}u_H^2 \simeq 254\,kJ\,mol^{-1}$ also lies well above the thermochemical threshold for reaction. Absolute values for the reactive cross-section $S(E_{CM})$, were estimated by using the photodissociation

$$HNO_3 \xrightarrow{hv} OH(v, j, f) + NO_2, \quad \phi = 1 \tag{7}$$

as a calibrant for the sensitivity of the laser induced fluorescence determination of the $OH(v, j, f)$ radical concentration. New values for $S(E_{CM})$ at other collision energies could be determined by varying the initial photolysis wavelength; since the reactive collisions are effectively monoenergetic, the measurements will also give the energy dependent rate constant, $k(E_{CM}) = S(E_{CM}) \times u_{rel} \simeq S(E_{CM}) \times u_H$.

As a bonus, analysis of the LIF spectrum of the OH gives the internal quantum state distribution(s) at the selected collision energy(ies). The fine structure spin–orbit states, for example, are found to be equally populated, but not the Λ-doublet components (despite their being much more closely spaced). Their unequal population reflects a preference for the unpaired π-electron in the OH product to lie in the plane of rotation, thus establishing a preferred stereochemistry for the reaction.

(iv) Vibrational state selection

Vibrational state selection first became a serious possibility with the discovery of non-equilibrium energy disposal in atomic and free radical reactions, the key to chemical laser systems, and subsequently with the advent of the chemical laser systems themselves. Their radiation can be used to pump absorbing molecules into the first, and sometimes higher excited vibrational levels providing a most versatile alternative to the 'pre-reaction' technique[5] which uses an exoergic reaction to introduce vibrationally excited reagent molecules directly into the chemical system (see Fig. B2.2), e.g.

$$H + X_2 \quad \rightarrow HX(v) + X (X = F, Cl, Br, I) \tag{8}$$

$$H + O_3 \quad \rightarrow HO(v < 9) + O_2 \tag{9a}$$

$$H + NO_2 \rightarrow HO(v < 3) + NO. \tag{9b}$$

The 'pre-reaction' technique does, however, provide an excellent means of probing

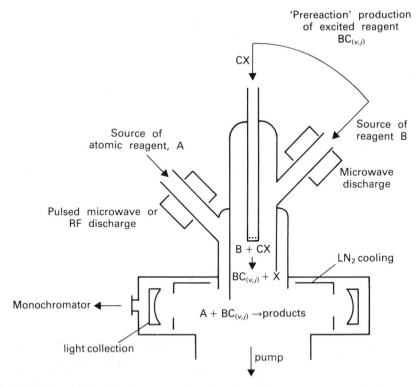

Fig. B2.2. Infra-red chemiluminescence system for studying the kinetics of vibrationally excited molecules by the 'pre-reaction' technique (after Polanyi J. C. and Woodall K. B. (1972) *J. Chem. Phys.*, **57**, 1574).

the influence of high reagent vibrational excitation on strongly endothermic reactions

e.g. $Br + HCl(v) \xrightarrow{k(|v; T)} HBr + Cl \quad \Delta H_0^{\ominus} = 65 \, kJ \, mol^{-1}.$ (10)

The relative state-selected rate constants, $k(|v; T)$, can be monitored via the depletion of the IR chemiluminescence from $HCl(v)$. Reaction 10 accelerates dramatically when $v > 2$, and the endothermicity can be overcome. The tunable infra-red laser pumping technique initially excites selected (v, j) levels in the absorbing reagent molecule but, unless the reagent is in a very low pressure environment or a molecular beam, any rotational disequilibrium is rapidly relaxed in collisions; the vibrational overpopulation is retained for a much longer period, however, since vibrational relaxation proceeds much less efficiently. The principal 'guns' in the IR laser artillery that are available for populating excited vibrational states in diatomic, or more complex reagents, are summarised in Table B2.1.

The convenience of employing hydrogen halide lasers to pump hydrogen halide reaction systems has encouraged intense research into the competition between near thermoneutral reaction (e.g. reaction 12a, 12b) and the alternative relaxation channel for $HX(v = 1)$ (e.g. reaction 12c; see Fig. B2.3). Typical elementary processes that have been studied include

$X + HX(v = 1) \quad \to XH + X$ (11a)

$\to X + HX(v = 0)$ (11b)

$H, D + HX(v = 1) \to H_2, \, HD + X$ (12a)

$\to HX, \, DX + H$ (12b)

$\to H, D + HX(v = 0)$ (12c)

$Y + HX(v = 1) \quad \to YH + X; \quad X = Hal, O$ (13a)

$\to Y + HX(v = 0).$ (13b)

Principal questions asked address the ability of *ab initio* calculations of the 'simple' reactive potential energy surfaces, and of the collision dynamics that proceed under their influence, to reproduce the observed rate constants and branching ratios. Reactive processes such as reactions 12a, 12b can be distinguished from relaxation processes such as reaction 12c by simultaneous monitoring of the populations in $HX(v = 1)$ and of the atomic products by IR chemiluminescence and atomic resonance fluorescence, respectively. For the system $D + HCl(v = 1)$, it is found that relaxation is far more probable than reaction, implying a large potential barrier in the atom transfer channel.[6]

Reactions involving vibrationally excited free radicals, e.g. $OH(v)$ may address the effect of vibrational energy in the attacking reagent (as opposed to the

Table B2.1. Tunable infra-red laser artillery

Laser medium	Tuning range/cm^{-1}
Optical parametric oscillator	7000–3000
F-centre	5000–3300
HF	3850–3300
Diodes (e.g. Pb salts)	3300– 330
DF	2780–2500
HCl	2770–2500
CO	2000–1400
CO_2	1100– 900
N_2O	1000– 900

molecule attacked) in an atom transfer reaction. More interestingly perhaps they may also address the influence of vibrational energy on free radical reactions that proceed via an intermediate complex, e.g.

$$O + OH(v = 1) \quad \rightarrow (HO_2)^\dagger \rightarrow OH(v = 0) + O \tag{14}$$

$$O(^3P) + CN(v) \begin{cases} \rightarrow CO(v') + N(^4S) & (15a) \\ \rightarrow \{OCN(^2\Pi)\}^\dagger \rightarrow CO(v'') + N(^2D) & (15b) \end{cases}$$

$$OH(v) + CO(v') \quad \rightarrow (HOCO)^\dagger \rightarrow H + CO_2. \tag{16}$$

More interestingly still the vibrational energy dependence of reaction 16 may be compared with the reverse endoergic reaction promoted by translational excitation of the H atom. Convenient routes for the introduction of vibrationally excited radicals are provided by photodissociation (e.g. reaction 7) or exothermic atom transfers (e.g. reactions 9a, 9b).

If vibrational state selection were restricted solely to diatomic molecules, its fascination—while intense for some—would be limited. Extension to polyatomic molecules has led to the tantalising vision of mode selective vibrational photochemistry.[7] Unfortunately, the competition between reaction and relaxation is now far more complex. Collisional relaxation may proceed via inter-mode energy transfer, which may be monitored by time resolved infra-red fluorescence in small polyatomic molecules. In highly excited or complex polyatomic molecules, where the density of ro-vibrational states is high, the mode composition of an initially selected level may be rapidly scrambled by intramolecular vibrational redistribution in the absence of collisions. A classic example of the effect of low levels of vibrational excitation in a polyatomic system is the experimental study of the branched reaction

$$NO(v = 1) + O_3(001, \text{ or } 010) \begin{cases} \rightarrow NO_2^\dagger(^2A_1) + O_2 & (17a) \\ \rightarrow NO_2^*(^2B_{2,1}) + O_2 & (17b) \end{cases}$$

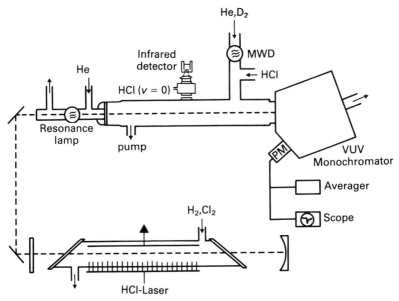

Fig. B2.3. Infra-red and resonance fluorescence detection system for monitoring the kinetics of both vibrationally excited molecular reagent and atomic reagent/product species. (Reproduced, with permission, from Arnoldi D. and Wolfrum J. (1976) *Ber. Bunsenges.*, **80**, 892.)

where O_3 is pumped into the (001) level by absorption from a pulsed CO_2 laser at 9.6 μm, or into the lower-lying (010) level by subsequent collisional relaxation, while $NO(v = 1)$ is populated by Zeeman tuning a rotational feature in the $v = 0 \rightarrow 1$ IR absorption into coincidence with a pulsed CO laser line at 5.3 μm. Both channels have a small activation energy and both are accelerated by vibrational excitation—as yet, however, there is no convincing evidence for mode or reagent selectivity.

High vibrational levels may be accessed specifically by overtone pumping using tunable pulsed or c.w. dye laser excitation and monitored by intracavity optoacoustic absorption spectroscopy. Alternatively, very high levels in the ground electronic states may be populated via a double resonance technique which involves stimulated emission pumping out of the first excited singlet state. Many of the high overtone experiments have focussed on local mode excitation of C–H or O–H vibrations in hydroperoxides, cyclic ethers and the like which readily undergo unimolecular decomposition or isomerisation. Their reactions have been monitored both by laser induced fluorescence of primary OH fragments and by final product analysis. The questions of possible mode selectivity and the validity of the RRKM assumption are generally addressed but interpretation is often clouded by the poor characterisation of the mode composition in the reacting molecules, and possibly by contributions from very weak electronic excitation underlying the high overtone absorption region.

The technique of stimulated emission pumping[8] uses a pulsed UV laser to populate bound rovibronic levels in the first electronically excited states of the molecule, followed by a second tunable pulsed dye laser operating at much longer wavelengths, which stimulates emission into selected, high-lying ro-vibrational levels in the ground electronic state. The technique has been used to populate very high levels in C_2H_2 ($\sim 28000 \text{ cm}^{-1}$) and in H_2CO ($\sim 10000 \text{ cm}^{-1}$) but its exploitation in state selected chemical kinetics remains to be developed.

(v) Rotational state selection

Rotational state selection remains a rare experimental procedure since collisional relaxation is so rapid in bulk systems. If a bimolecular reaction proceeds with a very high cross-section, relaxation prior to reaction can be avoided by pumping the reagent molecule into selected ro-vibrational levels by tuned infra-red laser absorption at very low pressures, or in a molecular beam. Examples are the influence of rotation in the reactions

$$Ca + HF(v = 1, j = 0-7) \rightarrow CaF + H \tag{18}$$

$$K + HF(v, j) \rightarrow KF + H. \tag{19}$$

Reaction 18 has been investigated under (Ca) beam–(HF) gas conditions at very low pressures, using tunable IR radiation from an optical parametric oscillator to excite the HF, whilst reaction 19 has been studied under crossed-beam conditions, with the excitation provided by a tuned HF chemical laser (see Fig. B2.4).

Rotational state selection is much more readily sustained in electronically excited diatomic or simple polyatomic molecules, where multiple collisions are prevented by rapid fluorescent decay. This has been utilised, for example, in a study of the electronic quenching of $OH(A^2\Sigma^+; v = 0)$ excited into the rotational states $N < 6$ by a pulsed tunable dye laser, frequency doubled into the ultra-violet; the quenching was found to depend both on the value of N as well as the nature of the collision partner. Collisional rotational relaxation in the available time scale may also be prevented by rapid unimolecular decomposition. Thus the rotationally state selected predissociation of H_2O, D_2O, excited into selected rovibronic levels of the $\tilde{C}^1B_1$ Rydberg state by tuned two-photon excimer laser absorption at 248 nm, has provided a uniquely detailed insight into the rotationally state dependent dynamics of a simple molecular photodissociation. Similarly, high overtone pumping of H_2O_2 and HOOD into vibrational levels lying just below the dissociation limit in the ground electronic state has exposed the contribution made by rotational excitation in making up the residual energy deficit. Finally, crude rotational state selection or, rather, narrowing may be achieved by the cooling promoted through supersonic jet expansion of the reagent.

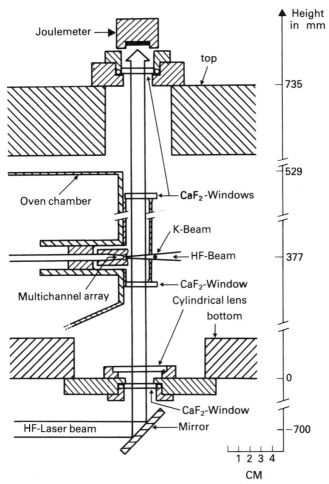

Fig. B2.4. Infra-red laser pumping of HF in a crossed molecular beam system. (Reproduced, with permission, from Hoffmeister M., Potthast L. and Loesch H. J. (1983) *Chem. Phys.*, **78**, 369.)

B2.2 Chemical reactions in colliding molecular beams

(i) Introduction

The ultimate experiment in molecular reaction dynamics involves the reaction of fully quantum state selected reagents under fully specified collision conditions and the measurements of the branching probabilities into each of the possible product channels within the constraints of energy and angular momentum conservation. This ideal is being approached by the application of molecular beam techniques; the present section explores the interface between molecular beam reaction dynamics and gas kinetics, though few gas kineticists have taken up the study of chemical reactions using molecular beam reagent sources. It is legitimate to ask 'Why not?' Is it the complexity, the expense or the limited scope of such experiments that is responsible; or are the results of such experiments simply not pertinent to the 'real world' of complex chemical reactions in bulk systems? The sceptic would probably answer 'Yes' in each case, but he would be wrong because if it were otherwise the lecture, on which this section is based, would not have been included in the Summer School; indeed its inclusion was at the behest of a majority of 'non-beamists'. So why go to the trouble? The key lies in the possibility of studying

(a) elementary bimolecular reactive (and inelastic) processes under single collision conditions so that the utilisation of energy and momentum in the reagents and their disposal in clearly identified primary products can be directly charted, and

(b) unimolecular reactions under collision free conditions, thus allowing the kinetics of intramolecular energy transfer processes and the branching into alternative product channels to be monitored, without the complication of secondary collisions involving neighbouring molecules or the confining walls of the reaction chamber.

(ii) The molecular beam experiment[9]

Three basic types can be identified, the crossed-beam arrangement, the beam-gas arrangement and, most recently, the jet-expansion experiment. The first is the classic molecular beam scattering experiment (see Fig. B2.5) developed initially by Moon, Taylor and Datz, Herschbach, Bernstein and subsequently by a host of others.[9] Its aim is the measurement of velocity and angle resolved differential scattering cross-sections for velocity selected, perhaps internally state-selected and possibly even spatially selected, aligned or oriented reagent beams. Its principal target is the generation of polar scattering maps (see section A3.2) and its principal product is detailed information on the reactive (or inelastic) collision dynamics.

Beam-gas systems sacrifice resolution of the relative velocities and translational energies of the collision partners to achieve a greatly enhanced collision frequency; the loss of resolution is minimal for fast reagent beams and/or heavy target molecules. A principle aim of this type of experiment is to observe the internal quantum state distributions in the primary products (usually via chemiluminescence or laser induced fluorescence measurements) as a function of the collision energy. The end-product is a set of relative and possibly state-resolved excitation functions for the reactive collisions, $S_{reac}(E_{CM})$ or $S_{reac}(n'|n; E_{CM})$.

What gas kinetics experiments can be done in a single beam jet-expansion since there have to be collision processes of some kind to promote reaction? The first possibility is the study of collisional energy transfer (principally rotational relaxation) during the expansion using LIF to monitor its progress. The second involves

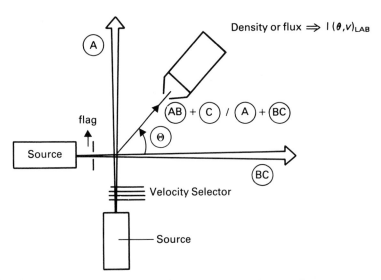

Fig. B2.5. Crossed molecular beams scattering experiment—schematic diagram.

laser beam–molecular beam interactions downstream of the expansion region and beyond the beam-skimmer. Experiments of this kind, originally developed by Wilson, use photofragment (translational) spectroscopy to probe the dynamics of molecular photodissociation.[10] Subsequently Lee has applied the same technique to probe the dynamics of infra-red multiple photon dissociation, IRMPD (more of this later). Most recently, and most excitingly, the ability of jet-cooled expansion sources to generate weakly bound molecular complexes has been exploited to study the reactive 'collisions' of atomic fragments, produced by pulsed laser dissociation of reagents pre-oriented by the intermolecular force fields—shades of the primitive days of steric factors. In addition, the 'collisions' are initiated at a narrowly defined range of impact parameters—a range that is otherwise impossible to control.[11]

(iii) Molecular beam techniques[12]

Sources

Reagent beams may be effusive, with a thermal Maxwell Boltzmann velocity distribution; or superthermal, generated by supersonic nozzle expansion of a gas into a vacuum (see Fig. B2.6) or mechanically accelerated by a high speed rotor blade (see Fig. B2.7); or 'seeded', generated by nozzle beam expansion in a carrier gas stream. Their velocity characteristics were briefly described in section B2.1. Here we may note in addition that the beam intensities (typically $\geqslant 10^{17}$ molecule $sr^{-1} s^{-1}$) provided by a nozzle beam expansion are at least three orders of magnitude greater than those obtainable from an effusive source. In addition, heavy reagent beams tend to be concentrated along the beam axis when swept along by a light carrier gas, since their lateral velocities are necessarily much slower. The insertion of a skimmer—a pierced, inverted cone (Fig. B2.6) in the region where molecular flow takes over from the high gas density continuum

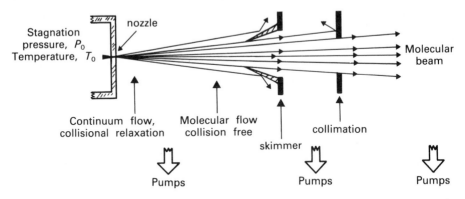

Fig. B2.6. Nozzle molecular beam source (schematic diagram). Each successive compartment is differentially pumped in order to sustain a background pressure in the final scattering chamber $< 10^{-7}$ Torr. Typical stagnation pressures, P_0, lie in the range > 350 Torr.

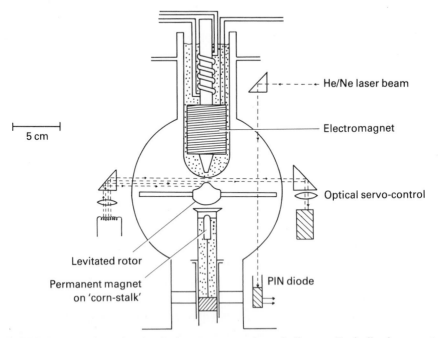

Fig. B2.7. Rotor accelerated molecular beam source—schematic diagram. Typically, the magnetically levitated rotor spins in a low pressure of the reagent gas at $\leqslant 250\,000$ r.p.m. to deliver a pulsed molecular beam travelling at $\leqslant 2.5$ km s^{-1}.

flow, consequently deflects much of the lighter gas away from the collision zone, which lies further downstream. Effusive sources are generally employed only where high temperatures are required to vaporise the atomic or molecular reagent.

Nozzle beams may be operated either continuously, or pulsed. In general, pulsing offers many advantages since it reduces the gas load and so allows the possibility of higher stagnation pressures, it helps velocity measurement *via* time of flight analysis and it is the method of choice when coupled with pulsed laser beam pump (or probe) techniques. The translational and rotational cooling associated with the nozzle beam expansion may allow stabilisation of weakly bonded dimers, complexes or reagent clusters—not necessarily an advantage if the aim is a study of the reaction dynamics of single, isolated molecules of course, but, as mentioned above, offering the possibility of initiating bimolecular reaction in van der Waals complexes at preset collision geometries.[11]

Beams of atomic or molecular fragments can be generated by pyrolysis of the gas stream, e.g. CH_3 from $(CH_3)_2N_2$; or by striking a microwave or radio frequency discharge in the gas flow, e.g. $O(^3P)$ from O_2/He; or by secondary reaction (e.g. OH from H/NO_2 reaction in a discharge flow); or by photolysis, either behind or at the nozzle tip (e.g. CN from BrCN). Electronically excited reagent beams can be generated by electron impact or by striking a d.c. discharge at the nozzle tip, e.g. Ar, Kr, Xe $(^3P_{2,0})$, $O(^1D_2)$ or $Ca(^1P_1)$, or by tunable dye laser absorption, e.g. $I_2[B^3\Pi(O_u^+)(v', j')]$. Vibrationally and rotationally excited molecular reagent beams can be generated by tunable IR laser pumping; the system shown in Fig. B2.4 was constructed for studying the reaction of $HF(v, j)$ with K under crossed-beam conditions.

As an alternative to excitation, individual rotational states can be selected from the initial thermal population distribution by passing the reagent beam through deflecting multipole electrical or magnetic fields. For example, diatomic or symmetric top polar molecules can be rotationally selected and focussed by quadrupolar or hexapolar electric fields. They may also be aligned or, by subsequent application of a d.c. field, oriented to point preferentially either 'heads' or 'tails' with respect to the approaching collision partner. Introductory descriptions of the techniques have been presented by Brooks[13] and by Bernstein.[9] Reagent spatial alignment or alternatively electronic orbital alignment can each be promoted by absorption of polarised laser light; two examples illustrate the technique:

(a) Photodissociation of a molecular halogen via a 'parallel' transition, e.g. $I_2[B^3\Pi(O_u^+) \leftarrow X^1\Sigma(O_g^+)]$ will lead to preferential alignment of the surviving molecules since these will have their rotation axes lying parallel to the ε-vector of the photolysis beam.

(b) The excitation of $Ca(3^1P_1 \leftarrow 3^1S_0)$ by a pulsed laser polarised either parallel or perpendicular to the beam axis, or to the relative velocity vector in a crossed molecular beam experiment, will align the electron density in the $3p$ orbital respectively parallel (σ, $M_L = 0$) or perpendicular (π, $M_L = \pm 1$) to that axis.

The application of techniques such as these allows the dynamical stereochemistry of elementary chemical reactions to be explored—a field which is likely to progress rapidly in the next few years. For the more traditional gas kineticist the further good news is the rapid growth of free radical molecular beam sources which should enable experiments on 'gas kinetically important' reagents such as O, F,Cl, OH, CH_2, NH_2, CH_3 . . . to become commonplace.

Detectors

The availability of suitably discriminating and sensitive detectors for the very low intensities of reaction products scattered from the collisions of intersecting reagent beams was the key to the original breakthrough of molecular beam chemistry. The earliest variety was the ionisation detector which provided a sensitive and rapid response to the arrival of alkali metal atoms or their halide salts on their surface. They ushered in the pioneering 'alkali age' of molecular beam reactive scattering.[14]

The current range of detectors can be classified (Table B2.2) under two main headings. Type A provides angle and velocity resolved scattering data from which the momentum distributions in the scattered products can be derived, while Type B provides spectrally resolved data which allow determination of internal quantum state distributions. The two types are, of course, complementary.

The 'universal' detector in molecular beam scattering experiments is the quadrupole mass spectrometer. Its universality is its big advantage, in that all products can be weighed, but against that is the requirement of stringent high vacuum conditions to minimise background interference and the need for careful mass and velocity analysis of fragment ion intensities when the monitored mass peak may be associated with more than one scattered species. Time resolved mass spectrometric detection limits the product energy resolution to translation; the total internal energies can only be derived *via* energy conservation. In favourable cases, the velocity resolution can be sufficient to allow some measure of internal, vibrational state resolution (see section A3.2) after employing appropriate deconvolution procedures to correct for the velocity/collision energy spread in the colliding reagent beams.

In general, internal state distributions are best determined by optical detection techniques. Tunable dye laser induced fluorescence (Fig. B2.8) offers very high levels of internal state resolution and, with polarised lasers, a state-selective probe of product rotational polarisation as well. In (rare) favourable cases it can even be possible to estimate angle resolved velocity distributions through analysis of the Doppler profiles of individual rovibronic features in the LIF spectrum. There is the obvious and possibly frustrating restriction to fluorescent species, though laser induced resonance enhanced multiple photon ionisation (REMPI) may provide an alternative. This technique, while sensitive, also has the unfortunate disadvantage of being non-linear so that quantitative population distributions may be

Table B2.2. Techniques for molecular beam detection/characterisation

Type A Angle velocity resolution

Ioniser/quadrupole mass filter	universal detector	number density detector
Surface ionisation	specific for alkali, alkali halides	flux detector
Electron multiplier	electronically excited reagents/ products (Auger electron emission)	flux detector

Type B Internal state resolution

Laser induced fluorescence
or Multiple photon ionisation
	can give (v, j) distributions
	fine structure (e.g. Λ-doublet
	  distributions)
	spin–orbit distributions
	velocity distribution via
	  Doppler spectroscopy
	polarisation state
		number density detector

Spontaneous fluorescence/ chemiluminescence	can give electronic branching ratios (v, j) distributions polarisation state	number density detector
Bolometer	measures energy flux	flux detector

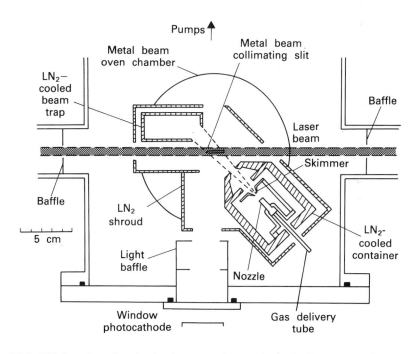

Fig. B2.8. LIF detection of molecular beam reaction products. An illustration of the system employed by Cruse, Dagdigian and Zare to monitor $BaX(v', j')$ generated in the reactions $Ba + HX \rightarrow BaX + H$, $X = F$, Cl, Br, I. (Reproduced, with permission, from (1973) *Faraday Discuss. Chem. Soc.*, **55**, 278.

difficult to extract; however, the rapid development of tunable vacuum UV laser sources through four wave mixing techniques, in which the non-linearities are introduced prior to the ionisation process, may avoid the problems by allowing single photon ionisation instead of REMPI. In the special case of chemilumines- cent reactions, when the primary products are spontaneously fluorescent, the laser can be dispensed with altogether. Analysis of the spectrally resolved and possibly polarised emission as a function of the collision energy can provide data for evaluation of the excitation functions into competing channels (via intensity measurements), the energy disposal and branching ratios (via spectral analysis) and the product rotational alignments (via polarisation measurements referenced to the collision velocity vector).

(iv) Some case histories

The typical crossed molecular beam experiment (see Fig. B2.9) records the angle and velocity resolved flux of reactively scattered products, often of a bimolecular

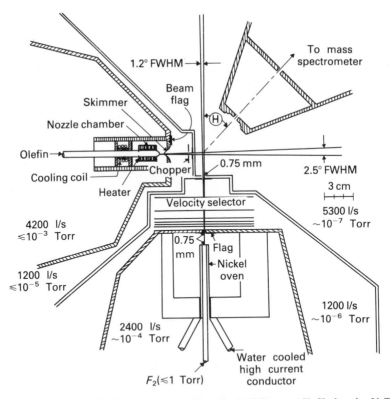

Fig. B2.9. A crossed molecular beam system: employed by J. M. Parsons, K. Shobotake, Y. T. Lee and S. A. Rice to study the unimolecular reaction of chemically activated $(C_2H_4F)^\dagger$, generated by fluorine atom addition to ethene

$$F + C_2H_4 \rightarrow (C_2H_4F)^\dagger \rightarrow C_2H_3F + H$$

(Reproduced with permission, from (1973) *Faraday Discuss. Chem. Soc.*, **55**, 345.)

reaction, initially in the laboratory frame and finally, after the appropriate trans-formation, in the centre of mass frame. Typically the transformed scattering patterns are presented as flux–velocity contour maps: some typical examples for the archetypal cases of forward scattering (stripping dynamics), backward scatter-ing (rebound dynamics) and symmetrical scattering (collision complex dynamics) were described and discussed in section A3.2. The techniques are well described in references such as 9 and 12. Rather than describe these again, we choose instead to outline two examples of unimolecular reactions where the application of molecu-lar beam conditions has resolved uncertainties or ambiguities in the analysis of conventional gas kinetic data obtained from bulk experiments. The examples are drawn from the particularly exemplary laboratory of Y. T. Lee at Berkeley and involve infra-red multiple photon dissociation.[15]

When the unimolecular decomposition of polyatomic molecules by IR multi-ple photon absorption is probed in a bulk environment the interpretation can be clouded by a number of possible complications. Do the activated molecules suffer bimolecular reactions? How do secondary reactions of free radical products affect the final product distributions? Do the primary products themselves decompose in the intense laser field? Complications such as these can be avoided or resolved if the primary products are detected in a collision free environment following irradiation of the molecular reagent beam by an intense pulsed IR laser. Angle resolved, time of flight analysis of the scattered primary fragments using mass spectrometric detection identifies the primary products and provides their transla-tional energy distributions. These may be compared with the predictions of RRKM theory (Chapter A.4) to give some guidance on the degree of energy randomisation in the activated molecule prior to its decomposition. The dissocia-tive lifetime may be estimated by assuming a kinetic competition, with saturation setting in when the absorbing molecule has been pumped to a mean energy $E > E^0$ at which the rate of decomposition matches the rate of further up-pumping to levels well above threshold (cf. Fig. 4 in the Introductory chapter describing conventional collisionally activated decomposition at low pressures). Interpreta-tion of the mass spectrally resolved data needs some caution since a fragment ion at a given mass may arise from more than one neutral molecular or fragment source.[16] Before the data can be accepted it is necessary to establish first a unique cracking pattern for the ions generated by electron impact ionisation of the detected neutral species, second, a proper correlation of the velocity distributions of the (suspected) primary fragments and third, that the two primary fragment velocities and masses satisfy the requirements of linear momentum conservation. We consider two examples: the 3-centre elimination of HCl from CHF_2Cl and its 4-centre elimination from CH_3CCl_3.[15]

CHF_2Cl

A pulsed CO_2 laser tuned to the R(26) line at $1082\,cm^{-1}$ lies within the rovibronic absorption contour of the asymmetric C—F stretching mode, when the absorption is shifted into the laser range by heating the nozzle molecular beam source. Quadrupole mass spectrometric detection monitors the time of arrival of HCl (via the Cl^+ and HCl^+ ion peaks) and CF_2 (via the CF^+ ion peak). The Cl^+ and HCl^+ ions have identical time of flight distributions, are distinct from that of CF^+, and the two distributions satisfy the requirements of linear momentum conservation. Thus the 3-centre elimination

$$F_2C\overset{H}{\underset{Cl}{\diagup}} + nh\nu \rightarrow \left(F_2C\cdot\overset{\cdot\,H}{\underset{\cdot\,Cl}{\vdots}}\right)_{E>E_0} \rightarrow F_2C + HCl, \quad \Delta H^{\ominus} = 201\ \text{kJ mol}^{-1} \qquad (20)$$

must constitute the major decomposition channel as expected. The alternatives

$$CF_2HCl \rightarrow CFCl + HF, \quad \Delta H^{\ominus} = 270\ \text{kJ mol}^{-1} \qquad (21a)$$

$$\rightarrow CHF_2 + Cl, \quad \Delta H^{\ominus} = 343\ \text{kJ mol}^{-1} \qquad (21b)$$

are far more endothermic. The translational energy distributions peak slightly above zero kinetic energy release and the experimental data are consistent with the predictions of statistical RRKM calculations, if the activated molecules are assumed to have a unimolecular lifetime $\sim 1.5\ \text{ns}$ together with a reverse exit barrier of $\sim 25\ \text{kJ mol}^{-1}$.

CH_3CCl_3

Irradiation of the molecular reagent beam by a pulsed CO_2 laser tuned to the R(12) line at $1073\ \text{cm}^{-1}$ initially excites both the C—C stretch and CH_3 rocking modes. Subsequent IRMPD leads to velocity correlated fragment ions at mass ratios corresponding to CH_2CCl^+ (from CH_2CCl_2) and HCl^+ (from HCl). These major products are consistent with decomposition via the 4-centre elimination

$$CH_3CCl_3 + nh\nu \rightarrow \left[\begin{matrix} H_2-C\cdots H \\ \vdots \quad \vdots \\ Cl_2-C\cdots Cl \end{matrix}\right]_{E>E_0} \rightarrow CH_2CCl_2 + HCl, \quad \Delta H^{\ominus} = 52\ \text{kJ mol}^{-1} \ (22)$$

again the least endothermic of the possible decomposition channels. A broad and relatively weak component in the time of flight distribution of the HCl^+ fragment was attributed to secondary IRMPD of the primary product CH_2CCl_2. Once again tolerable agreement with RRKM calculation could be obtained if it were assumed that the average energy of the decomposing molecules was controlled by the balance between energy fluence and the rate of unimolecular decomposition, to give an estimated unimolecular lifetime of $\sim 10\ \text{ns}$. The translational energy distribution among the primary products was consistent with a reverse energy barrier of $\sim 20\ \text{kJ mol}^{-1}$.

The equally interesting question of the internal energy partitioning and distribution between and within the primary products cannot be addressed by the time of flight analysis, of course. However, the complementary technique of laser pump–laser probe spectroscopy can be employed for CF_2HCl, where CF_2 can readily be probed by LIF. The collision free regime can still be attained, even in a bulk environment provided the pump–probe delay time lies well within the mean collision interval.

Finally we point out that a further example of the use of crossed molecular beams to elucidate the mechanism of a reaction that is of great interest in conventional gas kinetics is discussed in Chapter B4. The work was, once again, performed in Lee's laboratory and concerns the bimolecular reaction of ground state oxygen atoms with ethene. The crucial question concerns the products formed on the fragmentation of the initial adduct; it appears that this fragmentation changes if the adduct undergoes collisions and Lee's experimental results provided a clear picture of the route in the absence of collisions.[16]

B2.3 Suggestions for further reading

1. State-to-state kinetics

There are many, many references to methods of reagent state preparation for the study of state-to-state kinetics. As in other sections of this book, the discussions and compilation presented by Bernstein R. B. (1982). In *Chemical Dynamics via Molecular Beam and Laser Techniques*, Oxford, Clarendon Press provide an excellent starting point.

A selection of other leading references includes:

Levine R. D. and Bernstein R. B. (1974) *Molecular Reaction Dynamics*, Oxford University Press, Oxford.

Polanyi J. C. and Schreiber J. L. (1974) In *Physical Chemistry, An Advanced Treatise*, Vol. IIA (Ed. by W. Jost), p. 383. Academic Press, New York.

Levy M. R. (1979) *Progr. React. Kinet.* **10**, 1.

Kinetics of State Selected Species (1979) *Faraday Discuss. Chem. Soc.*, **67**.

A. Fontijn and M. A. A. Clyne (Eds) (1983) *Reactions of Small Transient Species*, Academic Press, London.

Smith I. W. M. (1980) *Kinetics and Dynamics of Elementary Gas Reactions*, Butterworths, London.

Smith I. W. M. (1980) Chemical Reactions of Selectively Energised Species. In *Physical Chemistry of Fast Reactions*, (Ed. by I. W. M. Smith), Plenum Press, New York.

Kneba M. and Wolfrum J. (1980) *Ann. Rev. Phys. Chem.*, **31**, 47.

Geddes J. (1980) *Contemp. Phys.*, **23**, 233.

Leone S. R. (1984) *Ann. Rev. Phys. Chem.*, **35**, 109.

Buelow S., Noble M., Radhakrishnan G., Reisler H., Wittig C. and Hancock G. (1986) *J. Phys. Chem.*, **90**, 1015.

2. Molecular beams

Ross J. (Ed.) (1966) Molecular Beams. *Adv. Chem. Phys.*, **10**.

Hartman H. (Ed.) (1968) *Chemische Elementarprozesse*, Springer-Verlag, Berlin.

Kinsey J. L. (1972) *Physical Chemistry, MTP Int. Rev.*, (Ed. by J. C. Polanyi) *Ser.* 1, **9**, 173.

Fluendy M. A. D. and Lawley K. P. (1973) *Chemical Applications of Molecular Beam Scattering*, Chapman and Hall, London.

Lawley K. P. (Ed.) (1975) Molecular Scattering, Physical and Chemical Applications. *Adv. Chem. Phys.*, **30**.

Toennies P. (1976) *Ann. Rev. Phys. Chem.*, **27**, 225.

Kinsey J. L. (1979) *Ann. Rev. Phys. Chem.*, **28**, 349.

Brookes P. R. and Hayes E. F. (Eds.) (1979) *State to State Chemistry*, ACS Symposium Ser. No. 56, Washington, D.C.

Davidovits P. and McFadden D. L. (Eds.) (1979) *The Alkali Halide Vapours*, Academic Press, New York.

Lee Y. T. and Shen Y. R. (1980) *Physics To-day*, **33**, 52.

3. A historical perspective on molecular beams

The development of molecular beam chemical reaction dynamics:

Fraser R. G. J. (1934) In *Free Radicals, Trans. Faraday Soc.*, **30**, 182.—the idea.

Bull T. H. and Moon P. B. (1954) *Discuss. Faraday Soc.*, **17**, 54.—the first demonstration.

Taylor E. H. and Datz S. (1955) *J. Chem. Phys.*, **23**, 1711.—the first crossed beam experiment.

Ramsey N. F. (1956) *Molecular Beams*, Clarendon Press, Oxford—the first source book.

An interesting perspective on the development of the field of reaction dynamics may be obtained from the Faraday Discussions of the Royal Society of Chemistry:

The Study of Fast Reactions, **17** (1954).
Inelastic Collisions of Atoms and Simple Molecules, **33** (1962).
Molecular Dynamics of the Chemical Reactions of Gases **44** (1967).
Molecular Beam Scattering, **55** (1973).
Kinetics of State Selected Species, **67** (1979).
Intramolecular Kinetics, **75** (1983).
Dynamics of Elementary Gas Phase Reactions, **84** (1987).

B2.4 References

1 Levy D. H. (1980) *Ann. Rev. Phys. Chem.*, **81**, 197; Anderson J. B., Andres R. P. and Fenn J. B. (1965) *Adv. Atom. Mol. Phys.*, **1**, 345.

2 Rettner C. T. and Simons J. P. (1979) *Faraday Discuss. Chem. Soc.*, **67**, 329.

3 Oldershaw G. A. (1977) Specialist Periodical Report. In *Gas Kinetics and Energy Transfer*, (Ed. by P. G. Ashmore and R. J. Donovan), Vol. 2, p. 96. Chem. Soc., London.

4 Wolfrum J. (1986) *J. Phys. Chem.*, **90**, 375.

5 Polanyi J. C. and Woodall K. B. (1972) *J. Chem. Phys.*, **57**, 1574.

6 Arnoldi D. and Wolfrum J. (1976) *Ber. Bunsenges Phys Chem.*, **80**, 892.

7 Crim F. F. (1984) *Ann Rev. Phys. Chem.*, **35**, 657; Zewail A. H. (1980) *Phys. To-day*, **33**, 27.

8 Reisner D. E., Vaccaro P. H., Kittrell C., Field R. W., Kinsey J. L. and Dai H. L. (1982) *J. Chem. Phys.*, **77**, 573.
9 Bernstein R. B. (1982) *Chemical Dynamics via Molecular Beam and Laser Techniques*, Clarendon Press, Oxford.
10 Reveiwed by Simons J. P. (1977) Specialist Periodical Report. In *Gas Kinetics and Energy Transfer*, (Ed. by P.; G. Ashmore and R. J. Donovan), Vol. 2, p. 56. Chem. Soc., London.
11 Radhakrishnan G., Buelow S. and Wittig C. (1986) *J. Chem. Phys.*, **84**, 727.
12 See (1966) *Adv. Chem. Phys.*, **10**, (Ed. J. Ross), *Molecular Beams* and books by N. F. Ramsey, M. A. D. Fluendy and K. P. Lawley, and R. B. Bernstein listed above; also reference 1.
13 Brooks P. R. (1976) *Science*, **193**, 11.
14 Herschbach D. R. (1973) *Faraday Discuss. Chem. Soc.*, **55**, 233.
15 Sudbø Aa. S., Schutz P. A., Shen Y. R. and Lee Y. T. (1973) *J. Chem. Phys.*, **69**, 2312; Schutz P. A., Sudbø Aa. S., Krajnovich D. J., Kwok H. S., Shen Y. R. and Lee. Y. T. (1979) *Ann. Rev. Phys. Chem.*, **30**, 379.
16 Buss R. J., Baseman R. J., He G. and Lee Y. T. (1981) *J. Photochem.*, **17**, 389.

B2.5 Problems

State selection of reagents and molecular beam techniques

B2.1. The slightly exoergic reaction:

$$K + HBr \rightarrow KBr + H, \quad \Delta H_0^{\ominus} \sim -18\,kJ\,mol^{-1}$$

has been studied using crossed molecular beams. Analysis of the rotational energy of the scattered KBr via deflection in an inhomogeneous electric field led to an estimate of the average rotational energy, $\langle E_r \rangle_{KBr}$, of KBr of $\sim 5\,kJ\,mol^{-1}$. It was also possible to measure the second moment of the angular distribution function $P(\hat{j}'_{KBr} . \hat{k})$, where $\hat{k}$ is a unit vector along the reagent relative velocity. The product rotational alignment, $\langle P_2(\hat{j}'_{KBr} . \hat{k}) \rangle \approx -0.35$.

Discuss these results and use them to estimate a value for the average relative cross-section, given that the average relative velocity of the reagents is $\sim 1500\,ms^{-1}$ and the rotational constant, $B_{KBr} = 0.08\,cm^{-1}$.

In the strongly exoergic reaction:

$$K + Br_2 \rightarrow KBr + Br, \quad \Delta H_0^{\ominus} = -188\,kJ\,mol^{-1}$$

a similar set of measurements gave $\langle E_r \rangle_{KBr} \simeq 23\,kJ\,mol^{-1}$ and $\langle P_2(\hat{j}' . \hat{k}) \rangle \simeq -0.13$. Discuss the differences between the reactions and comment on any difficulties that may arise from the fact that the electric deflection technique samples products scattered over a narrowly specified range of laboratory angles.

$$[P_2(\cos\theta) = \tfrac{1}{2}(3\cos^2\theta - 1)].$$

B2.2. In the slightly endothermic reaction:

$$K + HCl(v = 0) \rightarrow KCl + H.$$

The reaction cross-section, $S_{reac}(E_t)$ increases with collision energy, E_t, as follows:

$E_t/kJ\,mol^{-1}$	9	15	30	50
$10^{16}S_{reac}(E_t)/cm^2$	0.5	1.25	2.0	2.2.

Vibrational excitation of HCl ($v = 1 \leftarrow v = 0$) increases the cross-section by an order of magnitude. Estimate the threshold energy for the reaction and discuss the shape of the reaction potential energy surface. Assuming that the cross-section corresponds to a Boltzmann distribution over rotational states in HCl ($v = 0$) at 300 K, estimate the rate constant, $k(300)$ for the reaction.

B2.3. The first electronic transition in CF_3I is associated with the electron promotion:

$$\tilde{A}^1E,\ \tilde{a}^3E[(a_1)\,^2(e)\,^3(a_1^*)^1] \leftarrow \tilde{X}^1A_1[(a_2)\,^2(e)\,^4]$$

where the promoted electron [I, $5p(e)$] is transferred into the antibonding orbital [C—I, (a_1^*)]. The transition generates a broad continuum absorption centred at ~ 250 nm.

Draw up a correlation diagram for the dissociations:

$$CF_3I(\tilde{X}, \tilde{A}, \tilde{a}) \rightarrow CF_3(\tilde{X}^2A_1) + I(^2P_{3/2}, {}^2P_{1/2})$$

assuming strong spin–orbit coupling, and hence identify the spin–orbit state(s) which would need to be preferentially populated in the absorption continuum in order to generate an iodine photodissociation laser pumped by KrF laser radiation at 248 nm. What would be the minimum $[^2P_{1/2}]/[^2P_{3/2}]$ ratio for laser action?

Note:

(a) Species of spin function of C_{3v} symmetry:

S	0	$\tfrac{1}{2}$	1
Γ	A_1	$E_{1/2}$	$A_2 + E$

(b) Resolution of I($^2P_{3/2}$) and I($^2P_{1/2}$) under C_{3v} symmetry:

$$3/2 \rightarrow E_{1/2} + E_{3/2}; \quad 1/2 \rightarrow E_{1/2}.$$

(c) Some useful direct products:

$$E \times E = A_1 + A_2 + E; \quad E_{1/2} \times E_{1/2} = A_1 + A_2 + E$$

$$E_{1/2} \times E_{3/2} = E + E.$$

B2.6 Answers to problems

State selection of reagents and molecular beam techniques

B2.1. $K + Br \rightarrow KBr + H, \quad \Delta H_0^{\ominus} \sim -18 \, \text{kJ mol}^{-1}.$

We are given $\bar{u}_{rel}$, $\langle P_2(\hat{\boldsymbol{j}}' \cdot \hat{\boldsymbol{k}}) \rangle$, $\langle E_r \rangle_{KBr}$ and B_{KBr}, the rotational constant for KBr.

(a) The departing atom is light so that, from angular momentum considerations, $\boldsymbol{L} \sim \boldsymbol{j}'_{KBr}$ (the orbital angular momentum of the incoming particles equals that of the departing molecule; HBr has a large B value). This assertion may be checked from $\langle P_2 \rangle \simeq -0.35$, which shows that $\langle \cos^2 \theta \rangle \approx 0.1$ or $\theta \sim 70°$, where θ is the angle between $\hat{\boldsymbol{j}}'_{KBr}$ and $\hat{\boldsymbol{k}}'$—there is very strong perpendicular alignment.

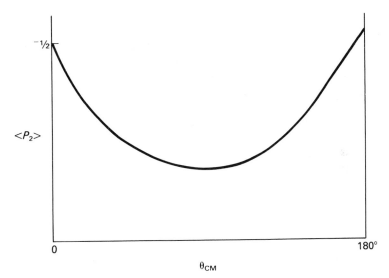

θ_{CM}

(b) Since $\boldsymbol{L} \sim \boldsymbol{j}'_{KBr}$

$$L = \mu_{K,HBr} \times b \times u_{rel} = \sqrt{j(j+1)}\hbar.$$

But $\langle E_{rot} \rangle_{KBr} = \langle j(j+1) \rangle \hbar^2 / (2I_{KBr}) = \langle j(j+1) \rangle B_{KBr} hc$

$\therefore b = [\langle E_{rot} \rangle_{KBr} / (hcB)]^{1/2} \hbar / \mu_{K,HBr} \cdot u_{rel}$

$\simeq 1.2 \times 10^{-10} \, \text{m}$

$\therefore S_{reac} \simeq 4.5\,A^2$.

$$K + Br_2 \rightarrow KBr + Br, \quad \Delta H_0^{\circ} = -188\,kJ\,mol^{-1}.$$

There are no kinematic constraints, so that $\langle P_2 \rangle$ is much smaller. There is, therefore, little correlation between L and j' and E_{rot} is determined largely by the dynamics of the strongly exoergic potential energy surface.

The alignment, $\langle \cos^2 \theta \rangle$, will depend on the scattering angle. For scattering at θ_{CM} near zero or $180°$, $\langle \cos^2 \theta \rangle \rightarrow 0$ and $\langle P_2 \rangle \rightarrow -\frac{1}{2}$. For scattering at $0° < \theta_{CM} < 180°$, $\langle \cos^2 \theta \rangle \neq 0$. Hence $\langle P_2 \rangle$ will depend on the range of scattering angles that are monitored.

B2.2. $K + HCl(v = 0) \rightarrow KCl + H$.

(a) Since $S_{reac}(E_t)$ increases with E_t, we can try a line-of-centres model:

$$S_{reac}(E_t) = S_0\{(E_t - E^0)/E_t\}$$

Plot $S_{reac}(E_t) \cdot E_t$ vs E_t:

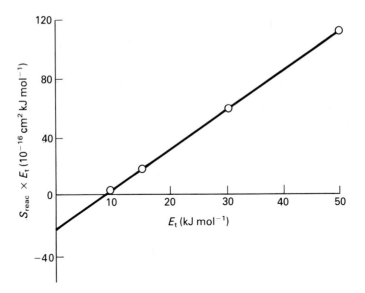

From the slope and intercept

$$S_0 = 2.6 \times 10^{-16}\,cm^2, \quad E^0 = 7\,kJ\,mol^{-1}.$$

(b) The energy difference $[HCl(v = 1) - HCl(v = 0)] \sim 34\,kJ\,mol^{-1}$.

The translational enhancement is therefore very much less than the vibrational enhancement, which suggests a late barrier.

(c) For a line-of-centres model,

$$k(T) = \{8k_B T/(\pi\mu)\}^{1/2}\,S_0 \exp(-E^0/RT)$$

$$\therefore k(300) = 9 \times 10^{-13}\,cm^3\,molecule^{-1}\,s^{-1}.$$

B2.3. The correlations are:

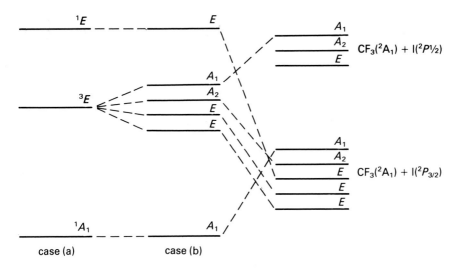

Note that:

(a) We need to populate the A_1 component of the ^3E state in order to correlate with $I(^2P_{1/2})$.

(b) For a population inversion,

$$\frac{[^2P_{1/2}]}{[^2P_{3/2}]} > \frac{g_{1/2}}{g_{3/2}} = \frac{1}{2}$$ at least 33% must be produced in the $^2P_{1/2}$ state.

Chapter B3
Analysis of Experimental Data

M. J. PILLING

B3.1 Introduction

The advent of digital methods of recording experimental data and the use of computers to analyse such data have significantly expanded the range of kinetic problems we can address. These developments have also placed a greater emphasis on the analysis techniques we must employ if we are to do justice to the intrinsic quality of the experimental data. This emphasis is further enhanced by the increasing application of kinetic data in simulations of complex reaction systems and the need for compilers of such complex schemes to evaluate the rate data available in the literature. This chapter examines methods of analysing experimental rate data and of presenting the resulting rate parameters.

The aim of experimental kineticists should be to obtain data under well defined and reproducible conditions, to analyse the data in an unbiased manner and so to produce best estimates of the required rate parameters together with estimates of the associated uncertainty. There are two typical analysis problems:

(a) Signal vs. time, where the signal is related to the concentration and the analysis is required to test a proposed decay mechanism and estimate the rate constant(s).

(b) Rate constant vs. some variable, e.g. temperature or vibrational quantum number, where the analysis is required to estimate Arrhenius parameters or to test a reaction dynamics model.

The following sections briefly outline some of the types of analysis which may be employed.

B3.2 Normally distributed errors[1]

It is easier to appreciate some of the basic principles of error analysis by considering first the probability distribution of a single quantity, y, which is subject to normally distributed random errors. The probability density, $P(y)$, for y is then

$$P(y) = \frac{1}{(2\pi\sigma^2)^{1/2}} \exp[-(y-\mu)^2/2\sigma^2] \tag{1}$$

where μ is the mean and σ^2 the variance. $P(y)dy$ is the probability that, in a single

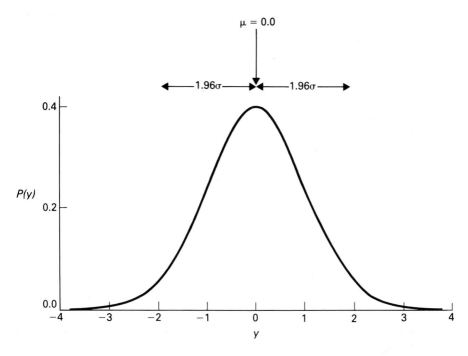

Fig. B3.1. Probability density distribution, $P(y)$, for normally distributed random errors. μ is the mean of the distribution and there is a 95% probability that y lies between $\mu \pm 1.96\,\sigma$, where σ^2 is the variance. In this figure, the mean is zero and the variance unity.

determination, an experimental measurement will lie between y and $y + dy$. There is a 95% probability that y lies between $\mu \pm 1.96\,\sigma$ (Fig. B3.1).

Our aim as experimentalists is to determine μ and σ. However, both μ and σ are not open to direct experimental determination because of the random nature of a single experimental result. The best we can do is to repeat the experiment a number of times, say n, which gives a sample of results with a characteristic sample mean and variance, $\bar{y}$ and s^2 respectively:

$$\hat{\mu} = \bar{y} = \sum_{i=1}^{n} y_i/n \tag{2}$$

$$\hat{\sigma}^2 = s^2 = \sum_{i=1}^{n} (y_i - \bar{y})^2/n \tag{3a}$$

(the $\hat{}$ notation is used to signify an estimate).

In the limit of a very large number of determinations, it is possible to show that the sample mean and variance converge to the true mean and variance. In smaller samples, it is often found that s^2, as defined by equation 3a, underestimates the true variance and it is customary to use the estimate:

$$s^2 = \sum_{i=1}^{n} (y_i - \bar{y}_i)^2/(n-1). \tag{3b}$$

The final result of our set of experiments is $\bar{y}$, our estimate of μ. We also wish to associate confidence limits with this estimate and so we need to calculate the variance of the mean σ_μ^2 which may be estimated from s^2

$$\sigma_\mu^2 = \sigma^2/n \simeq s^2/n. \tag{4}$$

It is generally recommended that 95% confidence limits are employed, corresponding to $\pm 1.96\,\sigma_\mu$. If n is small, $(s^2/n)^{1/2}$ should be multiplied by the Student's t value for the relevant number of degrees of freedom, to allow for the poorness of our estimate of σ_μ. The number of degrees of freedom is the number of determinations minus the number of estimated parameters, or $(n-1)$ in the present context. Table B3.1 lists t for several values of the number of degrees of freedom, v. As $n \to \infty$, $t \to 1.96$.

Table B3.1. Student's t values for 95% confidence limits

v	1	2	3	5	10	12	∞
t	12.71	4.30	3.18	2.57	2.23	2.18	1.96

The experimental estimate should be quoted as $[\bar{y} \pm t(s^2/n)^{1/2}]$, taking care to note explicitly that 95% confidence limits have been employed.

In the above discussion, we have assumed that the errors in the data points, y_i, have identical statistical properties. If they do not, then the data should be weighted (see section B3.5). It should also be noted that our confidence limits refer only to random errors and take no account of systematic errors, which should also be assessed (see section B3.9).

B3.3 Linear regression[1,2]

As noted in the Introduction (section B3.1), in many kinetic problems we are interested in the dependence of an observable, y, on a variable x. Many of the expected relationships between y and x can be expressed in linear form:

$$y = \beta_0 + \beta_1 x \tag{5}$$

e.g. $\ln k = \ln A - \dfrac{E_{act}}{R} \cdot \dfrac{1}{T}$

where $y = \ln k$ and $x = 1/T$. We wish to use our experimental data to estimate the parameters β_0 ($= \ln A$) and β_1 ($= -E_{act}/R$). This goal is achieved by minimising the sum, S, of the squares of the residuals, $\varepsilon_i = y_i - (\beta_0 + \beta_1 x_i)$, where once again i labels the different data points, and runs from i to n

$$S = \sum_{i=1}^{n} \varepsilon_i^2 = \sum_{i=1}^{n} [y_i - (\beta_0 + \beta_1 x_i)]^2. \tag{6}$$

In order to find the optimum values of β_0 and β_1, we minimise S with respect to β_0 and β_1 simultaneously:

$$\frac{\partial S}{\partial \beta_0} = -2 \sum_{i=1}^{n} (y_i - \beta_0 - \beta_1 x_i) = 0. \tag{7}$$

$$\frac{\partial S}{\partial \beta_1} = -2 \sum_{i=1}^{n} x_1(y_i - \beta_0 - \beta_1 x_i) = 0. \tag{8}$$

Equations 7 and 8 are then solved to give estimates, b_0 and b_1, of β_0 and β_1, which are given by

$$b_1 = S_{xy}/S_{xx} \tag{9}$$

$$b_0 = \bar{y} - b_1 \bar{x} \tag{10}$$

where

$$S_{xy} = \Sigma x_i y_i - (\Sigma x_i)(\Sigma y_i)/n \tag{11}$$

$$S_{xx} = \Sigma x_i^2 - n(\bar{x})^2 \tag{12}$$

and $\bar{x}$ and $\bar{y}$ are the sample mean values of x and y, given by $\bar{x} = (\Sigma x_i)/n$, $\bar{y} = (\Sigma y_i)/n$.

We generally make the assumption that the errors are normally distributed about the best fit or regression line, i.e. the residuals, ε_i, conform to a Gaussian distribution, with variance σ^2. We do not know σ^2 and would need to make a very large number of measurements to determine it accurately, but, once again, we can estimate it from our data:

$$\sigma^2 \simeq s^2 = \sum_i (y_i - b_0 - b_1 x_i)^2/(n-2) \tag{13}$$

s^2 is termed the estimated variance and $(n-2)$ is the number of degrees of freedom. The estimated variances in b_0 and b_1 are given by:

$$\hat{\sigma}^2(b_0) = \frac{s^2 . \Sigma x_i^2}{n \, S_{xx}}; \quad \hat{\sigma}^2(b_1) = \frac{s^2}{S_{xx}}.$$

The standard deviations in b_0 and b_1 are defined as the square roots of their variances. Once again, we include Student's t in our estimates of the confidence limits:

$$\hat{\beta}_0 = b_0 \pm ts \sqrt{\frac{\Sigma x_i^2}{n \, S_{xx}}}$$

$$\hat{\beta}_1 = b_1 \pm ts/S_{xx}^{1/2}.$$

Note that the analysis presented in this section made the assumption that each value of $\ln k_i$ had equal error and no attempt has been made to weight the data. The procedures necessary when weighting is required are discussed in section B3.5.

B3.4 Propagation and combination of errors[1,3]

Frequently, we determine a rate parameter from measured changes in several variables:

$$y = f(z_1, z_2, \ldots . z_j, \ldots . z_m)$$

or we need to estimate the uncertainty in y when it is not linearly related to z.

The variance in y can be calculated from those in z_j as follows, provided the errors in the z_j are uncorrelated

$$\sigma_y^2 = \sum_{j=1}^{m} \left\{ \left(\frac{\partial f}{\partial z_j} \right) \sigma_j \right\}^2 \tag{14}$$

e.g. $y = C z_1 z_2$

$$\sigma_y^2 / y^2 = (\sigma_1 / z_1)^2 + (\sigma_2 / z_2)^2.$$

Other useful combinations, which are readily derived using equation 14, are:

(a) $y = az_1 + bz_2$

$$\sigma_y^2 = a^2 \sigma_1^2 + b^2 \sigma_2^2$$

(b) $y = \ln z_1$

$$\sigma_y^2 = \sigma_1^2 / z_1^2$$

(c) $y = \exp(z_1)$

$$\sigma_y^2 = \sigma_1 y.$$

If the errors are correlated, then we must take account of the covariance, e.g. for two variables

$$y = f(z_1, z_2),$$

$$\sigma_y^2 = \sigma_1^2 \left(\frac{\partial f}{\partial z_1} \right)^2 + \sigma_2^2 \left(\frac{\partial f}{\partial z_2} \right)^2 + 2\sigma_{12}^2 \left(\frac{\partial f}{\partial z_1} \right) \left(\frac{\partial f}{\partial z_2} \right) \tag{15}$$

where σ_{12}^2 is the covariance of z_1 and z_2:

$$\sigma_{12}^2 = \lim_{n \to \infty} \sum_{i=1}^{n} (z_{1i} - \hat{z}_1)(z_{2i} - \hat{z}_2). \tag{16}$$

An important application of such a relationship is in the use of an Arrhenius expression to calculate the best estimate of a rate constant at a specific temperature, since A and E are correlated. A corollary of this observation is that covariance should really be quoted in Arrhenius expressions as well as variances.

B3.5 Weighting of data[1,2]

Not all data points are determined with equal precision and it is important to weight their contribution in the minimisation of the sum of squares. In Arrhenius plots, the $k_i(T)$ values have associated variances, σ_i^2, determined either from the error analysis of the decay data or from repeat determinations at a single temperature. The weight, ω_i, is set equal to σ_i^{-2} and the weighted sum of squares

$\Sigma_i \omega_i (y_i - \hat{y}_i)^2$ minimised, where $\hat{y}_i$ is the best fit value in the regression analysis (i.e. $\hat{y}_i = b_0 + b_1 x_i$). In analysing data, careful attention should be given to weighting the data especially for second-order plots. Figure B3.2 shows such a plot for methyl radical recombination. Because the reciprocal concentration is plotted, the largest values of y correspond to the lowest concentrations and so have the largest uncertainties. The sum of squares in an unweighted fit would, therefore, be primarily determined by these low concentration points and the minimisation would be unreliable. Weighting overcomes this problem since points with large uncertainty will not be given undue weight in the sum of squares.

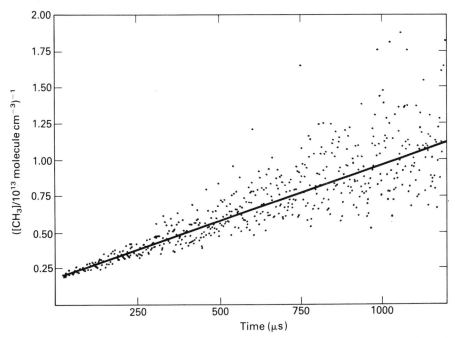

Fig. B3.2. Second-order plot for methyl radical recombinations. The plot is of the reciprocal concentration vs. time and shows how the scatter of the data points about the best fit line increases with time.

Generally speaking, if the variance of y_i is σ_i^2, then the weighting factor is σ_i^{-2}. It is important to adjust the weighting factors properly if the variable in the linear form is a function of the experimental observable (see section B3.4); e.g. for a first order plot, $y_i = \ln c_i$, where c is the concentration and $\omega_i = c_i^2/[\sigma(c_i)]^2$, whilst for a second-order plot, $y_i = 1/c_i$ and $w_i = c_i^4/[\sigma(c_i)]^2$, where $[\sigma(c_i)]^2$ is the variance of c_i.

It is not always possible to estimate the weighting factors from first principles. In photon counting experiments, the number of counts, N_i, follows a Poisson distribution and $[\sigma(N_i)]^2 = N_i$. With analogue signals, such as those found in absorption experiments, such a simple relationship does not apply and a more heuristic approach has to be adopted. One such approach, which works well if the weighting factors do not vary too strongly with x, is to fit initially to the unweighted data and then to fit a quadratic through the moduli of the residuals, $|\varepsilon_i| = |y_i - \hat{y}_i|$.[4] The quadratic gives a reasonable representation of how the residuals vary, on average, with x and relative weighting factors can then be set equal to $(\hat{\varepsilon}_i)^{-2}$. This procedure is open to question with second-order fits and it is then often better to use a non-linear least-squares technique (see section B3.6).

The relevant equations for a weighted linear least squares fit follow from section B3.2.[2] We now minimise $\sum\limits_{i=1}^{n} \omega_i (y_i - \hat{y}_i)^2$ with respect to β_0 and β_1 and find

$$S_{xy} = \Sigma w_i x_i y_i - (\Sigma w_i x_i)(\Sigma w_i y_i)/(\Sigma w_i) \tag{17}$$

$$S_{xx} = \Sigma w_i x_i^2 - (\Sigma w_i x_i)^2/(\Sigma w_i) \tag{18}$$

and

$$\bar{x} = (\Sigma w_i x_i)/(\Sigma w_i), \quad \bar{y} = (\Sigma w_i y_i)/(\Sigma w_i).$$

There are two estimates of the variances of b_0 and b_1 which can be made with weighted fits. Internal estimates may be made from the variances of the individual data points:

$$[\hat{\sigma}(b_0)]^2_{int} = \Sigma w_i \Sigma w_i x_i^2 / S_{xx} \tag{19}$$

$$[\hat{\sigma}(b_1)]^2_{int} = (\Sigma w_i)^2 / S_{xx}. \tag{20}$$

Internal variances are not determined by the deviations of the points from the straight line, but simply reflect a weighted average of the individual variances of the data used to construct the straight line.

External variances reflect the scatter about the best fit line (cf. equation 13):

$$s^2 = \Sigma w_i (y_i - \hat{y}_i)^2 / (n - 2) \tag{21}$$

$$[\hat{\sigma}(b_0)]^2_{ext} = \frac{s^2 \Sigma w_i x_i}{S_{xx} \Sigma w_i} \tag{22}$$

$$[\hat{\sigma}(b_1)]^2_{ext} = s^2 / S_{xx}. \tag{23}$$

$\hat{\sigma}^2_{ext}$ and $\hat{\sigma}^2_{int}$ should be similar (see reference 3) but it is often found that $\hat{\sigma}^2_{ext}$ is significantly larger than $\hat{\sigma}^2_{int}$. As an example, we take a set of rate constants, k_i, determined from a set of first-order decay curves which are used to construct an Arrhenius plot. The variances of the rate constants may be estimated from the first-order fits and the variances of A and E may be calculated internally or externally. An external estimate of the variance which significantly exceeds the internal estimate would arise if there were additional errors which were not picked up by an analysis based simply on the scatter of the data in single decay curves, e.g. random errors in temperature or in concentration measurements. One should then either re-estimate the individual weighting factors by recognising the additional sources of error, or use σ^2_{ext} as a better estimate of uncertainty.

Finally, our discussion has assumed that any experimental uncertainty is confined to y_i. This is frequently the case, but is not always so. Procedures for experiments in which uncertainties in both x_i and y_i are significant may be found in reference 5.

B3.6 Non-linear least squares fits[1,2,4]

Not all rate laws are amenable to linear regression. For example, for a mixed first-, second-order decay:

$$A \xrightarrow{k_1} \text{Products}$$

$$A + A \xrightarrow{k_2} A_2.$$

Then

$$[A] = \left\{ \left([A]_0^{-1} + \frac{2k_2}{k_1} \right) \exp(k_1 t) - \frac{2k_2}{k_1} \right\}^{-1}. \tag{24}$$

Under such circumstances, non-linear least squares fitting methods should be adopted. The method involves examining the variation in the quantity χ^2 defined by:

$$\chi^2 = \sum_i [\omega_i (y_i - \hat{y}_i)^2]$$

over parameter space (up to three parameters, k_1, k_2 and $[A]_0$ in the above example) and finding the minimum. In general terms, if $y = f(\beta_1, \beta_2 \dots \beta_j \dots \beta_m)$ then we need to minimise χ^2 with respect to each β parameter simultaneously:

$$\frac{\partial \chi^2}{\partial \beta_j} = \frac{\partial}{\partial \beta_j} \sum_{i=1}^{n} [\omega_i (y_i - \hat{y}_i)^2] = 0. \tag{25}$$

Figure B3.3 shows a schematic diagram of a χ^2 surface. In applying any fitting procedure we need to convince ourselves that there is a unique minimum, otherwise several sets of parameters provide potential solutions to our problem. It may be useful to examine the χ^2 surface, or sections through it, if there are more than two variable parameters, to demonstrate the feasibility of a unique fit.

An excellent account of a good approach to the method of minimising χ^2 may be found in chapter 11 of Bevington's book.[1] The chapter includes several useful FORTRAN subroutines, laid out in detail with plenty of comment statements, and the reader is strongly encouraged to consult this chapter.

An efficient approach to the problem is embodied in the so-called Marquardt algorithm. Initial guesses are made of all the β_j parameters and the method starts by employing a gradient search routine in which the line of steepest descent is followed as a means of approaching the minimum. This technique works very well

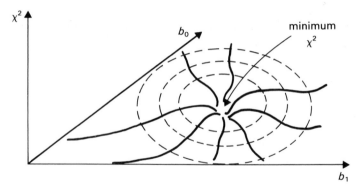

Fig. B3.3. Schematic χ^2 surface for a two parameter fit. The aim is to find the minimum in χ^2; surfaces with multiple minima lead to fitting problems.

if the set of parameters differs significantly from that at the minimum, but gets increasingly inefficient as the minimum is approached and the gradient tends to zero. An alternative method is to linearize the fitting function by expanding it as a Taylor series in the parameters, β_j, and truncating after the linear terms. In this form, the function may be fitted using linear least squares techniques; such an approach is likely to work best in the region near the minimum in the χ^2 surface, i.e. exactly where the gradient search is least efficient. If the two techniques are employed at the same point in parameter space, not surprisingly they suggest different directions for approaching the minimum. The Marquardt algorithm interpolates between these two directions and also calculates a step size so that the next point in parameter space may be determined and the minimum approached efficiently.

B3.7 Matrix techniques[2]

Regression analysis is ideally suited to a matrix formulation. For example, a set of data for fitting to a linear function (equation 5) may be expressed in the form:

$$Y = XB + \varepsilon$$

where Y is a column vector of dependent variables, y_i, X is a matrix containing the independent variables x_i, B is the vector of parameters and ε is an error vector:

$$
\begin{pmatrix} y_1 \\ y_2 \\ \cdot \\ \cdot \\ y_i \\ \cdot \end{pmatrix}
=
\begin{pmatrix} 1 & x_1 \\ 1 & x_2 \\ \cdot & \cdot \\ \cdot & \cdot \\ 1 & x_i \\ \cdot & \cdot \end{pmatrix}
\begin{pmatrix} \beta_0 \\ \beta_1 \end{pmatrix}
+
\begin{pmatrix} \varepsilon_1 \\ \varepsilon_2 \\ \cdot \\ \cdot \\ \varepsilon_i \\ \cdot \end{pmatrix}
.
$$

The technique is discussed in detail by Draper and Smith[2] in their chapter 2. The vector of best fit parameters, b

$$b = \begin{pmatrix} b_0 \\ b_1 \end{pmatrix}$$

where, as in section B3.3, b_0 and b_1 are the best estimates of β_0 and β_1 respectively, is given by:

$$b = (X'X)^{-1}X'Y \tag{26}$$

where X' is the transpose of X, whilst the vector of best fit values of the dependent variable, $\hat{Y}$, is given by

$$\hat{Y} = Xb. \tag{27}$$

The variances in b_0 and b_1, $\sigma^2(b_0)$ and $\sigma^2(b_1)$ and the covariance, $\sigma^2(b_0, b_1)$ are derived from the variance–covariance matrix of b:

$$V(b) = \begin{pmatrix} \sigma^2(b^0) & \sigma^2(b_0, b_1) \\ \sigma^2(b_0, b_1) & \sigma^2(b_1) \end{pmatrix}.$$

Draper and Smith[2] show that

$$V(b) = (X'X)^{-1}\sigma^2$$

where, as before, σ^2 is estimated from the sums of squares of the residuals.

$$\sigma^2 \sim s^2 = (Y'Y - b'X'Y)/(n-2).$$

Matrix techniques are particularly valuable in efficient computer programs; for example, the subroutines in the Marquardt algorithm, referred to in the previous section, are formulated in this way.

B3.8 Goodness of fit[1,2]

It is not sufficient simply to obtain best estimates of the parameters in our fitting procedure, it is also important to ensure that our data are compatible with the model we have employed. For example, a radical decay determined primarily by recombination and analysed according to a pure second-order rate law, would show deviations from such a law, and overestimates of the rate constant, if traces of oxygen were present.

The easiest and most widely employed method of testing the goodness of fit is to examine the residuals, ($y_i - \hat{y}_i$). Figure B3.4a shows a non-linear second-order fit for methyl radicals in the presence of oxygen together with a plot of the residuals—note that the residuals are not randomly scattered but show a distinct negative bias at short and at long times. Figure B3.4b shows the same data fitted according to a mixed first-, second-order decay (equation 24). The residuals are

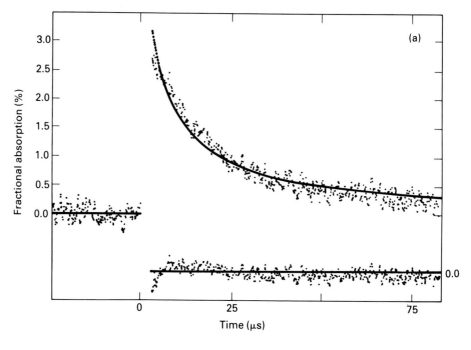

Fig. 3.4a.

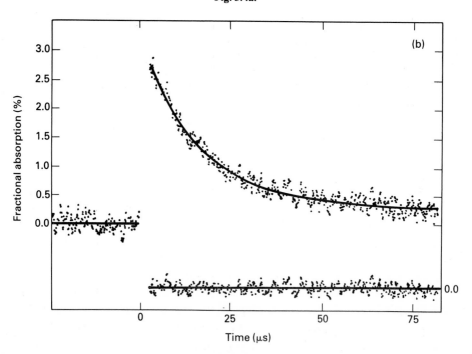

Fig. 3.4b.

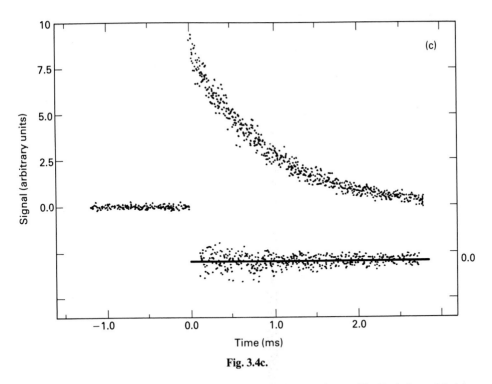

Fig. 3.4c.

Fig. 3.4. Demonstration of the use of residuals in assessing goodness of fit. Each figure (a)–(c) shows a decay curve and a residuals plot. (a) and (b) show data for methyl radical recombination in the presence of oxygen. In (a) the data have been fitted to a pure second-order decay and the residuals scatter unevenly around the zero line. The same data have been analysed according to a mixed-order decay, equation (24), in (b) and the randomly scattered residuals demonstrate a good fit. (c) shows data for the decay of hydrogen atoms in the presence of ethene. The residuals are unweighted and the larger scatter at short times demonstrates the importance of weighting the data.

now randomly scattered, demonstrating that the mixed-order model is a good one. Figure B3.4c shows data for the decay of H atom resonance fluorescence in the presence of C_2H_4, fitted according to a first-order model. The residuals are randomly scattered, demonstrating goodness of fit, but the variance of the short time data is greater than that of the long time data. This effect arises because the noise increases with signal strength. The plot illustrates the point made earlier about the importance of weighting data—in the present example, an unweighted fit would give too much emphasis to the short time data.

It is not always convenient to employ a visual test and numerical statistics have many advantages. For example, the reduced χ_v^2, where $\chi_v^2 = \chi^2/\nu$ (ν is the number of degrees of freedom) should be close to unity and large χ_v^2 values indicate a poor fit. Additional statistical tests have been described by Bevington[1] (see the F test in his chapter 10) and by Draper and Smith.[2]

B3.9 Systematic errors[1,3]

It is important to assess sources of error additional to those revealed from a statistical analysis of, say, a decay curve. Contributions from obvious examples, such as temperature and concentration, may be incorporated into the final result via equation 14. Quite frequently, reactions additional to that being studied may occur and their effect on the decay of the monitored species should be assessed via numerical simulation of a full reaction scheme. If the effect is well within experimental error, then no further action is required, but significant contributions from additional reactions may require the experimentalist to refine the approximate analytical result using a full numerical analysis of the experimental data (see Chapter C4). It is then necessary to include uncertainties in rate constants employed in the numerical analysis in the quoted overall uncertainty for the measured rate parameters.

There may also be potential sources of error whose magnitude cannot be assessed. For example, in the determination of radical heats of formation (Chapter B4) the entropy of the radical is required and, for larger species, the necessary vibrational frequencies are unavailable and must be estimated. Whilst it is easy to calculate the effect specific uncertainties in these frequencies will have on $\Delta H_f^{\ominus}$, it is very difficult to obtain a realistic, numerical estimate of the uncertainties themselves. In such a situation, it is important that the final result is presented along with the estimated entropy, with the statement that potential errors in the entropy have not been included in the quoted uncertainty.

Finally, it should always be remembered that kinetic systems may contain unanticipated complexities. A vigilant, painstaking approach is necessary in which conditions are varied as widely as possible to allow such complexities an opportunity to reveal themselves. There is no substitute, however, for measurements in different laboratories by different techniques, before rate parameters can be considered to be well established.

B3.10 References and suggestions for further reading

1 Bevington P. R. (1969) *Data Reduction and Error Analysis for the Physical Sciences*, McGraw-Hill, New York.
   A very useful handbook explaining how to apply statistical analysis, with several subroutine listings. Not so good on the basic theory.
2 Draper N. R. and Smith H. (1981) *Applied Regression Analysis*, 2nd edn, Wiley, New York.
   A more detailed, but accessible, discussion of basic theory. Good on matrix techniques.
3 Cvetanovic R. J., Singleton D. L. and Paraskevopoulos G. (1979) *J. Phys. Chem.*, **83**, 50.
   A discussion of error analysis in a kinetics context.
4 Tulloch J. M., Macpherson M. T., Morgan C. A. and Pilling M. J. (1982) *J. Phys. Chem.*, **86**, 3812.
   Application of non-linear least squares techniques to an analysis of decay data.
5 Irvin J. A. and Quickenden T. I. (1983) *J. Chem. Ed.*, **60**, 711.
   Weighting procedures when there are uncertainties in both x and y.

B3.11 Problem

Analysis of experimental data

The following rate data refer to the reaction

$$H + C_2H_4 \rightarrow C_2H_5$$

at the high pressure limit:

T/K	$10^{12}k^\infty/cm^3 \text{ molecule}^{-1}s^{-1}$
198	0.200 ± 0.016
216	0.298 ± 0.050
234	0.407 ± 0.040
258	0.631 ± 0.086
283	0.943 ± 0.110
285	0.860 ± 0.230
298	1.130 ± 0.108
320	1.45 ± 0.110
400	2.83 ± 0.15
511	4.27 ± 0.33
604	7.69 ± 0.37

The quoted error limits refer to ± 2 standard deviations.
Determine:

(a) the best estimates for the Arrhenius parameters A and E,
(b) the variance in A and E,
(c) 95% confidence limits for A and E,
(d) the best estimate of k at 550 K
(for 9 degrees of freedom $t = 2.262$).

B3.12 Answer to problem

Analysis of experimental data

Let $x = 10^3 K/T$
    $y = \ln k$.

Need linear regression on $y = b_1 x + b_0$

where $b_1 = -(E/R) \times 10^{-3}$ and $b_0 = \ln A$.

The weightings are given by $(k_i^2/\sigma_i^2)^\S$.

§Note only relative weightings are necessary for our purposes.

x_i	$-y_i$	$w_i =$ $\left(\dfrac{k_i}{\sigma_i}\right)^2 \times 10^{-2}$	$w_i x_i$	$w_i x_i^2$	$-w_i y_i$	$-x_i y_i w_i$	$-y_i$	$-\hat{y}_i$	$w_i(y_i - \hat{y}_i)^2$
5.051	29.24	6.25	31.57	159.45	182.75	923.1	29.24	29.35	0.0756
4.630	28.84	1.42	6.58	30.44	40.95	189.6	28.84	28.89	0.0036
4.274	28.53	4.14	17.69	75.63	118.11	504.8	28.53	28.51	0.0017
3.876	28.09	2.15	8.33	32.30	60.39	234.1	28.09	28.08	0.0002
3.534	27.69	2.94	10.39	36.72	81.41	287.7	27.69	27.71	0.0012
3.509	27.78	0.56	1.96	6.89	15.56	54.6	27.78	27.68	0.0056
3.356	27.51	4.38	14.70	49.33	120.49	404.4	27.51	27.51	0.0000
3.125	27.50	4.33	13.53	42.29	119.08	372.1	27.50	27.27	0.2291
2.500	26.59	14.24	35.60	89.00	378.64	946.6	26.59	26.59	0.0000
1.957	26.18	6.70	13.11	25.66	175.41	343.3	26.18	26.00	0.2171
1.656	25.59	16.88	27.95	46.29	431.96	715.3	25.59	25.68	0.1367
Σ		63.99	181.41	594.0	1724.7	4975.6			0.7708

For a weighted fit

$$b_0 = \frac{1}{\Delta}\{\Sigma w_i x_i^2 \, \Sigma w_i y_i - \Sigma w_i x_i \Sigma w_i x_i y_i\}$$

$$b_1 = \frac{1}{\Delta}\{\Sigma w_i \Sigma w_i x_i y_i - \Sigma w_i x_i \Sigma w_i y_i\}$$

$$\Delta = \Sigma w_i \Sigma x_i^2 w_i - (\Sigma w_i x_i)^2$$

Thus

$$b_0 = \frac{-594.0 \times 1724.7 + 181.41 \times 4975.6}{63.99 \times 594.0 - (181.41)^2} = -23.89$$

$$b_1 = \frac{-63.99 \times 4975.6 + 181.41 \times 1724.7}{63.99 \times 594.0 - (181.41)^2} = -1.080.$$

Thus, the best estimates of the Arrhenius parameters are:

$$A = \exp b_0 \quad = \quad 4.21 \times 10^{-11} \, \text{cm}^3 \, \text{molecule}^{-1} \, \text{s}^{-1}$$

$$E = -10^3 R b_1 \quad = \quad 8.98 \, \text{kJ} \, \text{mol}^{-1}.$$

Both external (i.e. calculated from the scatter of the data points) and internal (i.e. calculated from the standard deviations of the data points) standard deviations should be calculated (see reference 3). We shall calculate the former explicitly, to demonstrate which data points show the largest scatter (see Table).

$$\sigma_{\text{ext}}^2 = 2.00/9 = 0.0856$$

$$(\sigma_{b_1})_{\text{ext}} = 0.293 \left\{\frac{63.99}{5100}\right\}^{1/2} = 0.033$$

$$(\sigma_{b_0})_{\text{ext}} = 0.293 \left\{\frac{594}{5100}\right\}^{1/2} = 0.100.$$

$$(\sigma_{b_1})^2_{int} = \Sigma \frac{1}{\sigma_i^2} \Big/ \left\{ \Sigma \frac{1}{\sigma_i^2} \Sigma \left(\frac{x_i}{\sigma_i}\right)^2 - \Sigma \left(\frac{x_i}{\sigma_i}\right)^2 \right\}$$

$$(\sigma_{b_0})^2_{int} = \Sigma \left(\frac{x_i}{\sigma_i^2}\right) \Big/ \left\{ \Sigma \frac{1}{\sigma_i^2} \Sigma \left(\frac{x_i}{\sigma_i}\right)^2 - \Sigma \left(\frac{x_i}{\sigma_i}\right)^2 \right\}$$

where σ_i^2 is the (standard deviation)^2 in y_i i.e. $= (w_i^{-1}) \times 10^2$

$$\therefore \ (\sigma_{b_1})^2_{int} = \frac{10^{-2}\Sigma w_i}{\Delta}; \quad (\sigma_{b_0})^2_{int} = \frac{10^{-2}\Sigma w_i x_i}{\Delta}$$

$$(\sigma_{b_1})_{int} = 0.011 \quad (\sigma_{b_0})_{int} = 0.034$$

Thus the internal standard deviations are significantly smaller than the external ones, suggesting that the standard deviations of the individual data points are too small—there may be an additional source of error, which shows up in the Arrhenius plot, but is not evident from the scatter of the data points used to determine the individual k's. We shall use the larger standard deviations in our estimates, but this discrepancy casts some doubt on our weighting procedures. Propagating the standard deviations:

$$\sigma_A = A \cdot \sigma_{b_0} = 0.42 \times 10^{-11} \, cm^3 \, molecule^{-1} s^{-1}$$

$$\sigma_E = 10^3 \sigma_{b_1} R = 0.27 \, kJ \, mol^{-1}.$$

There are 11 data points, and 2 parameters have been determined—9 degrees of freedom, so that t, for 95% confidence limits, is 2.262.

$$\therefore \ A = (4.2 \pm 1.0) \times 10^{-11} \, cm^3 \, molecule^{-1} s^{-1}$$

$$E = (9.0 \pm 0.6) \times 10^3 \, J \, mol^{-1} \, K^{-1}$$

quoted errors refer to 95% confidence limits.

To determine the best estimate of k at 550 K:

$$\hat{k}_{550} = 5.91 \times 10^{-12} \, cm^3 \, molecule^{-1} s^{-1}.$$

Variance $(\ln \hat{k}_0)^2$

$$= \left\{ \frac{\Sigma w_i (y_i - \hat{y}_i)^2}{\Sigma w_i (n-2)} + \frac{(x_0 - \bar{x})^2}{\Sigma w_i (x_i - \bar{x})^2} \frac{\Sigma w_i (y_i - \bar{y}_i)^2}{(n-2)} \right\}$$

where $\bar{x}$ is the (weighted) mean.

$$\bar{x} = 2.834; \quad \Sigma w_i (y_i - \hat{y}_i)^2 = 0.771$$

$$\Sigma w_i = 63.99; \quad (x_0 - \bar{x})^2 = 1.032$$

$$\Sigma w_i (x_i - \bar{x})^2 = 79.33$$

$$\therefore \ Var (\ln \hat{k}_0) = 6.7 \times 10^{-3}$$

$$\sigma(\ln \hat{k}_0) = 0.08$$

$$\therefore \qquad \sigma(\hat{k}_0) = 0.47.$$

At 550 K, best estimate is $(5.9 \times 1.1) \times 10^{-12} \, cm^3 \, molecule^{-1} s^{-1}$ (95% confidence limits).

Chapter B4
Case Studies

M. J. PILLING

B4.1 Introduction

In this chapter, we shall examine the application of some of the techniques discussed in Chapter B1 in the elucidation of the mechanisms of two types of elementary reaction and in the determination of thermodynamic parameters for free radicals. All of the reactions we shall examine are relevant to the complex systems which will be described in Section C of this book, so this chapter not only illustrates the application of experimental and theoretical techniques, but also forms a link between the study of elementary reactions and their incorporation in reaction schemes of complex systems.

Throughout this chapter, it is important to recognise that our understanding of reactions and our ability to parameterise rate constants and their temperature dependencies is in a state of constant, evolutionary (we hope!) flux. The picture we shall present is simply a snapshot taken during this process and should not be taken as the final word; it may, indeed, already be out of date, in some aspects, by the time this book is published. The main aim of the chapter is not to present a finished complete view of the set of chosen reactions, but simply to illustrate how the thinking of kineticists evolves as a result of research employing a variety of different techniques.

B4.2 Addition of oxygen atoms to unsaturated hydrocarbons

The problem with which we are concerned in this section is the determination of the product channels resulting from the addition of ground state oxygen atoms (O^3P) to ethene and ethyne. For a long time it has been thought that there are, potentially, two major channels for each reaction:

$$O + C_2H_2 \rightarrow {}^3CH_2 + CO \tag{1a}$$

$$\rightarrow H + HCCO \tag{1b}$$

$$O + C_2H_4 \rightarrow CH_3 + HCO \tag{2a}$$

$$\rightarrow H + CH_2CHO \tag{2b}$$

where ${}^3CH_2$ is the ground triplet state of methylene, $CH_2({}^3B_1)$.

Both reactions may be thought of as involving addition to form a triplet biradical, channel (a) requiring a subsequent 1,2 H-atom shift, followed by C—C bond cleavage, channel (b) requiring C—H cleavage. Several 'bulb' experiments have demonstrated the occurrence of both channels. Jones and Bayes[1] detected $^3CH_2$ by photoionization mass spectrometry in a flow tube study of $O + C_2H_2$ and also observed that HCCO grows to a steady state at short times, suggesting that it is an 'early' product in the complex chemical sequence which occurs following the initial elementary reaction. End product analysis of $O + C_2D_2 + H_2$ showed the presence of HD, arguing for D atom formation via reaction 1b.[2] A classic series of experiments was conducted by Cvetanovic[3] in which the products from the addition of oxygen atoms to olefins were probed by end product analysis. This work suggested that (a) is the major channel in reaction 2.

The detailed, quantitative results of these studies differed, but there was undoubtedly strong general support for channels 1a and 2a. These types of experiment have been criticised, not always justifiably, because they are 'indirect'—products are analysed, in a variety of ways, and the initial radical yields must be deduced using an overall reaction mechanism, which may be subject to debate. Gutman's research group was the first to try to probe the elementary reactions more directly. They employed a crossed molecular beam system, coupled with detection by photoionization mass spectrometry[4,5] (see section B1.2). They used 'diffuse' molecular beams, so that the reactions were not studied under truly 'single-collision' conditions, but the experiments were undoubtedly more able to probe the elementary reaction directly than had previously proved possible. For $O + C_2H_2$, they found major ion signals from both CH_2 and HCCO, suggesting, once again, that both channels 1a and 1b operate.[4] For reaction 2 their experiments favoured channel (a) and the vinoxy radical was not detected.[5]

Theory provides an interesting, alternative view of the likely reaction path. The routes, including the symmetries of the electronic states involved, are shown in Scheme I for reaction 1. Harding[6] calculated the reaction path energies for $O + C_2H_2$ assuming that both channels occur via the $^3A''$ state. His calculations, incorporated configuration interaction and polarisation and were based on a double ζ basis set. His results were unequivocal, but in direct conflict with experiment; he found that the 1,2 H-shift gave channel (a) an activation barrier nearly $70 \, kJ \, mol^{-1}$ greater than that for channel (b). His examination of the configurations of the transition states also suggested a larger A factor for channel (b). Similar arguments can be made for reaction 2.

Convincing confirmatory evidence for this interpretation was provided by the work of Lee and his group at Berkeley.[7] They employed a crossed molecular beam apparatus with mass spectrometric detection and angular resolution under genuine single collision conditions. From $O + C_2H_4$ they observed peaks at m/e values of 42 (from cracking C_2H_3O) and 15 (from CH_3). However the angular distribution of the mass 15 peak was identical to that of mass 42, suggesting that it was

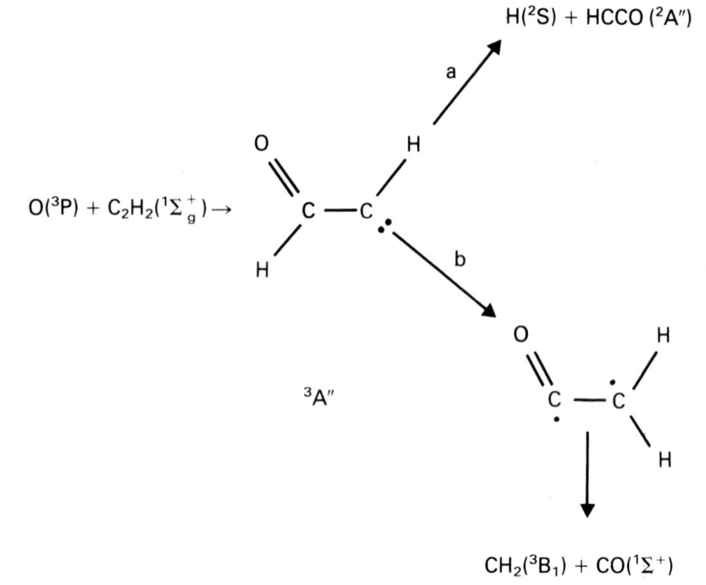

$$H(^2S) + HCCO\ (^2A'')$$

$$O(^3P) + C_2H_2(^1\Sigma_g^+) \rightarrow$$

$$^3A''$$

$$CH_2(^3B_1) + CO(^1\Sigma^+)$$

Scheme 1.

produced entirely from the fragmentation of C_2H_3O in the mass spectrometer. Were any CH_3 to be produced by reaction 2a then it would be expected to show an angular distribution different from that for the C_2H_3O from reaction 2b. The conclusion from these experiments is clear: if reaction 2a occurs under single collision conditions, then it is a very minor channel. Further confirmation came, at about the same time, from Kleinermans and Luntz, who detected vinoxy radicals by LIF.[8]

A possible rationalisation of these seemingly conflicting results was provided by Hunziker et al.[9] in their molecular modulation study of reaction 2. They utilised the sensitivity of the technique to detect the vinoxy radical, CH_2CHO, via its absorption spectrum at ~ 300 nm, which they observed for the first time, following the mercury photosensitised decomposition of N_2O (to give O^3P) in the presence of ethylene. From the yield of vinoxy, they deduced a value of 0.36 for $k_{2a}/(k_{2a}+k_{2b})$, independent of pressure for their experimental conditions ($P \geqslant 40$ Torr). Using absorption spectroscopy at 563 nm they also detected HCO with a fractional yield of ~ 0.55. From a consideration of the subsequent radical reactions, given these yields, Hunziker deduced molecular product yields in excellent agreement with those of Cvetanovic,[3] whose experiments were conducted many years before, but under similar conditions.

The key to the interpretation given by Hunziker is the difference in the conditions probed by 'bulb' and 'beam' experimentalists: the initial adduct in reaction 2, $CH_2CH_2O^*$, experiences several collisions in the former experiments and none in the latter. A simplified, schematic version of the proposed mechanism

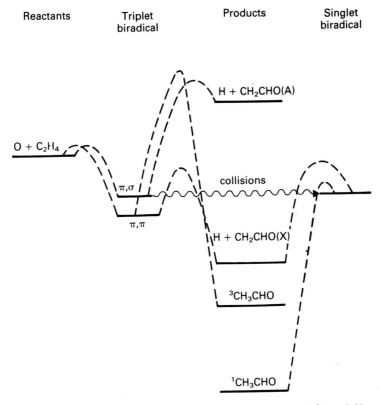

Fig. B4.1. Schematic representation of the proposed mechanism of the $O(^3P) + C_2H_4$ reaction. Adapted from reference 9.

is shown in Fig. B4.1. There are two possible types of triplet biradical which may be formed in the initial addition, labelled (π, σ) and (π, π), depending on the types of orbital occupied by the electrons. The (π, π) undergoes facile H elimination to produce the ground state vinoxy radical (X) whilst the (π, σ) correlates with an excited vinoxy (A), via a high activation barrier. Note also that the barrier to forming triplet acetaldehyde, via a 1,2 H-atom shift, is high. Thus the (π, σ) species has nowhere to go and, in the absence of collisions, simply falls apart to regenerate the reactants. As the pressure is increased, however, it is suggested that collisions induce a transition to the singlet biradical, which can also lose H or can undergo a facile 1,2 H-atom shift to generate singlet acetaldehyde, which has so much internal energy that it dissociates to generate the products of channel 2a. The absence of any pressure effects in Hunziker's experiments demonstrate that the collision induced intersystem crossing is complete even at his lowest pressure of 40 Torr.

Thus a seemingly inconsistent set of experimental results can be accommodated within a single rational mechanism, which *may* be correct! Experiments continue and recent papers which may be read with profit are those of Clemo *et al.*[10] in which the dynamics of reaction 2 are probed in crossed beam experiments and of Vinckier *et al.*[11] which demonstrates the care which must be exercised in interpreting the results of a flow tube study of reaction 1.

B4.3 Kinetics and mechanisms of HO_2 disproportionation

HO_2 is an important radical sink in the troposphere and a central species in combustion processes. An important characteristic is its lack of reactivity with molecular species and the consequent importance of its reactions with atoms and radicals. The disproportionation reaction to generate hydrogen peroxide:

$$HO_2 + HO_2 \rightarrow H_2O_2 + O_2 \tag{3}$$

is one of the most important of these reactions, both in the atmosphere and in combustion.

For a long time, the reaction was thought to occur via a simple abstraction mechanism, possibly with a small activation barrier. One of the earliest indications that the reaction might not be so straightforward was the observation that the reaction rate is enhanced by the presence of water vapour.[12] There followed a series of investigations in which the temperature dependence of the rate constant was probed by molecular modulation spectroscopy,[13] pulse radiolysis/absorption spectroscopy[14] and flash photolysis/absorption spectroscopy.[15] Rather surprisingly, k_3 was found to decrease strongly with temperature. A further relevant experiment was the demonstration, in a discharge flow/laser magnetic resonance experiment, that the rate constant shows a linear dependence on pressure.[16] These observations suggest that the reaction proceeds via an adduct, $H_2O_4^*$, which rapidly decomposes back to the reactants unless it is stabilised (hence the pressure dependence) and which has a shorter dissociative lifetime at high temperatures, so that it is more difficult to stabilise as the temperature increases (hence the 'negative activation energy').

The HO_2 radical has a small absorption cross-section in the UV and so is difficult to monitor by UV spectrophotometry. It is also difficult to generate in flash photolysis experiments and needs quite a complex photochemical generation scheme. Despite these difficulties Sander and coworkers,[17] in a series of investigations, completed an impressive and thorough analysis of the pressure and temperature dependence of k_3. They used conventional flash photolysis, coupled with multipass optics. Figure B4.2 shows the surprising results. The pressure and negative temperature dependences are indeed confirmed, but the rate constant does not tend to zero as $P \rightarrow 0$, as would be expected for a simple association type of mechanism. Instead there is a non-zero intercept and the rate constant may be expressed in the form:

$$k_3 = k_3^{II} + k_3^{III}[M]$$

where k_3^{II} is a second-order and k_3^{III} a third order rate constant. The flash photolysis technique employed was not easily extended to very low pressures, but further experiments, at such pressures, by Sander[18] using discharge flow/mass spectrometry and by Thrush and Tyndall,[19] using flash photolysis coupled with diode laser spectroscopy, confirmed the conclusions.

How can one explain these observations and what is the significance of k_3^{II} and k_3^{III}? Patrick et al.[20] attempted to answer this question based on the scheme shown in Fig. B4.3. Both k_3^{II} and k_3^{III} decrease with temperature and they argued that k_3^{II} could not, therefore, be ascribed to a simple abstraction. Instead, the zero pressure contribution was assumed to arise from the vibrationally excited H_2O_4 molecules formed by the initial association, which rearrange to form the products rather than

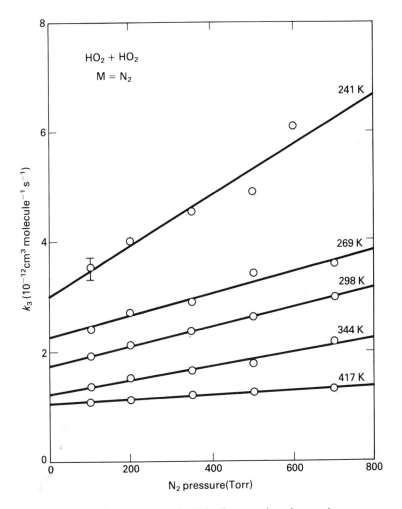

Fig. B4.2. Dependence of the rate constant for HO_2 disproportionation on nitrogen pressure and temperature. Reproduced, with permission, from reference 17.

redissociate to the reactants. The pressure dependent contribution arises from stabilisation of H_2O_4, which may be reactivated to form H_2O_2 and O_2 or $2HO_2$.

A simple steady-state analysis[20] of the scheme given in Fig. B4.3 results in:

$$k_3 = \frac{k_a k_r + k_r k_{-s}[M][H_2O_4]/[HO_2]^2}{k_s[M] + k_{-a} + k_r}$$

so that, if $[M] = 0$,

$$k_3 = k_3^{II} = k_a k_r / (k_{-a} + k_r)$$

whilst, if $[M] \neq 0$, but $(k_{-a} + k_r) \gg k_s[M]$

$$k_3 = k_3^{III} = \frac{k_r k_{-s}}{(k_{-a} + k_r)} \frac{[H_2O_4]}{[HO_2]^2} \cdot [M]$$

and the reaction shows a pressure dependent component. Patrick examined the scheme in more detail, using RRKM theory to incorporate energy dependent rate constants, along the lines of those contained in equation 39 of Chapter A4. H_2O_4 has been observed at low temperatures in a matrix and its dissociation energy, E_1^0, has been estimated as $\sim 34\,kJ\,mol^{-1}$. Attempts to fit the experimental data using unimolecular rate theory met with reasonable success, with $E_2^0 \sim 10\,kJ\,mol^{-1}$ provided the structure of H_2O_4 is very loose—much looser than the species observed by matrix isolation. The question of the full interpretation of the experimental data, however, remains an open one. In particular, its extrapolation

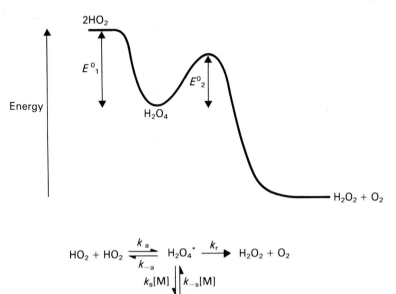

Fig. B4.3. Schematic potential energy diagram for HO_2 disproportionation. Adapted from reference 20.

outside the experimental range is a perilous procedure and the value of k_3 in combustion systems, where reaction 3 is very significant, is subject to considerable uncertainty.

B4.4 Heats of formation of free radicals

Thermodynamic parameters for free radicals are important in determining heat release in combustion processes, in calculating radical concentrations in high temperature systems where local equilibrium may be maintained and in estimating rate constants, either from the reverse rate constant via the equilibrium constant, or from similar reactions via linear energy or linear free energy relations. Many smaller radicals are spectroscopically well characterised so that entropies and specific heats may be precisely calculated. In cases where frequencies are not known, they can often be estimated by approximate or *ab initio* techniques. Benson's book on 'Thermochemical Kinetics' is a classic account of approximate methods.[21]

Heats of formation (or, equivalently, bond dissociation energies) present a more difficult problem. Benson[21] describes ways in which $\Delta H_f^\ominus$ may be estimated and great strides are being made with *ab initio* calculations, but the most reliable technique remains experimental measurement. A recent review of the field for hydrocarbon free radicals may be found in reference 22, where the classic techniques for determining bond dissociation energies and radical heats of formation are described. An example may be found in problem B1.3. Many of the techniques rely on measurements of forward and backward rate constants and hence of the equilibrium constant. An important example is provided by reaction $(4, -4)$

$$I + i\text{-}C_4H_{10} \underset{-4}{\overset{4}{\rightleftharpoons}} t\text{-}C_4H_9 + HI \tag{4}$$

in which k_4 was determined from the thermal $i\text{-}C_4H_{10}/I_2$ system and k_{-4} from a very low pressure photolysis experiment. The ratio of the rate constants was combined with the entropies of the reactants to obtain $\Delta H_4^\ominus$ and hence, since all the other heats of formation are known, $\Delta H_f^\ominus(t\text{-}C_4H_9)$. A common problem, however, has been that estimates using a different reaction give significantly different radical heats of formation. For example, measurements of k_5 and k_{-5}

$$((CH_3)_3C)_2 \underset{-5}{\overset{5}{\rightleftharpoons}} 2t\text{-}C_4H_9 \tag{5}$$

give a value for $\Delta H_f^\ominus(t\text{-}C_4H_9) \sim 12\,\text{kJ}\,\text{mol}^{-1}$ higher than does k_4. A detailed account of the controversy surrounding $\Delta H_f^\ominus(t\text{-}C_4H_9)$ may be found in the article by McMillen and Golden.[22] A major drawback of this kind of approach is that it relies on measurements of forward and backward rate constants which are made on complex chemical systems in different laboratories, often over different temperature ranges and, frequently, with poorly assessed uncertainties.

A more direct technique has been employed by Benson and coworkers to determine radical heats of formation, including $\Delta H_f^\ominus(\text{t-C}_4\text{H}_9)$.[23,24] It was referred to in Problem B1.3. In a typical investigation bromine atoms from a microwave discharge were flowed into a very low pressure reactor and their concentration measured, using a quadrupole mass spectrometer, as they left the reactor through a small pinhole. The decrease in [Br] on adding isobutane enabled k_6 to be determined, whilst the equilibrium constant, K_6, was determined from the changes on flowing in both HBr and isobutane

$$\text{Br} + \text{i-C}_4\text{H}_{10} \underset{-6}{\overset{6}{\rightleftharpoons}} \text{t-C}_4\text{H}_9 + \text{HBr}. \tag{6}$$

The results obtained by Benson and his coworkers supported the lower heat of formation, but there is no general agreement on the value which should be adopted.

A time-resolved technique has also been developed in which a pulsed method is used to generate a radical and then the time dependence of the approach to equilibrium is observed:

$$\text{R} + \text{X} \underset{-7}{\overset{7}{\rightleftharpoons}} \text{RX} \tag{7}$$

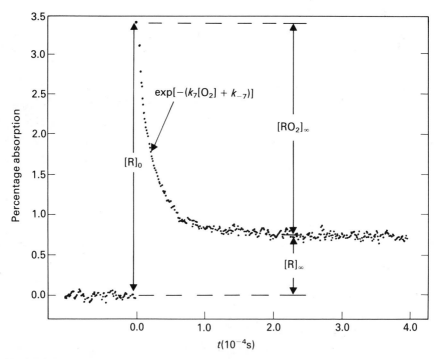

Fig. B4.4. Time dependence of the radical absorbance, following its initial production by laser flash photolysis in the presence of oxygen and its subsequent equilibration with the peroxy radical, RO_2, via reaction $(7, -7)$. The zero-time absorbance is proportional to the initial radical concentration and the long-time limiting absorbance to the equilibrated radical concentration $[R]_\infty$. By stoichiometry, their difference is proportional to $[RO_2]_\infty$. The data were obtained for reaction $(8, -8)$.[26]

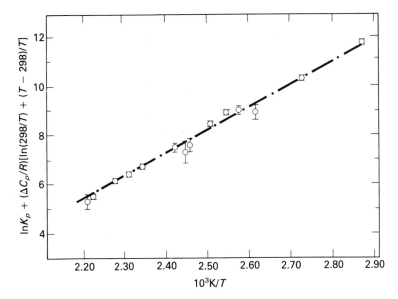

Fig. B4.5. Van't Hoff plot for reaction (8, −8):[26] the lowest two temperature points were obtained by photoionisation mass spectrometry[25] and the others by absorption spectroscopy.[26] Note the need to include a ΔC_p term in the ordinate—a simple plot of ln K_p vs. $1/T$ is not accurate enough (c.f. problem B4.1). Reproduced, with permission, from reference 26.

where R and RX are free radicals and X a molecule. The basis of the technique is shown in Fig. B4.4. The approach to equilibrium is exponential, with a first-order rate constant $(k_7[X] + k_{-7})$, whilst the concentrations of R at zero time and in the long time limit are related by:

$$([R]_0 - [R]_\infty)/[R]_\infty = k_7[X]/k_{-7}$$

(assuming R is the photochemically generated radical and $[RX]_0 = 0$). Thus k_7, k_{-7} and the equilibrium constant, K_7 may be found from the form of the approach to equilibrium. This simple analysis does not always apply in practice and there are generally additional first-order loss terms (e.g. pump-out of the reactants from the reaction vessel), so that the decay becomes biexponential—the details may be found in the references cited below—nonetheless the principle is unchanged.

The technique was pioneered by Bayes and collaborators,[25] using laser flash photolysis/photoionisation mass spectrometry on the reaction:

$$C_3H_5 + O_2 \underset{-8}{\overset{8}{\rightleftharpoons}} C_3H_5O_2. \tag{8}$$

These experiments were supplemented with higher temperature measurements using laser flash photolysis/absorption spectroscopy,[26] so that K_8 could be determined over a wide range of temperatures and $\Delta H_8^\ominus$ estimated from the slope of the van't Hoff plot, Fig. B4.5. Gutman[27] has applied the technique to several alkyl radical/alkylperoxy equilibria, which are of great importance in low temperature combustion (see p. 265). The values obtained by Gutman indicate that the R—O_2 bond strength increases by 20 kJ mol^{-1} as the size of the R group increases from CH_3 to t-C_4H_9. This is a very significant change as far as combustion is concerned and is contrary to estimates based on group additivity.[21]

The direct measurement of equilibration has also been applied to the determination of alkyl radical heats of formation. Problem B4.1 concerns the estimation of $\Delta H_f^{\ominus}(C_2H_5)$ from data obtained by flash photolysis/resonance fluorescence.[28] The problem illustrates the approach which must be adopted if a wide range of temperatures cannot be employed. $\Delta S^{\ominus}$ for the reaction then has to be calculated from vibrational frequencies for the radicals involved. These calculations can be a significant source of error and are, in some cases, the weak point of the 'direct' approach.

B4.5 References

1 Jones I. T. N. and Bayes K. D. (1973) *Proc. Roy. Soc. London, Ser. A.*, **335**, 547.
2 Williamson D. G. (1971) *J. Phys. Chem.*, **75**, 4053.
3 Cvetanovic R. J. (1955) *Can. J. Chem.*, **33**, 1684.
4 Kanofsky J. R., Lucas D., Pruss F. J. and Gutman D. (1974) *J. Phys. Chem.*, **78**, 311.
5 Pruss F. J., Slagle I. R. and Gutman D. (1974) *J. Phys. Chem.*, **78**, 663.
6 Harding L. B., (1981) *J. Phys. Chem.*, **85**, 10.
7 Buss R. J., Baseman R. J., He G. and Lee Y. T. (1981) *J. Photochem.*, **17**, 389.
8 Luntz A. and Kleinermans K. (1981) *J. Phys. Chem.*, **85**, 1966.
9 Hunziker H. E., Kneppe H. and Wendt H. R. (1981) *J. Photochem.*, **17**, 377.
10 Clemo A. R., Duncan G. L. and Grice R. (1982) *J. Chem. Soc., Faraday Trans. 2*, **78**, 1231.
11 Vinckier C., Schaekers M. and Peeters J. (1985) *J. Phys. Chem.*, **89**, 508.
12 Hamilton E. J. and Lii R.-R. (1977) *Int. J. Chem. Kinet.*, **9**, 875.
13 Cox R. A. and Burrows J. P. (1979) *J. Phys. Chem.*, **83**, 2560.
14 Lii R.-R., Gorse Jr. R. A., Sauer Jr. M. C. and Gordon S. (1979) *J. Phys. Chem.*, **83**, 1803.
15 Patrick R. and Pilling M. J. (1982) *Chem. Phys. Letters*, **91**, 343.
16 Thrush B. A. and Wilkinson J. P. T. (1979) *Chem. Phys. Letters*, **66**, 441.
17 Kircher C. C. and Sander S. P. (1984) *J. Phys. Chem.*, **88**, 2082.
18 Sander S. P. (1984) *J. Phys. Chem.*, **88**, 6018.
19 Thrush B. A. and Tyndall G. S. (1982) *J. Chem. Soc., Faraday Trans 2*, **78**, 1469.
20 Patrick R., Barker J. R. and Golden D. M. (1984) *J. Phys. Chem.*, **88**, 128.
21 Benson S. W. (1976) *Thermochemical Kinetics*, 2nd edn, Wiley-Interscience, New York.
22 McMillen D. F. and Golden D. M. (1982) *Ann. Rev. Phys. Chem.*, **33**, 493.
23 Baghal-Vayjooe M. H., Colussi A. J. and Benson S. W. (1979) *Int. J. Chem. Kinet.*, **11**, 147.
24 Islam T. S. A. and Benson S. W. (1984) *Int. J. Chem. Kinet.*, **16**, 995.
25 Ruiz R. P., Bayes K. D., Macpherson M. T. and Pilling M. J. (1981) *J. Phys. Chem.*, **85**, 1622.
26 Morgan C. A., Pilling M. J., Tulloch J. M., Ruiz R. P. and Bayes K. D. (1982) *J. Chem. Soc., Faraday Trans 2*, **78**, 1323.
27 Slagle I. R., Ratajczak E. and Gutman D. (1986) *J. Phys. Chem.*, **90**, 402.
28 Brouard M., Lightfoot P. D. and Pilling M. J. (1986) *J. Phys. Chem.*, **90**, 445.

B4.6 Problem

Case studies

B4.1. The equilibrium

$$H + C_2H_4 \underset{k_{-1}}{\overset{k_1}{\rightleftharpoons}} C_2H_5$$

was studied directly at 800 K by monitoring the hydrogen atom resonance fluorescence signal, following the laser flash photolysis of C_2H_4. Analysis of the transient decay to equilibrium gave $k_1 = 2.7 \times 10^{-12}\,\text{cm}^3\,\text{molecule}^{-1}\,\text{s}^{-1}$ and $k_{-1} = 162\,\text{s}^{-1}$ at 100 Torr total pressure.

(a) Show that

$$\Delta H_{298}^{\ominus} = T\Delta S_{298}^{\ominus} - RT \ln K_p + T \int_{298}^{T} \Delta C_p \mathrm{d}\ln T - \int_{298}^{T} \Delta C_p \mathrm{d}T$$

where $K_p = k_1/(k_{-1}RT)$.

(b) Given that the change in molar heat capacities for reaction $(1, -1)$ is given by

$$\Delta C_p = [-15.0 + 8.4(T/1000\,\text{K})]\,\text{J mol}^{-1}\,\text{K}^{-1}$$

and the change in molar entropies at 298 K by

$$\Delta S_{298}^{\ominus} = -86.0\,\text{J mol}^{-1}\,\text{K}^{-1},$$

evaluate $\Delta H_{298}^{\ominus}$ and hence the standard heat of formation of the ethyl radical given that $\Delta H_{f,298}^{\ominus}(\text{H}) = 217.9\,\text{kJ mol}^{-1}$ and $\Delta H_{f,298}^{\ominus}(C_2H_4) = 52.5\,\text{kJ mol}^{-1}$.

B4.7 Answer to problem

Case studies

B4.1. Setting $\Delta C_p = a + bT$, the heat of reaction at any temperature is given by:

$$\Delta H(T) = \Delta H_{298} + \int_{298}^{T} (a + bT)\mathrm{d}T$$

$$= \Delta H_{298} + a(T - 298) + (b/2)(T^2 - 298^2)$$

$$RT \ln K_p(T) = RT \ln K_p(298) + T \left\{ \frac{\Delta H_{298}}{298} - \frac{\Delta H_{298}}{T} + \right.$$

$$\left. \int_{298}^{T} \left(\frac{a}{T} - \frac{298a}{T^2} + (b/2) - \frac{298^2 b}{2T^2} \right) \mathrm{d}T \right\}$$

$$= T\Delta S_{298} - \Delta H_{298} + T\left\{ a \ln\left(\frac{T}{298}\right) + \frac{298a}{T} - a + (b/2)T - (b/2)298 \right.$$

$$\left. + \frac{298^2 b}{2T} - \frac{298b}{2} \right\}$$

$$\Delta H_{298} = T\Delta S_{298} - RT \ln K_p(T) + aT \ln\left(\frac{T}{298}\right) - a(T - 298) + b\left(\frac{T^2}{2} - 298T\right.$$
$$\left. + \frac{298^2}{2}\right)$$

Alternatively

$$RT \ln K_p(T) = RT \ln K_p(298) + T\left\{\frac{\Delta H_{298}}{298} - \frac{\Delta H_{298}}{T}\right\} +$$

$$T\int_{298}^{T} \frac{1}{T^2}\left(\int_{298}^{T} \Delta C_p \, dT\right) dT$$

$$\therefore \quad \Delta H_{298} = T\Delta S_{298} - RT \ln K_p(T) - \int_{298}^{T} \Delta C_p \, dT + T\int_{298}^{T} \frac{\Delta C_p \, dT}{T}$$

which leads to the same result.
   For $T = 800$ K

$$T\Delta S = -68.80 \text{ kJ mol}^{-1}.$$

At 800 K,

$$RT \ln K_p = 8.314 \times 800 \ln\left\{\frac{2.97 \times 10^{-12} \times 6.023 \times 10^{20}}{162 \times 0.0821 \times 800}\right\}$$

$$= 79.99 \text{ kJ mol}^{-1}.$$

$$aT \ln (T/298) \qquad\qquad = -11.87 \text{ kJ mol}^{-1}$$

$$a(T - 298) \qquad\qquad\quad = -7.52 \text{ kJ mol}^{-1}$$

$$b\left(T^2/2 - 298T + \frac{298^2 b}{2}\right) = +\ 1.05 \text{ kJ mol}^{-1}$$

$$\Delta H_{298}^{\ominus} = 152.1 \text{ kJ mol}^{-1}$$

$$\Delta H_{f,298}^{\ominus}(C_2H_5) = 118.3 \text{ kJ mol}^{-1}.$$

Section C
Kinetics of complex
reactions

The effect that applied areas of kinetics have had on the development of our understanding of elementary reactions was stressed in the introduction to Section B. The early impetus was provided by the high temperature processes occurring in hydrocarbon pyrolysis, explosions and flames, and led, in the 1920's and 1930's to progress in theories of chain reactions in gases. At the same time, photochemistry was developing and providing a means of studying reactions involving atoms and radicals at lower temperatures, where atmospheric chemistry served as the applied focus for research. From the 1950's, steady-state experiments increasingly made way for more direct, time-resolved methods, but the impact of these three major areas of applied interest, atmospheric chemistry, combustion and pyrolysis, remained. They were joined in the 1960's by a new field, derived from the emerging study of reaction dynamics, that of chemical lasers, in which the new dimension of state-resolved kinetics is of primary interest.

In Section C we present a discussion of these four areas concentrating entirely on chemical kinetic aspects. In any application, elementary reactions are components of a model which contains other important features such as hydrodynamics and heat and mass transfer. The kineticist's concerns are the construction of kinetic models and the provision and evaluation of rate data. The other elements are not our primary interest and are not discussed in detail in this book, although the informed kineticist should be aware of them and we hope that advantage will be taken, in this respect, of the bibliographies provided.

Atmospheric chemistry is an enormous subject and Chapter C1 is, of necessity, selective. The stability of the stratospheric ozone layer, which has received wide popular exposure, depends on a complex chemistry involving a series of cycles in which 'odd oxygen' [$O(^3P)$ and O_3] is converted to O_2; the ability of small concentrations of free radicals to catalyse this conversion lies at the heart of the anthropogenic impact on the ozone layer. Chapter C1 also discusses the chemistry of the polluted and unpolluted troposphere, where reactions of OH assume particular importance.

Combustion is associated with a wide variety of kinetic phenomena, whose elucidation, coupled with the commercial and environmental interest in energy utilisation, has led to a recently accelerating research effort on hydrocarbon oxidation. At low temperatures, cool flames, which are sometimes oscillatory in nature, are of primary interest and can be described in terms of the degenerate branching theory proposed originally by Semenov. The behaviour can be related to the formation and reactions of peroxy radicals, and the tendency of different fuels to show cool flame behaviour—and to undergo pre-ignition in a petrol engine—is correlated with the structure of the radicals formed during oxidation. Chapter C2 discusses low temperature combustion in some detail. In many ways, the chemistry becomes simpler at higher temperatures, because the larger radicals decompose rapidly and the kinetics are dominated by atoms and small free radicals; this latter region is treated more schematically.

Chapter C3 differs from all the other chapters in the book in that it is not directly concerned with elementary reactions. An understanding of flames is developed by combining detailed modelling with experimental probing, whose aim is to determine, *inter alia*, temperature profiles and species concentrations. The use of lasers to probe flames without perturbing them has developed rapidly in recent years. Because many of the techniques are the same as those used to measure elementary reaction rates, and often depend on an understanding of energy transfer and relaxation processes, kineticists have played an important part in this development.

The coupling of kinetic data for elementary reactions with models of complex systems owes much to advances in numerical techniques for integrating coupled differential equations. If we were still limited to the use of the steady-state approximation, or to analytical or simplified numerical solutions of time-dependent equations, then the chemical complexity that can now be built into models would be impossible. Chapter C4 outlines the basis of numerical integration, including algorithms for so-called 'stiff' systems, where the characteristic time constants of the component steps differ widely. The chapter also makes reference to sensitivity analysis, in which the sensitivity of a species concentration to specific rate parameters is determined.

The final two chapters return to the part played by elementary reactions in complex systems, but stress the need for a detailed understanding of rate constants. Chapter C5 discusses hydrocarbon cracking and demonstrates that, although the basic mechanism has been appreciated for half a century, detailed modelling is difficult because of the need to extrapolate rate data obtained at lower temperatures to industrial conditions. For several important dissociation and association reactions this involves a fall-off analysis along the lines discussed in Chapter A4, whilst H-atom abstraction reactions appear to have a significant tunnelling contribution. Chapter C6 returns to state-to-state reactions in a discussion of chemical lasers. The gain of a chemical laser depends on energy disposal and on the extent to which a chemical reaction generates products in an inverted population distribution. The categorisations available through the application of information theory, already encountered in Chapter A3, are of benefit in assessing potential chemical lasers and are discussed at some length, and the chapter concludes with an account of hydrogen halide lasers.

Section C is intended to demonstrate to the reader the approach needed to develop a detailed chemical kinetic model of a complex reaction system. The reactions to be included in the model must be assessed and rate parameters, if they are available, evaluated and extrapolated to the region of interest. If the rate constants are not available, they must either be measured or estimated. The model must then be compared with experiment, its deficiencies recognised and refinements incorporated. The examples chosen to demonstrate this approach are illustrative rather than exhaustive. Others—such as detonations, interstellar chemistry or plasma etching—could well have been chosen instead. Nor have we included any discussion of the important field of data evaluation, in which available rate data are assessed in detail, thus saving the modeller a great deal of effort in an area in which he may not be well equipped. Finally, a word of exhortation to try the problems which will, we hope, complete the development of a practical understanding of the analysis of complex reaction systems.

Chapter C1
Atmospheric Chemistry

R. A. COX

C1.1 Introduction

The origins of the study of atmospheric kinetics may be found in the early experiments in photochemistry. Photodissociation of gases produced active free radical fragments and photochemical studies provided an insight into their complex chemistry. The development of the flash photolysis technique by Norrish, Porter and Thrush[1] in the 1950's provided, for the first time, the possibility of direct study of the kinetics of radical reactions by spectroscopic methods. Also at this time the early developments in the field of aeronomy, i.e. the chemistry and photochemistry of the upper atmosphere[2] and in the photochemistry of air pollution in California[3] were being made.

By the early 1960's the body of information on the reactivity of atoms and small radicals containing H, C, O, N, S, Halogen, etc., was mainly qualitative and techniques were mainly indirect. Inference of reaction mechanisms was based on measurement of stable products and rate constants were obtained from competitive kinetic studies in chemical systems that were usually quite complex.

In the following years, the development of non-intrusive spectroscopic techniques for the specific detection of transient atomic and radical species in the gas phase led to vast improvements in the techniques for direct measurement of the rates of fast reactions of these species (see Chapter B1 and reference 4). This activity has been further stimulated by efforts to obtain a more detailed understanding of such practical problems as photochemical smog and the modification of the earth's ozone shield.

The development of an understanding of the complex phenomena associated with atmospheric chemistry requires a marrying of results of studies of the phenomena themselves, of laboratory simulation of associated chemistry, of kinetic data for elementary reactions and of numerical simulation studies. Chemical change results from fast reactions of atomic and molecular free radicals present in 'local' steady state. The total concentration of radical species is controlled by a balance between production and loss and the relative concentrations of the various radical species is determined by fast interconversion reactions. The kinetic distinction between fast and slow processes offers the possibility of simplifying the mathematical analysis through the steady state approximation and the grouping of fast reactions into subsets. This approach continues to be useful in gaining an insight into the kinetic behaviour of complex systems, even when more explicit numerical integration techniques are employed to model the system. Furthermore, a full description of atmospheric phenomena requires that the chemistry is incorporated into models describing the physical environment, air motion and transport. Such models currently demand large computing capacity and simplification of the chemistry submodel is required before linking with the physical models. Improvement of computing capability in this respect can be expected in the coming years.

In this chapter we review the chemistry of photochemical smog[5] and of stratospheric[6] and tropospheric ozone.[7] We outline in detail the major chemical reactions involved and stress the interconnections between the families of radical species.

C1.2 Chemistry of photochemical smog

The phenomenon of photochemical smog was discovered in the Southern California air basin around Los Angeles in the late 1940's. It is characterised by a reduction in visibility due to the build up of light scattering aerosols and nitrogen dioxide and, more importantly, by elevated concentrations of ozone and other oxidising species, which cause irritation of the eye and respiratory tract. It has subsequently been shown to be a widespread urban phenomenon in developed countries when appropriate meteorological conditions prevail. Study of this phenomenon in the 1950's showed that it was caused by the action of sunlight on part-per-million concentrations of nitrogen oxides (NO and NO_2) and hydrocarbons, which originated primarily from automobile emissions. Laboratory simulation experiments, involving the photoirradiation with simulated sunlight of mixtures of NO_x and simple hydrocarbons diluted in air at ambient pressure and temperature, played an important part in the early efforts to understand this system. These studies were conducted in large glass or metal reaction vessels of volume $1000\,dm^3$ or more; various analytical techniques, including wet chemical analysis, gas chromatography and dispersive infra-red spectroscopy were employed to follow concentrations of reactants and products over periods of hours. These were termed 'smog chamber' experiments.

Figure C1.1 shows how the concentrations of several significant species vary with time during the irradiation of propene/NO mixtures in a laboratory smog chamber.[8] System behaviour is characterised by oxidation of the hydrocarbon to aldehydes, CO, etc., with accompanying oxidation of NO to NO_2. After a while, NO_2 maximises and O_3 starts to build up when NO has declined to a low concentration. As will be seen from Fig. C1.2, a similar pattern was found in the

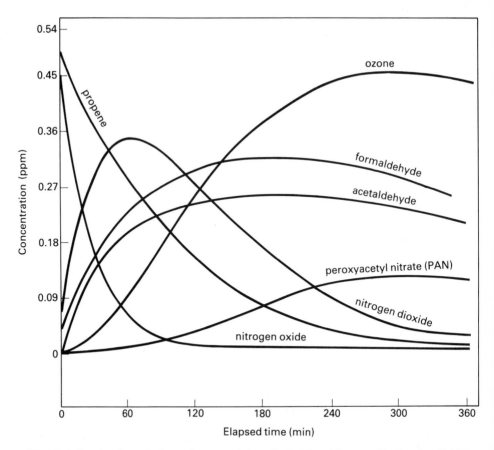

Fig. C1.1. Results of a typical experiment carried out in the Riverside evacuable chamber. Initial concentration of reactants are: nitrogen oxide, 0.45 ppm; nitrogen dioxide, 0.05 ppm; propene, 0.50 ppm. Reproduced from reference 8 with permission.

diurnal variations of the concentrations of these species in urban air, and this observation encouraged a large number of empirical studies aimed at 'mapping' the complex relationships between the precursor concentrations and ozone. This approach, whilst useful in defining the chemical problem and demonstrating that different hydrocarbons exhibit different reactivity towards ozone formation, failed to provide real insight into the mechanism of this complex system. The major advances in understanding the system were made when the photochemistry and kinetics of the elementary reactions were examined, for example by Leighton in 1961,[3] and later using computer modelling techniques by Demerjian, Kerr and Calvert.[9]

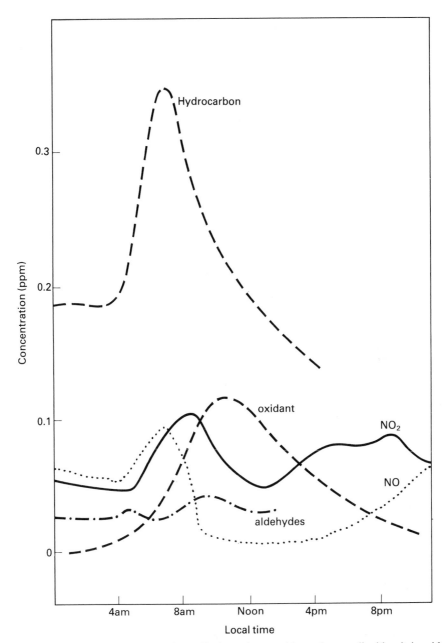

Fig. C1.2. Plot of average concentrations of hydrocarbons, oxidant, nitrogen dioxide, nitric oxide and aldehydes in the atmosphere as a function of time of day in downtown Los Angeles. Data were generated by the Los Angeles County Air Pollution control district for days when eye irritation was present. Data for hydrocarbons, aldehydes and ozone (1953–54); NO and NO₂ (1958). Reproduced from Kerr J. A., Demerjian K. and Calvert J. G., (1972) *Chem. Brit.* **8**, 252, with permission.

(i) Photostationary state relationship between NO, NO$_2$ and O$_3$

The relationships between NO, NO$_2$ and O$_3$ concentrations in Fig. C1.1 can be understood in terms of the following reaction system:

$$NO_2 + h\nu(\lambda < 400\,nm) \rightarrow NO + O(^3P) \tag{1}$$

$$O(^3P) + O_2 + M \rightarrow O_3 + M \tag{2}$$

$$NO + O_3 \rightarrow NO_2 + O_2. \tag{3}$$

Nitrogen dioxide photolyses rapidly via reaction 1 in sunlight and produces ground state O(^3P), which forms ozone by recombination with O$_2$. The rapid reaction 3 reforms NO$_2$, so no net conversion results from reactions 1–3; a photo-stationary state is established where the concentration of O$_3$ is governed by the NO$_2$/NO ratio, i.e.

$$[O_3] = \frac{k_1[NO_2]}{k_3[NO]}. \tag{i}$$

The ratio k_1/k_3 is determined by light intensity and temperature: $k_3 = 2.2 \times 10^{-12} \exp(-1430/T)\,cm^3\,molecule^{-1}\,s^{-1}$. k_1 is typically $\sim 10^{-2}\,s^{-1}$ in midday sunlight and the photostationary state is therefore established on a time scale of minutes in the lower atmosphere.

Net production of ozone from the above system occurs when some process other than reaction 3 oxidises NO to NO$_2$ thus altering the ratio NO$_2$/NO. In the atmosphere this occurs via the reaction of HO$_2$ or organic peroxyradicals, RO$_2$, with NO:

$$RO_2 + NO \rightarrow NO_2 + RO \quad (R = H\ or\ organic\ radical). \tag{4}$$

The key to a detailed understanding of ozone production in photochemical smog is therefore a knowledge of factors governing the concentration of RO$_2$ and HO$_2$ radicals in hydrocarbon oxidation in air at low temperatures.

(ii) Mechanism of low temperature photo-oxidation of hydrocarbons

The broad features of the mechanism of hydrocarbon oxidation in air have now been established as a result of elementary reaction rate studies and chamber type studies over the past 10–15 years. Most volatile hydrocarbon species emitted into the atmosphere are reactive towards the OH radical. The product radical reacts with O$_2$ to form a peroxyradical which then undergoes the O atom transfer reaction with NO, reaction 4, forming an alkoxy type radical which reacts further to produce a carbonyl compound, R'CO, as the first stable oxidation product together with HO$_2$. HO is regenerated by the rapid reaction of HO$_2$ with NO. The overall scheme may be written:

$$OH + RH \rightarrow R + H_2O \tag{5}$$

$$R + O_2 \rightarrow RO_2 \tag{6}$$

$$RO_2 + NO \rightarrow RO + NO_2 \tag{4}$$

$$RO + O_2 \rightarrow HO_2 + R'CO \tag{7}$$

$$HO_2 + NO \rightarrow OH + NO_2. \tag{8}$$

This sequence leads to the oxidation of 2 molecules of NO to NO_2, thereby leading to net production of 2 molecules of ozone from reactions $1 + 2$.

The rate of RO_2 production is clearly dependent on the rate of OH attack on the hydrocarbon and therefore on the nature of the hydrocarbon species. Recognition of this fact has stimulated measurement of the rate constants for the elementary OH + hydrocarbon reactions. There is now a large body of kinetic data for reaction of OH with a wide range of hydrocarbons which has been reviewed by a number of authors.[10,11] It has been found that the magnitudes of the OH rate constants correlate well with the 'hydrocarbon reactivity' scale based on ozone formation potential obtained from empirical smog chamber studies.[12]

The detailed nature of the oxidation mechanism depends on the nature of the hydrocarbon species, and has only been established for some of the simpler species. Some of the key features of the mechanism are exemplified in the scheme for oxidation of n-butane outlined in Fig. C1.3. The initial attack by OH is 85%

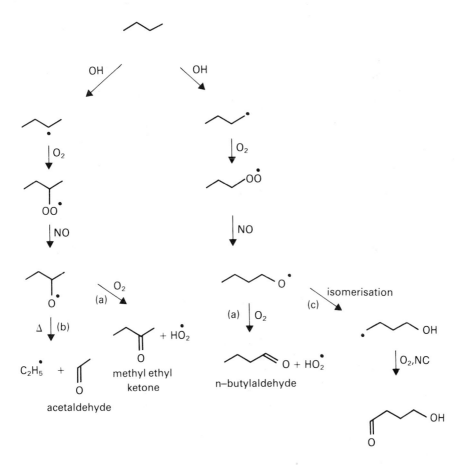

Fig. C1.3. Degradation pathways in the atmospheric photo-oxidation of n-butane.

on the secondary H atoms and 15% on the primary H atoms in accordance with the known reactivity of OH with these types of bonds in the homologous series of alkanes. After addition of O_2 to the C_4H_9 radical and reaction of the peroxyradical with NO, two types of butoxy radical are formed, primary and secondary.

Three competitive reaction patterns for alkoxy radicals have been identified:
(a) reaction with O_2 to form a carbonyl compound and an HO_2 radical,
(b) thermal decomposition to form an aldehyde or ketone with fewer carbon atoms, together with an alkyl radical,
(c) isomerisation by internal H-abstraction from a carbon atom remote from the C—O· group.
Process (a) only occurs in primary and secondary alkoxy radicals since the tertiary radical has no α-H atom. For secondary radicals (a) and (b) can compete and for tertiary radicals decomposition is the exclusive pathway. The isomerisation reaction (c) requires a structure with a straight chain of at least four carbon atoms. The process seems to be dominant for n-alkoxy radicals derived from nC_5 and nC_6 hydrocarbons and thus agrees with the predictions of thermochemical kinetics calculations. [13] From the example given in Fig. C1.3 it is clear that the relative rates of the alkoxy radical reactions decide the nature of the initial carbonyl products during oxidation.

The attack of OH on alkenes and aromatics at ambient temperatures occurs by addition rather than abstraction. At atmospheric pressure in air, the OH–alkene adducts add O_2 to form stable peroxy radicals, analogous to alkyl radicals. The alkoxy radical produced in the oxidation of alkenes predominantly cleaves to produce carbonyl fragments. In ethylene oxidation, a small amount of glycolaldehyde is formed,[14] showing that reaction with O_2 is competitive with decomposition of the 2-hydroxy ethoxy radical:

The oxidation of aromatic hydrocarbons, initiated by O_2, leads to extensive break up of the benzene ring. The detailed mechanism of this process is not well understood at the present time but it is clearly different from, and more complex than, that of aliphatic hydrocarbons. The peroxy radical formed by addition of O_2 to simple aromatics is much less stable and the oxidation via transfer in the $RO_2 + NO$ reaction is not a major reaction path. The most recent work[15] suggests break up of the ring via internal addition forming an endoperoxy radical, e.g. for toluene:

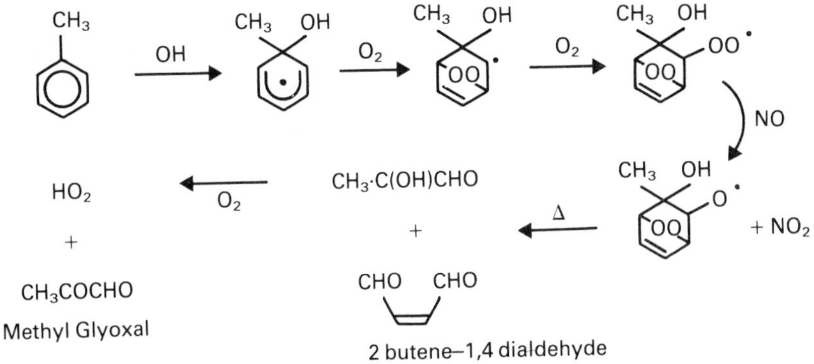

The variety of products identified in the OH + toluene reaction shows that other parallel oxidation pathways are also occurring.[16]

(iii) FTIR instrumented smog chambers—product analysis

The elucidation of atmospheric photo-oxidation mechanisms has benefitted from the development of a number of experimental techniques over the past decade. In particular the application of high resolution Fourier transform infra-red spectroscopy coupled with long pathlength absorption cells has allowed sensitive *in situ* diagnostic measurements in photochemically initiated reacting systems. Figure C1.4 shows a schematic illustration of a fully instrumented smog chamber used at the University of California at Riverside[8] which incorporates an FTIR spectrometer. The multi-reflective optical system gives the infra-red beam a long path in the gas mixture providing adequate sensitivity for detection at the sub ppm concentration level. The essential components are an IR source, an interferometer and a detector coupled to a computer. The latter is used for storage of the interferogram, Fourier transformation and subsequent arithmetic manipulation of the spectra. Typically spectra can be acquired over time scales of ~ 10 s upwards with adequate resolution for spectral stripping of reference spectra of unreacted mixtures and pure compounds. The great advantage of this *in situ* method is its ability to detect species such as HNO_3, $HCOOH$ and peroxy compounds, which have proved too unstable to diagnose and measure by more conventional methods.

(iv) Chemistry of secondary products—carbonyl compounds and PAN

The carbonyl compounds formed during the breakdown of hydrocarbons are further oxidised by attack of OH, e.g.

$$OH + CH_3CHO \rightarrow H_2O + CH_3CO \tag{9}$$

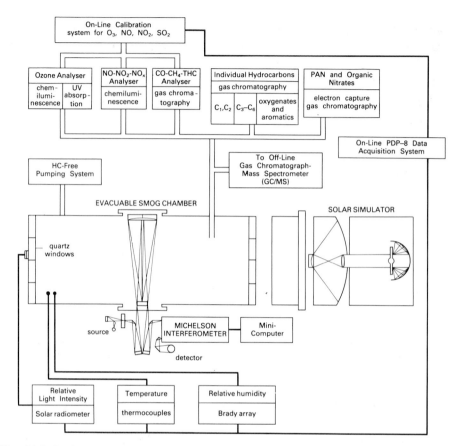

Fig. C1.4. On-line instrumentation, calibration, and data acquisition facilities in Riverside evacuable smog chamber. Reproduced from Winer A. M., (1980) In *Advances in Environmental Science and Technology* (Ed. by J. N. Pitts Jr. and R. L. Metcalfe), Vol. 10, p. 493. Copyright © J. Wiley & Son. Reprinted by permission of John Wiley and Sons, Inc.

and also by photodissociation, since carbonyl compounds possess absorption bands in the near UV region, e.g.

$$CH_3CHO + h\nu(\lambda < 330\,nm) \rightarrow CH_3 + HCO. \tag{10}$$

The radical fragments are subsequently oxidised, mainly to CO as end product, via intermediate formation of formaldehyde.

Some of the carbon ends up as acyl radicals, RCO (e.g. CH_3CO from reaction 9), which are of particular interest from the point of view of atmospheric chemistry, since the corresponding acylperoxy radicals, formed by addition of O_2, can form relatively stable peroxynitrates by an association reaction with NO_2, e.g., the well known first member of the series, peroxyactyl nitrate (PAN):

$$CH_3COO_2 + NO_2 + M \rightarrow CH_3COO_2NO_2 + M. \tag{11}$$

This compound is interesting since it was first detected by infra-red spectroscopy in smog chambers and in polluted air over Los Angeles. ('Compound X' in the early air pollution literature.[3]) PAN and related compounds are oxidising agents, they cause plant damage and are also apparently associated with eye irritation in smog.

The stability of the peroxynitrates is only apparent and results from its being in reversible equilibrium with its precursor species, the acetylperoxy radical and NO_2:[17]

$$CH_3COO_2NO_2 \rightleftarrows CH_3COO_2 + NO_2. \qquad (12,11)$$

In the presence of NO reaction 13 can compete with reaction 11

$$CH_3COO_2 + NO \rightarrow CH_3 + CO_2 + NO_2 \qquad (13)$$

the methyl radical produced being oxidised to formaldehyde and OH in the $RO_2 \rightarrow RO \rightarrow HO_2$ reaction sequence (reactions 6, 4, 7 and 8). Note that the overall process converts 3 molecules of NO to NO_2 and therefore results in ozone production in photoirradiation. At 298 K and 1 atm, the rate constant for decomposition of PAN to CH_3COO_2 radicals, k_{12}, is $4.5 \times 10^{-4} s^{-1}$; if all these radicals reacted with NO, PAN would decompose with a half life of half an hour. However since reaction of CH_3COO_2 with both NO and NO_2 is rapid, $(k_{11}/k_{13} \approx 0.5$ at 1 atm pressure and 298 K), and NO_2 is normally the dominant oxide of nitrogen in the atmosphere, decomposition of PAN will be slower than this. The actual rate at which PAN is removed is given as a function of the ratio $[NO]/[NO_2]$ by the following expression

$$-\frac{d}{dt}\ln[PAN] = k_{12}\left\{1 - \frac{1}{1 + (k_{13}[NO]/k_{11}[NO_2])}\right\} \qquad (ii)$$

where $k_{12} = 4.0 \times 10^{15} \exp(-13040/T) s^{-1}$ at 1 atm pressure. In daylight or in a photoirradiated smog chamber, when NO is always present from photodissociation of NO_2, prevailing PAN concentrations will reflect a balance between production and loss of CH_3COO_2. In the dark, if NO is not present, thermal decomposition does not provide a net sink and PAN will be apparently stable in the absence of other sinks.

Alkyl peroxy radicals and HO_2 form peroxynitrates by association with NO_2, but the reverse dissociations have lower activation energies than reaction 12 and occur much more rapidly at ambient temperature (for HO_2NO_2 at 298 K, $t_{1/2} \simeq 10$ s). Consequently, at a given $[NO]/[NO_2]$ ratio, the rate of decomposition as given by the corresponding equation (ii), is much more rapid. At the lower temperatures and pressures prevailing in the upper troposphere and lower stratosphere, peroxynitric acid is thermally more stable and plays an important role in the NO_x chemistry there.

(v) Radical sources and sinks

The rate of oxidation of hydrocarbons, oxidation of NO to NO_2 and O_3 formation, in smog chambers and in the surface atmosphere, depend on the total concentrations of OH, HO_2, RO_2 and RO radicals. The reactions considered so far have, in the main, simply converted one radical species to another. The total radical concentration depends on the magnitude of the radical production and removal terms.

The radical production reactions are predominantly photochemical. The most rapid photolysis reaction is the dissociation of NO_2 to produce ground state oxygen atoms. However since $O(^3P)$ reacts rapidly with O_2 via reaction 2, its steady state concentration is very small under atmospheric conditions and only an insignificantly small fraction of the O atoms react with organic species to produce OH or organic radicals which can propagate the oxidation sequence.

The most important source of OH radicals results from the production of electronically excited $O(^1D)$ atoms from the photolysis of ozone at wavelengths less than 310 nm. Although $O(^1D)$ is rapidly relaxed to the ground state by O_2 and N_2, a significant fraction reacts with water vapour to produce OH:

$$O(^1D) + H_2O \rightarrow 2OH. \tag{14}$$

Although reaction 14 is the major radical source in the lower atmosphere, it does not occur in the initial stages of smog chamber photo-oxidation studies, since the O_3 concentration is too low. The initial radical source in chamber studies of NO_x–hydrocarbon photo-oxidations has not been fully established, but the most recent work[18] seems to point to photolysis of nitrous acid, HONO, as a major radical source:

$$HONO + h\nu \rightarrow OH + NO. \tag{15}$$

HONO is believed to be formed in heterogeneous reactions involving NO, NO_2 and adsorbed water, e.g.

$$2NO_2 + H_2O \rightarrow HNO_3 + HNO_2. \tag{16}$$

In addition, photodissociation of HCHO and other carbonyl compounds provides an important source of HO_2 and organic radicals, through reaction of the primary photofragments with O_2, both in the atmosphere and in chamber experiments. Peroxides also photolyse, but rather slowly, to yield RO and OH radicals. A few thermal sources of radicals have been found, but they are of only minor importance except for the understanding of night-time atmospheric chemistry.[19] They mainly involve ozone reactions, e.g. with alkenes producing a small yield of HO_x radicals and with NO_2 to produce NO_3.

The dominant free radical sinks in air containing nitrogen oxides all involve reactions of radicals with NO_2. The most important one is the formation of nitric acid:

$$OH + NO_2 + M \rightarrow HNO_3 + M. \tag{17}$$

The reaction of peroxyradicals to form peroxynitrates has already been mentioned. These reactions only act as sinks if the peroxynitrate is stable and does not decompose rapidly to regenerate the radicals. Another sink involving nitrogen oxides is the alternative pathway for the reaction of large alkylperoxy radicals with NO to form alkyl nitrates, $RONO_2$,[20] which competes with the O-atom transfer pathway, reaction 4:

$$RO_2 + NO(+ M) \rightarrow [RO_2NO] \rightarrow RONO_2. \tag{4a}$$

As with HNO_3, the organic nitrate esters formed here do not react rapidly with OH or photolyse significantly at low altitude.

When NO concentrations are very low, for example in the unpolluted troposphere, the radical sinks mainly involve HO_2, i.e.

$$HO_2 + HO_2 \rightarrow H_2O_2 + O_2 \tag{18}$$

$$HO_2 + RO_2 \rightarrow ROOH + O_2. \tag{19}$$

The peroxides formed in these reactions photolyse only slowly with near UV radiation and therefore act as temporary radical sinks.

C1.3 Determination of relative rate coefficients in photochemical steady state systems

As our knowledge of the mechanisms of the reactions occurring in NO_x–O_3–hydrocarbon photo-oxidation has improved, it has been possible to design photochemical steady state experiments based on these systems for the determination of rate constants by competitive kinetic techniques. A particular advantage is the ability to make measurements under pseudo-atmospheric conditions. Most of the work has focussed on reactions of OH, but the technique has also been used to determine relative rate constants for HO_2[21] and NO_3[22] reactions.

(i) Relative rate constants for OH radical reactions

The simplest photochemical OH source that has been used in chamber studies in the photolysis of HONO[23] which, in air at 1 atm pressure, proceeds via the reactions:

$$HONO + h\nu \rightarrow OH + NO \tag{15}$$

$$OH + HONO \rightarrow H_2O + NO_2. \tag{20}$$

As the products NO and NO_2 build up OH is also removed by reaction 17 with NO_2 and reaction 21:

$$OH + NO + M \rightarrow HONO + M. \tag{21}$$

The steady state concentration of OH is given by

$$[OH] = \frac{k_{15}}{k_{20} + k_{17}[NO_2]/[HONO] + k_{21}[NO]/[HONO]}. \tag{iii}$$

Initially, when $[HONO] \gg [NO]$ and $[NO_2]$, the steady-state concentration of OH is simply (k_{15}/k_{20}) and independent of $[HONO]$. Experimentally convenient light sources give $k_{15} \sim 10^{-3} s^{-1}$ and $k_{20} \sim 10^{-11} cm^3 molecule^{-1} s^{-1}$. Consequently, HONO has a half life of ~ 10 min and $[OH] \sim 10^8 molecule\ cm^{-3}$.

The preparation and handling of HONO without unacceptable amounts of NO and NO_2 is difficult and a more convenient indirect OH source is the photolysis in the presence of air or O_2 of methyl nitrate,[24] which can be readily prepared in the pure state. OH is produced by the reactions:

$$CH_3ONO + h\nu \rightarrow CH_3O + NO \tag{22}$$

$$CH_3O + O_2 \rightarrow HCHO + HO_2 \tag{23}$$

$$HO_2 + NO \rightarrow NO_2 + OH. \tag{8}$$

The photolysis rate k_{22} is similar to that for HONO and since the reaction of OH with CH_3ONO is much slower than with HONO $[k(OH + CH_3ONO) \approx 1.5 \times 10^{-13} cm^3 molecule^{-1} s^{-1}]$, a higher steady state hydroxyl concentration can, in principle, be generated from this source. In practice, NO is added to the CH_3ONO–Air mixtures, to prevent build up of O_3 from the NO to NO_2 conversion in reaction 8, and so $[OH]$ is controlled by reaction 21 and later by reaction with NO_2 and HCHO. Even so OH concentrations of up to $3 \times 10^8 molecule\ cm^{-3}$ can be achieved, falling to lower values as reaction proceeds.

If a hydrocarbon is added to the OH–precursor–air mixture, then it will be attacked by OH and the oxidation sequence involving reactions 4–8 will be initiated, in which OH is regenerated. Thus the steady state concentration for OH will be unperturbed by the added substrate, provided no additional radical sink reactions are introduced by the new chemical pathways. If OH reaction is the only loss route the rate of removal of the substrate will be given by:

$$-d\ln[S]/dt = k_s[OH]. \tag{iv}$$

If a second reference organic compound is added, its decay rate will be given by:

$$-d\ln[R]/dt = k_r[OH] \tag{v}$$

where k_s and k_r are the rate constants for OH reaction with the substrate and the reference compound and $[OH]$ is the local steady state concentration of OH, which can be factorised out:

$$d\ln[S]/dt = \frac{k_s}{k_r}d\ln[R]/dt$$

and hence

$$\ln\frac{[S]_0}{[S]_t} = \frac{k_s}{k_r}\ln\frac{[R]_0}{[R]_t} \tag{vi}$$

where the 0 and t subscripts refer to concentrations of S and R at initial time t_0 and at elapsed reaction time t. Thus plots of the logarithm of the ratio of concentration at t_0 and t for the substrate against the corresponding ratio for the reference compound should yield straight lines with zero intercept and a slope k_s/k_r. Some typical plots for several alkanes using n-hexane as reference are shown in Fig. C1.5 from the work of Atkinson *et al.*[25] It will be noticed that determinations of the ratio k_s/k_r by this method do not require measurements of absolute concentration, relative measurements being sufficient. The main requirements are:

(a) A precise, specific and accurate sampling and analysis technique for substrate and reference compound. Both gas chromatography and *in situ* visible or infra-red spectroscopy have been employed.

(b) The chemical system must contain only one significant attacking radical. Problems can occur, for example, with alkenes if significant concentrations of ozone build up providing an additional loss reaction.

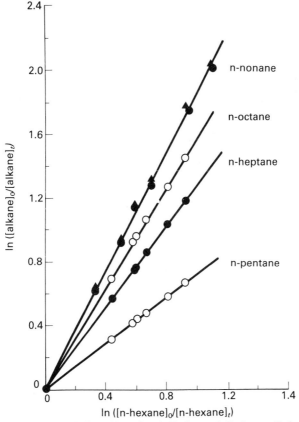

Fig. C1.5. Plots of $\ln([\text{alkane}]_0/[\text{alkane}]_t)$ against $\ln([\text{n-hexane}]_0/[\text{n-hexane}]_t)$ for n-pentane, n-heptane, n-octane and n-nonane. The differing symbols for n-nonane refer to two independent gas chromatographic analyses. Reproduced from reference 25. Copyright © J. Wiley & Son. Reprinted by permission of John Wiley & Sons, Inc.

(c) The substrate must be stable with respect to wall loss or other non-photochemical reactions, on the time scale of the experiments.

Since the chemical systems are highly reproducible and time variation of relative concentrations of a wide variety of organic species can be measured readily over the experimental time periods involved using modern techniques, the method is capable of producing highly accurate relative data. This has been well demonstrated by the work of Atkinson and co-workers at U.C. Riverside, who have used the technique to investigate in detail the relationships between structure and reactivity in room temperature OH + organic substrate reactions.[25,26] Some of their results are compared with OH rate constants measured by direct techniques in Table C1.1, which illustrates the high precision achievable by the indirect method.

The limitations of the technique lie in the difficulty of measuring the decay of substrates that react only slowly with OH and in the restrictions in the pressure and temperature ranges dictated by the nature of the experimental arrangements (large smog chambers or Teflon bags are used normally) and by the chemical system. Some of the latter limitations have been overcome by modifying the experimental technique to allow coverage of a range of pressures and temperatures, by the use, for example, of cooled, evacuable reaction chambers.

Extension of the technique to compounds of lower reactivity has been achieved in a number of ways including multiple, successive additions of CH_3ONO precursor and monitoring of product formation, instead of reactant removal, to determine $\ln [S]_0/[S]_t$. The disadvantage of the latter method lies in the requirement of absolute concentration measurements and also of the reaction stoichiometry. An interesting variation on this approach is to measure the rate of NO to NO_2 conversion at short reaction times in the photolysis of a flowing HONO–air mixture with known amounts of added substrate.[27] This method provides a value for the ratio k_s/k_{20}, but absolute concentration measurements and a knowledge of stoichiometry are still required.

Other variations of the technique include the use of thermal OH sources (e.g.

Table C1.1. Quality of rate constant data from relative rate technique for OH + organic compounds

| | $10^{12} k(cm^3 molecule^{-1} s^{-1})$ at 299 K* | |
Organic compound	relative rate technique	direct methods
neo-pentane	0.77 ± 0.05**	0.82
n-butane	—	2.58
c-hexane	7.57 ± 0.05	7.95
2-,3-dimethylbutane	6.26 ± 0.06	7.45
ethene	8.48 ± 0.39	7.85 ± 0.79
propene	—	25.0
2-methyl-2-butene	86.4 ± 3.3	87.3 ± 8.8
isoprene	96.0 ± 4.3	93 ± 9

*Using $n-C_6H_{14}$, $c-C_6H_{12}$ and C_3H_6 as reference compounds. ** ± 2σ.

$N_2O_4 + O_3$ mixtures) for investigation of OH rates with photolabile compounds such as CH_3ONO.[28] Compounds which undergo photolysis in competition with reaction with OH can be investigated using a non-photolabile reference compound. In this case substrate loss is described by:

$$-\frac{d \ln [S]}{dt} = k_s[OH] + k_p \qquad (vii)$$

where k_p is the photolysis constant, i.e. the product of the quantum yield, the absorption cross-section and the light intensity, $\phi\sigma I$. The integrated form of equation vii is:

$$\ln\left(\frac{[S]_0}{[S]_t}\right) = k_s \int_{t_0}^{t} [OH]\, dt + k_p(t - t_0) \qquad (viii)$$

and the form equivalent to equation vi is:

$$\frac{1}{t - t_0} \ln\left(\frac{[S]_0}{[S]_t}\right) = k_p + \frac{k_s}{k_r(t - t_0)} \ln\left(\frac{[R]_0}{[R]_t}\right) \qquad (ix)$$

where t is the elapsed experimental time. The ratio k_s/k_r and the photolysis constant, k_p, can be obtained from the slope and intercept, respectively, of a plot of $(t - t_0)^{-1} \ln\{[S]_0/[S]_t\}$ against $(t - t_0)^{-1} \ln \{[R]_0/[R]_t\}$.

(ii) Relative rate constants for HO_2 and NO_3 reactions

Measurement of the relative rates of HO_2 reactions have been studied using the thermal decomposition of peroxynitric acid, HO_2NO_2, as a source of HO_2.[21] At ambient temperature, steady state HO_2 concentrations of 10^{10}–10^{11} molecule cm^{-3} are maintained by the equilibrium:

$$HO_2NO_2 \rightleftharpoons HO_2 + NO_2. \qquad (24, 25)$$

Addition of a substrate which reacts with HO_2

$$HO_2 + S \xrightarrow{k_s} products$$

leads to a perturbation of the equilibrium and decay of HO_2NO_2 will occur at a rate equal to that of HO_2 reaction with the substrate, i.e.

$$-\frac{d}{dt}[HO_2NO_2] = k_s[HO_2][S] = k_s K^* \frac{[HO_2NO_2]}{[NO_2]} \cdot [S] \qquad (x)$$

where $K^* (= k_{24}/k_{25})$ is the equilibrium constant for reaction (24, 25).

It follows that k_s is related to k_d, the overall first order coefficient for loss of HO_2NO_2, by:

$$k_s = \frac{1}{K^*} \frac{[NO_2]}{[S]} \cdot k_d. \qquad (xi)$$

Quite large uncertainties are inherent in this method since the absolute concentrations of NO_2 and S are required and any regeneration of HO_2 in the reaction system or removal of HO_2NO_2 by secondary chemistry invalidates the derivation of k_s. Suitable experimental conditions for application of the substrate/reference compound decay technique for relative rate measurements, have not so far been reported.

A similar method has been used for NO_3 reactions, using the thermal decomposition of N_2O_5 as a source of NO_3.[22,29]

$$N_2O_5 \rightleftharpoons NO_3 + NO_2. \tag{26, 27}$$

In this case experimental conditions can be employed where reaction with NO_3 is the sole chemical loss process for the organic substrate. Use of a reference compound allows determination of the rate constant ratio k_s/k_r for NO_3 reaction, from measurement of the decay rate of the organic species. Rate constants can also be obtained from measurements of the rate of decay of N_2O_5 in the presence of an added substrate which reacts with NO_3.

C1.4 Stratospheric ozone chemistry

(i) Introduction

Ozone is present in the earth's atmosphere at all altitudes from the surface up to at least 100 km. The bulk of the ozone resides in the stratosphere with $[O_3]_{max} \approx 5 \times 10^{12}$ molecule cm^{-3} at about 25 km. In the mesosphere (> 60 km) O_3 densities are quite low and are not discussed in this chapter. Although concentrations of O_3 in the troposphere are also less than in the stratosphere, ozone plays a vital role in the atmospheric chemistry in this region and also affects the thermal radiation balance in the lower atmosphere.

Atmospheric ozone is formed by combination of atomic and molecular oxygen

$$O(^3P) + O_2 + M \rightarrow O_3 + M \tag{2}$$

where M is a third body required to carry away the energy released in the combination reaction. At altitudes above approximately 20 km, the net production of O atoms results almost exclusively from photodissociation of molecular O_2 by short wavelength ultraviolet radiation:

$$O_2 + h\nu\,(\lambda < 243\,\text{nm}) \rightarrow O + O. \tag{28}$$

At lower altitudes and particularly in the troposphere, O atom formation from the photodissociation of nitrogen dioxide by long wavelength ultraviolet radiation is more important:

$$NO_2 + h\nu\,(\lambda < 400\,\text{nm}) \rightarrow NO + O(^3P). \tag{1}$$

Ozone itself is photodissociated by both UV and visible light:

$$O_3 + h\nu \rightarrow O_2 + O \tag{29}$$

but this reaction, together with the combination reaction 2, only serves to partition the 'odd oxygen' species between O and O_3. The processes producing odd oxygen, reactions 28 and 1, are balanced by chemical and physical loss processes. Until the 1960's, chemical loss of odd oxygen was attributed only to the reaction:

$$O(^3P) + O_3 \rightarrow O_2 + O_2 \tag{30}$$

originally proposed by Chapman in 1930.[32] It is now known that ozone in the stratosphere is removed predominantly by catalytic cycles involving homogeneous gas-phase reactions of active free radical species in the HO_x,[33] NO_x[34] and ClO_x[35] families:

$$X + O_3 \rightarrow XO + O_2 \tag{31}$$

$$XO + O \rightarrow X + O_2 \tag{32}$$

net $O + O_3 \rightarrow 2O_2$

where the catalyst $X = H$, OH, NO, Cl and Br. Thus these species can control the abundance and distribution of ozone in the stratosphere with varying degrees of efficiency depending on the respective rate coefficients of reactions 31 and 32 and the local abundance of the radicals. Assignment of the relative importance and the prediction of the future impact of these catalytic species are dependent on a detailed understanding of the chemical reactions which form, remove and inter-convert the active components of each family. This in turn requires knowledge of the atmospheric life cycles of the hydrogen, nitrogen and halogen-containing precursor and sink molecules, which control the overall abundance of the HO_x, NO_x and ClO_x species. Recently, detailed descriptions of the basic chemical and photochemical processes which occur in the atmosphere, and which control ozone and other trace gas budgets, have been given in books by Brasseur and Solomon[36] and Wayne.[37]

Physical loss from the stratosphere is mainly by dynamical transport to the troposphere where further photochemically driven sources and sinks modify the ozone concentration. Ozone is destroyed heterogeneously at the surface of the earth and so there is an overall downward flux in the lower part of the atmosphere. Physical removal of ozone and other trace gaseous components can also occur in precipitation and on the surface of atmospheric aerosols. Since most of the precursor and sink molecules for the catalytically active species in ozone removal in the stratosphere are derived or removed in the troposphere, global tropospheric chemistry is a significant feature in the overall behaviour of atmospheric ozone.

Numerical simulation techniques are used to describe and investigate the behaviour of the complex chemical system controlling atmospheric composition, the models having elements of chemistry, radiation and transport. The chemistry in such models may include some 150 elementary chemical reactions and photochemical processes. Laboratory measurements of the rates of these reactions have progressed rapidly over the past decade and have given us a basic understanding of the kinetics of these elementary processes and the way they act in controlling ozone. This applies particularly in the upper stratosphere where the local chemical composition is controlled predominantly by photochemistry.

It has proved more difficult to describe adequately both the chemistry and the dynamics in the lower stratosphere. Here the chemistry is complicated by the involvement of temporary reservoir species such as $HOCl$, H_2O_2, HNO_3, HNO_4, N_2O_5 and $ClONO_2$ which 'store' active radicals and which strongly couple the HO_x, NO_x and ClO_x families. The long photochemical and thermal lifetimes of ozone and the reservoir species in this region give rise to a strong interaction between chemistry and dynamics (transport) in the control of the distribution of ozone and other trace gases. Seasonal variability and natural perturbations, due to volcanic injections of gases and aerosol particles, further complicate the description and interpretation of atmospheric behaviour in this region. Moreover most of the changes in the predicted effects of chlorofluoromethanes and other pollutants on ozone column density have resulted from changes in our view of the chemistry in the lower stratosphere. A great deal of importance must therefore be attached to achieving an understanding of the key factors in ozone chemistry in this region of the atmosphere.

Description of atmospheric chemistry in the troposphere is similarly complicated by dynamical influence and additionally by involvement of the precipitation elements (i.e. cloud, rain and snow) in the chemical pathways. The homogeneous chemistry of the troposphere is centred around the role of the hydroxyl radical in promoting oxidation and scavenging of trace gases released from terrestrial sources. Photochemical smog is a manifestation of this general process, occurring in polluted air. Although the mechanisms are more complex than in the stratosphere owing to the involvement of larger and more varied molecular species, the overall pattern of relatively rapid photochemical cycles involving a coupled carbon/hydrogen/nitrogen and oxygen chemistry is similar to that in the stratosphere.[38] The influence of hydrocarbon oxidation both on the odd hydrogen budget and on ozone itself, through coupling with NO_2 photochemistry, is particularly important. A comprehensive review of tropospheric trace gas chemistry has been given by Cox and Derwent.[7]

Provision of an evaluated photochemical and kinetics data base for modelling atmospheric chemistry and ozone perturbations, has been recognised as an important feature of atmospheric programmes for some years now. With the rapid growth in the amount of information and expertise available in recent years this has become even more important. Recent evaluations prepared by the NASA panel for data evaluation[39] and the CODATA Task Group for Chemical

Kinetics[40] provide useful sources of data for atmospheric chemistry and for low temperature kinetics generally. Atkinson and Lloyd have recently evaluated data for modelling ozone formation in the troposphere.[41] In the following sections some of the key elementary reactions controlling HO_x, NO_x and halogen chemistry are introduced and the role they play is outlined.

(ii) Odd hydrogen chemistry

The main reservoir species for HO_x in the stratosphere is water vapour, from which the active species are released by reaction with excited atomic oxygen:

$$O(^1D) + H_2O \rightarrow 2OH. \tag{14}$$

$O(^1D)$ is produced throughout the atmosphere by photodissociation of O_3 (reaction 29) at wavelengths $\lesssim 310\,nm$. Two important O_3 destruction cycles involving HO_x occur. In the upper stratosphere O_3 removal is predominantly caused by:

$$OH + O_3 \rightarrow HO_2 + O_2 \tag{33}$$

$$HO_2 + O \rightarrow OH + O_2. \tag{34}$$

In the lower stratosphere, where O is not so abundant, reaction 34 is replaced by:

$$HO_2 + O_3 \rightarrow OH + 2O_2. \tag{35}$$

There is today an excellent experimental data base for all the reactions of OH and HO_2 which have been identified as important in controlling the partitioning and abundance of OH and HO_2. Unexpected temperature, pressure and water vapour effects on the rate constants for some HO_2 reactions have now been confirmed; these need to be taken into account when applying laboratory results to the atmosphere and they also offer a challenge to the theoretical description of rate processes involving radical species. The reader should refer to the recent work cited in the NASA[39] and CODATA[40] evaluation for detailed discussion of these mechanisms.

Removal of HO_x occurs by the reactions:

$$OH + HO_2 \rightarrow H_2O + O_2 \tag{36}$$

$$HO_2 + HO_2 \rightarrow H_2O_2 + O_2 \tag{18}$$

reaction 36 being more important at higher altitudes.

(iii) Odd nitrogen chemistry

The main source of NO_x in the stratosphere is N_2O and the reactive species are again released by reaction with $O(^1D)$:

$$O(^1D) + N_2O \rightarrow 2NO. \tag{37}$$

The following catalytic cycle is the dominant ozone loss process in the middle stratosphere (25—35 km):

$$NO + O_3 \rightarrow NO_2 + O_2 \tag{3}$$

$$NO_2 + O(^3P) \rightarrow NO + O_2. \tag{38}$$

At lower altitudes, where [O] is lower, photodissociation of NO_2 nullifies the odd oxygen removal effect, since ground state oxygen atoms are produced:

$$NO_2 + h\nu \rightarrow NO + O(^3P). \tag{1}$$

The main reservoir species for NO_x is nitric acid, which is formed in a reaction that couples the HO_x and NO_x chemistry through the reaction:

$$OH + NO_2 + M \rightarrow HNO_3 + M. \tag{17}$$

Active nitrogen species are only slowly regenerated from HNO_3 by photolysis and by reaction with OH:

$$HNO_3 + h\nu \rightarrow OH + NO_2 \tag{39}$$

$$OH + HNO_3 \rightarrow H_2O + NO_3. \tag{40}$$

Reaction 17, followed by 40, is also an important loss process for odd hydrogen species in the lower stratosphere, since the net process leads to conversion of two hydroxyl radicals into water. The photolysis of NO_3 leads to reformation of NO_2 either directly:

$$NO_3 + h\nu(\lambda < 580\,nm) = NO_2 + O \tag{41}$$

or through the formation of NO followed by reaction 3:

$$NO_3 + h\nu(\lambda < 900\,nm) \rightarrow NO + O_2. \tag{42}$$

The minor stratospheric NO_x species, NO_3, N_2O_5 and HO_2NO_2 are all involved in cyclical reactions which affect the active species distribution and the local ozone budget.

(iv) Odd chlorine chemistry

The major source of stratospheric chlorine is the photodissociation of organic chlorine compounds such as methyl chloride, other chlorinated hydrocarbons and chlorofluorocarbons:

$$RCl + h\nu(\lambda \lesssim 215\,nm) \rightarrow R + Cl. \tag{43}$$

The ozone destruction cycle involves Cl and ClO

$$Cl + O_3 \rightarrow ClO + O_2 \tag{44}$$

$$ClO + O(^3P) \rightarrow Cl + O_2. \tag{45}$$

The effect is modified in the presence of NO_x by the reactions:

$$ClO + NO \rightarrow Cl + NO_2 \tag{46}$$

$$NO_2 + \boldsymbol{hv} \rightarrow NO + O(^3P) \tag{1}$$

which offset the odd oxygen loss in reaction 45. Thus the extent of ozone destruction at a particular altitude by ClO_x depends on the amount of NO_x present. A further important coupling between the chlorine and nitrogen oxides is the formation of chlorine nitrate:

$$ClO + NO_2 + M \rightarrow ClONO_2 + M. \tag{47}$$

This is a typical 'temporary reservoir species' (other examples are N_2O_5, HO_2NO_2, $HOCl$ and H_2O_2) which do not directly participate in odd oxygen destruction catalytic cycles but which serve to tie up active species, thereby reducing the rate of ozone destruction.

The main removal of active chlorine species occurs via the reaction:

$$Cl + CH_4 \rightarrow HCl + CH_3. \tag{48}$$

Reaction with other H-containing species such as HO_2, $HCHO$ and C_2H_6 also contribute to HCl formation.

HCl is the major reservoir species for stratospheric chlorine from which active Cl is released by reaction with OH:

$$OH + HCl \rightarrow H_2O + Cl. \tag{49}$$

Some HCl is transported downwards into the troposphere from where it is removed by rain, thereby completing the atmospheric chlorine cycle.

(v) Bromine chemistry

Active bromine species Br and BrO undergo a catalytic cycle for ozone destruction exactly analogous to that for Cl and ClO

$$Br + O_3 \rightarrow BrO + O_2 \tag{50}$$

$$BrO + O \rightarrow Br + O_2. \tag{51}$$

Bromine is much more efficient than chlorine at destroying O_3 since conversion of Br to inactive HBr is much slower than the corresponding reactions for Cl. H-abstraction by Br from the major H-containing species CH_4, C_2H_6 and H_2 is endothermic and is negligibly slow at stratospheric temperatures. Furthermore, reaction of OH with HBr is much faster than with HCl, and most of the Br present in the stratosphere is in the active form.

Bromine also couples with chlorine chemistry via the reaction

$$BrO + ClO \rightarrow Br + Cl + O_2 \tag{52}$$

which together with the $Br+O_3$ and $Cl+O_3$ reactions provides an additional catalytic cycle for the destruction of ozone. The corresponding reaction of ClO is $10\,kJ\,mol^{-1}$ endothermic and is relatively slow even at room temperature. The self reaction of ClO:

$$ClO + ClO \rightarrow products \tag{53}$$

occurs mainly by association to form the dimer $(ClO)_2$, which is probably stable at stratospheric temperatures. The reaction is nevertheless too slow, compared to the $ClO + NO_2$ reactions, to be important in the present-day stratosphere where NO_x abundance is considerably higher than ClO_x.

C1.5 Ozone in the clean troposphere

The troposphere differs from the stratosphere in that there is an abundance of water vapour, the air is mixed rapidly by convective transport and only long wavelength solar UV radiation ($\lambda > 290$) penetrates to these altitudes. Although only about 10% of all atmospheric O_3 is located in this region, it is significant for atmospheric chemistry.

Ozone photolysis is a major source of hydroxyl radicals in the troposphere through the reactions:

$$O_3 + h\nu(\lambda < 310\,nm) \rightarrow O(^1D) + O_2(^1\Delta) \tag{29}$$

$$O(^1D) + H_2O \rightarrow 2OH. \tag{14}$$

Attack by OH initiates the oxidation of many gases in the troposphere including hydrocarbons, halogenocarbons, organosulphur compounds, and many inorganic gases such as CO, NO_2, H_2S and SO_2. In this way OH acts as a scavenging agent preventing build up of C, N, S and halogen containing gases in the lower atmosphere. Control of the transfer of these gases to the stratosphere is important since they are the source gases for active species involved in the catalytic cycles governing ozone abundance.

In the background troposphere about 70% of the OH radicals react with CO and 30% with CH_4 to yield peroxy radicals:

$$OH + CO \rightarrow CO_2 + H \xrightarrow{O_2} HO_2 \tag{54}$$

$$OH + CH_4 \rightarrow H_2O + CH_3 \xrightarrow{O_2} CH_3O_2. \tag{55}$$

The HO_2 can be converted back to OH by the reactions:

$$HO_2 + O_3 \rightarrow OH + 2O_2 \tag{35}$$

$$HO_2 + NO \rightarrow OH + NO_2. \tag{8}$$

The cycle $54 + 35$ removes ozone according to the net overall reaction

$$O_3 + CO \rightarrow CO_2 + O_2. \tag{56}$$

If HO_2 is converted to OH in reaction 8 ozone is produced since NO_2 is rapidly photolysed to $NO + O(^3P)$, giving overall:

$$2O_2 + CO + h\nu \rightarrow CO_2 + O_3. \tag{57}$$

The methyl peroxy radical does not appear to react at a significant rate with ozone. The two reactions:

$$CH_3O_2 + NO \rightarrow CH_3O + NO_2 \tag{58}$$

$$CH_3O_2 + HO_2 \rightarrow CH_3OOH + O_2 \tag{59}$$

occur at competitive rates. The latter is a radical termination reaction and results in ozone removal only at the slow rate of removal of $O(^1D)$ by reaction with H_2O, since no catalytic cycle is involved.

The CH_3O radical produced in reaction 58 is oxidised to formaldehyde:

$$CH_3O + O_2 \rightarrow HCHO + HO_2. \tag{23}$$

Photodissociation of HCHO:

$$HCHO + h\nu \rightarrow H + HCO \tag{60}$$

then provides an additional source of HO_2 radicals, by reaction of the photofragments with O_2. Since NO is converted to NO_2 in reaction 58, oxidation of CH_4 results in net production of O_3 as well as augmented radical concentrations when NO is present. In the absence of NO, CH_3OOH is formed. This species can react either by photolysis or by reaction with OH:

$$CH_3OOH + h\nu \rightarrow CH_3O + OH \tag{61}$$

$$OH + CH_3OOH \rightarrow H_2O + HCHO + OH \quad (50\%) \tag{62a}$$

$$\rightarrow H_2O + CH_3O_2 \quad (50\%). \tag{62b}$$

Combining each of these reactions with reaction 59 produces three different pathways. Reactions 59 and 61 convert $CH_3O_2 + HO_2$ to $HCHO + OH + HO_2$; i.e. there is no net change in the concentrations of radicals. Reactions 59 and 62a replace reaction 59 alone by the alternative radical termination:

$$CH_3O_2 + HO_2 \rightarrow HCHO + H_2O + O_2. \tag{63}$$

Finally, reactions 59 and 62b give a CH_3O_2-catalysed removal of OH and HO_2:

$$OH + HO_2 \rightarrow H_2O + O_2. \tag{36}$$

The kinetics of the reactions forming and removing CH_3OOH are not well established at the present time, and there is a clear need for further laboratory measurements.

Laboratory simulation of ozone chemistry

Much of the detailed understanding of photochemical smog formation and chemistry in the lower atmosphere has evolved from simulation studies involving smog chambers and other complex systems. On the other hand the approach to chemistry in the clean troposphere and the stratosphere has involved laboratory studies of the kinetics of key elementary reactions in isolation (see Chapter B1). These reactions have been identified from chemical intuition, observations in the atmosphere and modelling studies. Whereas this approach must ultimately provide a complete and accurate picture of the complex chemical kinetics, there is always the possiblility of error and inaccuracy resulting from extrapolation of rate data from laboratory conditions and the problem of chemistry involving unidentified species or reactions. A case can be argued for the use of simulation experiments to check the photochemical mechanism invoked to describe ozone in the stratosphere and the clean troposphere. One of the difficulties is the design of experiments which offer a useful mimic of atmospheric conditions, since temperatures and pressures are usually far removed from ambient surface values and the concentrations of active species and overall reaction rates which can be conveniently studied in the laboratory are much higher than those occurring in the atmosphere. Nevertheless some useful clues regarding ozone chemistry have been obtained from simple laboratory photochemical experiments.

The interaction of HO_x chemistry with ozone was the subject of much debate during the 1960's, with various theories based on observations in the upper atmosphere and laboratory flash photolysis experiments on 'wet' ozone mixtures. A breakthrough was made in the simple laboratory photochemical experiments reported by De More in 1973.[30] The experiments involved the photolysis with short wavelength (185 nm) light of O_2–H_2O and O_3–H_2O–CO mixtures in a static reactor. Both ozone and HO_x radicals are produced and ozone reaches a steady state concentration controlled by the chain reaction:

$$OH + O_3 \rightarrow O_2 + HO_2 \tag{33}$$

$$HO_2 + O_3 \rightarrow 2O_2 + OH \tag{35}$$

which balances the production of ozone by the reaction

$$O_2 + h\nu \rightarrow 2O \xrightarrow{O_2} 2O_3 \quad rate = 2I_a^{O_2}$$

where $I_a^{O_2}$ is the rate of light absorption by O_2.

The production of HO_x radicals via

$$H_2O + h\nu \rightarrow OH + H \xrightarrow{O_2} HO_2 \quad rate = 2I_a^{H_2O}$$

is balanced by the termination reactions:

$$OH + HO_2 \rightarrow H_2O + O_2 \tag{36}$$

$$HO_2 + HO_2 \rightarrow H_2O_2 + O_2. \tag{18}$$

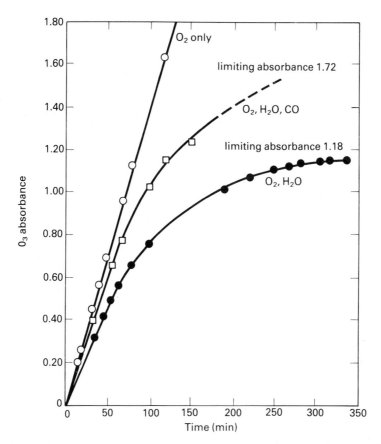

Fig. C1.6. Time dependence of O_3 concentrations for irradiation of various mixtures at 184.9 nm. For pure O_2, the production of O_3 is linear with time. With added water, a steady state is reached when the rate of O_3 destruction by OH and HO_2 radicals equals the rate of production. Addition of excess CO ($CO \gg O_3$) converts OH to HO_2, and a different steady state results. In the apparatus used, 1 absorbance unit corresponds to $67\,\mu m$ of O_3. Reproduced from reference 30 with permission. Copyright 1973 by the AAAS.

Table C1.2. Rate constant ratios for elementary processes in the photolysis of ozone–water vapour mixtures

Rate constant ratio	Experimental value*	Calculated from consensus values of rate constants for elementary reactions
$k_{35}/k_{18}^{1/2}$	$6.4 \times 10^{-8} \exp(-1220/T)$	$7.2 \times 10^{-8} \exp(-1203/T)$
k_{33}/k_{54}	$16.8 \exp(-1233/T)$	$68 \exp(-1000/T)^{\dagger}$
k_{36}	$(1.2-1.6) \times 10^{-10\ddagger}$	$(1.0 \pm 0.2) \times 10^{-10}$

*Units: cm^3, molecule, s. †Calculated using a temperature independent value of k_{54} for high pressure conditions. ‡1 atm pressure values determined relative to k_{18}, k_{35} and k_{33}.

Under steady state conditions $k_{33}[OH] = k_{35}[HO_2]$ and the ozone steady state is given by

$$[O_3]_s = I_a^{O_2}/k_{35}[HO_2]. \tag{xii}$$

On addition of CO, the ozone steady state concentration is modified because reaction 54:

$$OH + CO(+O_2) \rightarrow CO_2 + HO_2 \tag{54}$$

competes with reaction 33; in addition, reaction 36 becomes less important because $[OH]$ is depressed. With $[CO] \gg [O_3]$:

$$[O_3]_s = 2I_a^{O_2}/k_{35}[HO_2] \text{ and } [HO_2]_s^2 = I_a^{H_2O}/k_{18}. \tag{xiii}$$

Figure C1.6 shows a plot of the experimental data. Analysis of the steady state concentrations of O_3 and the rates at which they are attained gives values for the following ratios at 300 K

$$k_{35}/k_{18}^{1/2}, k_{33}/k_{54} \text{ and } k_{36}k_{35}/k_{18}k_{33}.$$

The values obtained turn out to be remarkably close to those predicted from the currently accepted values for the rate constants of the elementary rate constants measured in isolation. In later work[31] the same consistency was found over a range of temperatures, as shown in Table C1.2 where the temperature dependences from the relative rate studies are compared with those from current consensus values derived from measurements on the individual elementary reactions. The agreement confirms that the interpretation of the chemistry is fundamentally correct. Such experiments therefore provide a useful check on the validity of integrated chemical systems relevant to the atmosphere.

Conclusions

It is clear from the above discussion that the chemistry controlling the budget of ozone in the stratosphere and the troposphere is quite complex. Accurate knowledge of the detailed kinetics of the key reactions is central to the understanding of the atmospheric ozone balance and the effects of pollutant gases on ozone. At the present time we rely on photochemical models of the atmosphere to predict the potential effects of these pollutant gases and the availability of accurate chemical data is a crucial factor contributing to our confidence in the validity of these models.

C1.6 Suggestions for further reading

Photochemical smog chemistry

Leighton P. A. (1961) *Photochemistry of Air Pollution*, Academic Press, New York.
Demerjian K. A., Kerr J. A. and Calvert J. G. (1974) The Mechanism of Photochemical Smog Formation. In *Advances Environmental Science and Technology*, Vol. 4 (Ed. by J. N. Pitts, Jr. and R. L. Metcalfe, Assoc. Ed., A. C. Lloyd), J. Wiley and Sons, New York.
Findlayson B. J. and Pitts Jr. J. N. (1976) Photochemistry of Polluted Troposphere. *Science*, **192**, 111.

Tropospheric chemistry

Cox R. A. and Derwent R. G. (1981) Specialist Periodical Report. Gas Phase Chemistry of the Minor Constituents of the Troposphere. In *Gas Kinetics and Energy Transfer*, Vol. 4, p. 189. Royal Society of Chemistry, London.

Chamiedes W. L. and Davis D. D. (1982) Chemistry in the Troposphere. *Chem. Eng. News*, **60**, 39.

Relative rate measurements

Cox R. A., Derwent R. G. and Williams M. R. (1980) Atmospheric Photo-oxidation Reactions: Rates, Reactivity and Mechanism for Reaction of Organic Compounds with Hydroxyl Radicals. *Environ. Sci. Technol.*, **14**, 57.

Atkinson R., Carter W. P. L., Winer A. M. and Pitts Jr. J. N. (1981) An Experimental Protocol for the Determination of OH Radical Rate Constants with Organics Using Methyl Nitrate Photolyses as an OH Radical Source. *J. Air Pollution Contr. Assoc.*, **31**, 1090.

Stratospheric chemistry

McEwan M. J. and Phillips L. F. (1975) *Chemistry of the Atmosphere*, Chap. 4, Edward Arnold, London.

Brasseur G. and Solomon S. (1984) *Aeronomy of the Middle Atmosphere: Chemistry and Physics in the Stratosphere and Mesosphere*, D. Reidel, Dordrecht.

General

Wayne R. P. (1985) *Chemistry of Atmospheres*, Clarendon Press, Oxford.

Findlayson-Pitts B. J. and Pitts Jr. J. N. (1986) *Atmospheric Chemistry*, Wiley-Interscience, New York.

C1.7 References

1 Norrish R. G. W., Porter G. and Thrush B. A. (1953) *Proc. R. Soc. London, Ser. A*, **216**, 165.

2 Bates D. R. and Nicolet M. (1950) *J. Geophys. Res.*, **55**, 301.

3 Leighton P. A. (1961) *Photochemistry of Air Pollution*, Academic Press, New York.

4 Kaufman F. (1985) *Science*, **230**, 313–399.

5 Findlayson B. J. and Pitts Jr. J. N. (1976) *Science*, **192**, 111.

6 Brasseur G. and Solomon S. (1984) *Aeronomy of the Middle Atmosphere: Chemistry and Physics in the Stratosphere and Mesosphere*, D. Reidel, Dordrecht.

7 Cox R. A. and Derwent R. G. (1981) Specialist Periodical Report. *Gas Kinetics and Energy Transfer*, Vol. 4, p. 189. Royal Society of Chemistry, London.

8 Pitts Jr. J. N., Lloyd A. C. and Sprung J. L. (1975) *Chemistry in Britain*, 247.

9 Demerjian K. L., Kerr J. A. and Calvert J. G. (1974) In *Advances in Environmental Science and Technology* (Ed. by J. N. Pitts Jr. and R. L. Metcalfe), Vol, 4, p. 1. Wiley, New York.

10 Baulch D. L. and Campbell I. M. (1981) Specialist Periodical Report. *Gas Kinetics and Energy Transfer*, Vol. 4, p. 137. Royal Society of Chemistry, London.

11 Atkinson R. (1985) *Chem. Rev.*, **85**, 69.

12 Darnell K. R., Lloyd A. C., Winer A. M. and Pitts Jr. J. N. (1976) *Environ. Sci. Technol.*, **10**, 908.

13 Baldwin A. C., Barker J. R., Golden D. M. and Hendry D. G. (1977) *J. Phys. Chem.*, **81**, 2423.

14 Niki H., Maker P. D., Savage C. M. and Breitenbach L. P. (1981) *Chem. Phys. Letters*, **80**, 499.

15 Atkinson R., Carter W. P. L., Darnall K. R., Winer A. M. and Pitts Jr. J. N. (1980) *Int. J. Chem. Kinet.*, **12**, 779.

16 Dumdei B. E. and O'Brien R. J. (1984) *Nature*, **311**, 248.

17 Cox R. A. and Roffey M. (1977) *Environ. Sci. Technol.*, **11**, 900.

18 Carter W. P. L., Atkinson R., Winer A. M. and Pitts Jr. J. N. (1982) *Int. J. Chem. Kinet.*, **14**, 1071.

19 Cantrell C. A., Stockwell W. R., Anderson L. G., Busarow K. L., Perner D., Schmeltekkopf A., Calvert J. G. and Johnston H. S. (1985) *J. Phys. Chem.*, **89**, 139.

20 Atkinson R., Aschmann S. M., Carter W. P. L., Winer A. M. and Pitts Jr. J. N. (1982) *J. Phys. Chem.*, **86**, 4653.

21 Graham R. A., Winer A. M., Atkinson R. and Pitts Jr. J. N. (1979) *J. Phys. Chem.*, **83**, 1563.

22 Japar S. M. and Niki H. (1975) *J. Phys. Chem.*, **79**, 1629.

23 Cox R. A., Derwent R. G. and Williams M. R. (1980) *Environ. Sci. Technol.*, **14**, 57.

24 Atkinson R., Carter W. P. L., Winer A. M. and Pitts Jr. J. N. (1981) *J. Air Pollution Contr. Assoc.*, **31**, 1090.

25 Atkinson R., Aschmann S. M., Carter W. P. L., Winer A. M. and Pitts Jr. J. N. (1982) *Int. J. Chem. Kinet.*, **14**, 781.

26 Atkinson R., Aschmann S. M., Carter W. P. L. and Pitts Jr. J. N. (1982) *Int. J. Chem. Kinet.*, **14**, 839.

27 Cox R. A. (1975) *Int J. Chem. Kinet.*, Symp. Issue 1, 379.

28 Tuazon E. C., Carter W. P. L., Atkinson R. and Pitts Jr. J. N. (1983) *Int. J. Chem. Kinet.*, **15**, 619.

29 Atkinson R., Plum C. N., Carter W. P. L., Winer A. M. and Pitts Jr. J. N. (1984) *J. Phys. Chem.*, **88**, 1210.

30 De More W. B. (1973) *Science*, **180**, 735.

31 De More W. B. (1979) *J. Phys. Chem.*, **83**, 1113.

32 Chapman S. (1930) A Theory of Upper Atmospheric Ozone, *Memoirs Roy. Meteorological Soc.*, **3**, 103.

33 Hunt B. G. (1966) *J. Geophys. Res.*, **71**, 1385.

34 Crutzen P. J. (1970) *Quart. J. Roy. Meteorological Soc.*, **96**, 320.

35 Stolarski R. S. and Cicerone R. J. (1974) *Can. J. Chem.*, **52**, 1610; Wofsy S. C. and McElroy M. B. (1974) *Can. J. Chem.*, **52**, 1582.

36 Brasseur G. and Solomon S. (1984) *Aeronomy of the Middle Atmosphere*, D. Reidel, Dordrecht.

37 Wayne R. P. (1985) *Chemistry of Atmospheres*, Clarendon Press, Oxford.

38 Crutzen P. J. (1974) *Tellus*, **26**, 47.

39 NASA panel for evaluation (1985) DeMore W. B., Margitan J. J., Molina M. J., Watson R. T., Golden D. M., Hampson R. F., Kwylo M. J., Howard C. J. and Ravishankara A. R., *Chemical Kinetics and Photochemical Data for use in Stratospheric Modelling, Evaluation No. 7*, JPL Publication, 85-37, California.

40 Baulch D. L., Cox R. A., Hampson Jr. R. F., Kerr J. A., Troe J. and Watson R. T. (1980) *J. Phys. Chem. Ref. Data*, **9**, 295; (1984), **13**, 1259.

41 Atkinson R. and Lloyd A. C. (1984) Evaluation of Kinetic and Mechanistic Data for Modelling of Photochemical Smog, *J. Phys. Chem. Ref. Data*, **13**, 315.

C1.8 Problems

Atmospheric chemistry

C1.1. Figure C1.7 shows concentration time profiles for disappearance of reactants and the formation of NO, NO_2, PAN (peroxyacetyl nitrate) and formaldehyde in the near UV photolysis of $HONO$–CH_3CHO–Air mixtures containing initially 4 ppm each of HONO and CH_3CHO and 0.4 ppm each of NO and NO_2.

The other products formed in the system were CO, CO_2, methyl nitrate, H_2O and traces of CH_3ONO. The yields after 15 min irradiation are given in Table C1.3 and the time dependence of the principal reactants is given in Table C1.4.

Table C1.3. Reactant consumption and product yields in 360 nm photolysis of CH_3CHO–HONO–Air mixture

Photolysis time 15 min	molecule cm^{-3}
ΔHONO	-5.73×10^{13}
ΔCH_3CHO	-3.84×10^{13}
$\Delta(NO + NO_2)$	2.79×10^{13}
CO_2	1.91×10^{13}
PAN	1.94×10^{13}
HCHO	1.59×10^{13}
CO	0.24×10^{13}
CH_3ONO_2	0.08×10^{13}

1 ppm $= 2.46 \times 10^{13}$ molecule cm^{-3} at   1 atm, 300 K.

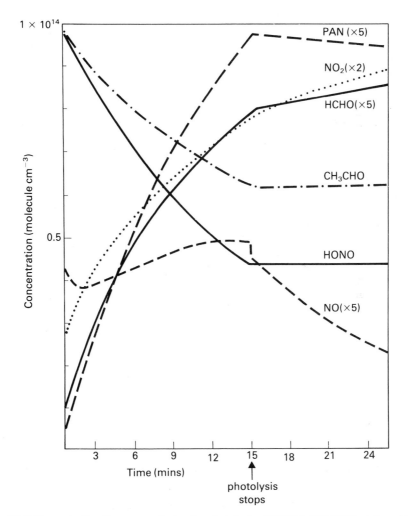

Fig. C1.7. Concentration time curves during the photolysis of HONO–CH_3CHO mixtures in air.

Table C1.4. Time dependence of nitrous acid and acetaldehyde in 360 nm photolysis of CH$_3$CHO–HONO–Air mixture

Time (min)	[HONO] (molecules cm^{-3} × 10^{-14})	[CH$_3$CHO]
0	1.0	1.0
3	0.833	0.881
6	0.700	0.793
9	0.541	0.722
12	0.501	0.644
15	0.427	0.616

The behaviour of the system on cessation of photolysis is characterised by rapid loss of NO, slow loss of PAN and an increase in other products.

The first-order rate constants for photolysis of HONO and NO$_2$ were $7.5 \times 10^{-4} s^{-1}$ and $2.5 \times 10^{-3} s^{-1}$, respectively. CH$_3$CHO photolysis can be neglected.

Devise a mechanism describing the observations. How could one determine the relative rate coefficients for the reactions of OH with HONO and CH$_3$CHO.

Describe the time dependence of ozone during photolysis.

C1.2. Outline the requirements, the advantages and the disadvantages of the competitive rate method for the determination of OH reaction rate constants with organic compounds using the 'atmospheric photo-oxidation' technique. Compare the accuracy and precision of this method with direct techniques.

Devise a possible method for determining rates of peroxyactyl reactions with organic substrates. What would be the possible sources of error in the methods adopted.

C1.3. When mixtures of Cl$_2$ and O$_3$ are irradiated with near UV light (350 nm), ozone is decomposed to O$_2$, with little or no overall change in Cl$_2$ concentration, provided concentrations are kept low. At high concentrations of reactants, Cl$_2$O$_6$ and Cl$_2$O$_7$ can be formed as products.

The time dependence of several species in a typical experiment is shown in Fig. C1.8. The kinetics of the decay of ozone are zero order and follow the rate law:

$$-d[O_3]/dt = \phi k_a[Cl_2]$$

over a wide range of concentrations, where k_a = photodissociation rate of Cl$_2$. Quantum yields, ϕ, are given in Table C1.5. These were obtained in the presence of 1 atm O$_2$, and they increase with temperature. In N$_2$ at room temperature, the value is slightly higher.

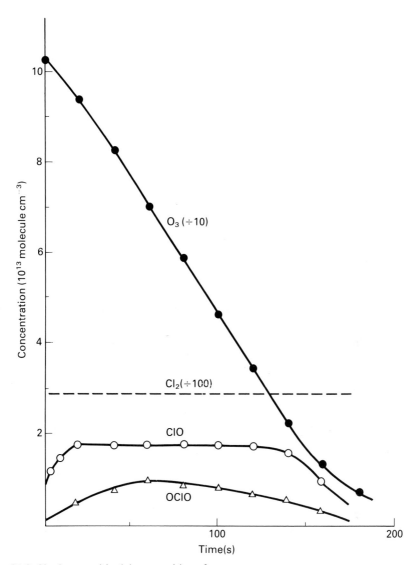

Fig. C1.8. Cl_2 photosensitised decomposition of ozone.

Intermediates which have been detected in this reaction are ClO and OClO. Cl, ClOO and O_2 ($^1\Delta$ or $^1\Sigma$) are also expected to be present. The time dependence of O_3, ClO and OClO is shown in Fig. C1.8. During photolysis ClO rises to a steady state value which is proportional to $(k_a[Cl_2])^{1/2}$ and which remains more or less constant during the decay of ozone. OClO rises to a maximum value, which only shows a very weak dependence on $k_a[Cl_2]$ but increases with increasing initial O_3 concentration.

Devise a mechanism for the Cl_2 photosensitised decomposition of ozone.

What is the time dependence of [Cl]? Derive an expression for the quantum yield for ozone removal in terms of the rate coefficients for the reactions of ClO with itself. (Neglect the minor route forming OClO). Comment on the origin of the temperature dependence of ϕ.

What are the possible origins of O_2^* and indicate possible reasons for a higher quantum yield in N_2?

What would be the effect of using radiation at 250 or 600 nm. How does this sytem differ from the effect of chlorine radicals on stratospheric ozone?

Table C1.5. Quantum yields for Cl_2 photosensitised decomposition of ozone

In 1 atm O_2

T/K	$\phi*(O_3)$
278.2	3.04
287.2	3.20
297.4	3.52
306.9	3.75
318.3	4.08

In 1 atm N_2

297.0	6.0

$*\phi = -d[O_3]/dt/k_a[Cl_2]$.

C1.4. Table C1.6 gives a simplified reaction scheme describing ozone photochemistry and methane oxidation in the background troposphere. Rate data for these reactions at 1 atm pressure and 300 K are provided, together with photodissociation coefficients for a solar zenith of 45°.

Derive an expression for the steady state concentration of OH in the lower atmosphere and evaluate [OH] for these conditions and with the following concentrations (molecule cm^{-3}) of the background trace gases:

$[O_3]$ $= 1.0 \times 10^{12}$

$[H_2O]$ $= 3.2 \times 10^{17}$

$[HCHO] = 6.0 \times 10^9$

$[CO]$ $= 2.5 \times 10^{12}$

$[CH_4]$ $= 4.1 \times 10^{13}$

$[NO_2]$ $= 1.7 \times 10^9$

$[NO]$ $= 0.8 \times 10^9$

Suggest the effects of altitude, season, clouds, CO and NO_x concentrations on [OH].

Could this system, or any parts of it, be simulated in the laboratory? How and what information could be gained?

Table C1.6. Mechanism for background O_3/NO_x chemistry in the troposphere

Reaction	k (300 K)/cm^3 molecule^{-1} s^{-1}
Source of OH	
$O_3 + hv \rightarrow O_2 + O(^1D)$	$J_1 = 5 \times 10^{-6} (s^{-1})$
$O(^1D) + H_2 \rightarrow OH + OH$	$k_2 = 2.8 \times 10^{-10}$
$O(^1D) + M \rightarrow O(^3P) + M$	$k_3 = 4.0 \times 10^{-11}$
$O(^3P) + O_2 + M \rightarrow O_3 + M$	$k_4 = 5.6 \times 10^{-34} [M]^*$
Source of HO$_2$	
$HCHO + hv \rightarrow H + HCO$	$J_5 = 2.0 \times 10^{-5} (s^{-1})$
$H + O_2 + M \rightarrow HO_2 + M$	$k_6 = 5.7 \times 10^{-32} [M]^*$
$HCO + O_2 \rightarrow HO_2 + CO$	$k_7 = 5.1 \times 10^{-12}$
Conversion of OH to HO$_2$	
$OH + CO \rightarrow CO_2 + H$	$k_8 = 2.2 \times 10^{-13}$
$OH + CH_4 \rightarrow CH_3 + H_2O$	$k_9 = 7.9 \times 10^{-15}$
$CH_3 + O_2(+M) \rightarrow CH_3O_2$	$k_{10} = 2.0 \times 10^{-12}$
$CH_3O_2 + NO \rightarrow CH_3O + NO_2$	$k_{11} = 7.6 \times 10^{-12}$
$CH_3O + O_2 \rightarrow HCHO + HO_2$	$k_{12} = 1.3 \times 10^{-15}$
Conversion of HO$_2$ to OH	
$HO_2 + NO \rightarrow NO_2 + OH$	$k_{13} = 8.3 \times 10^{-12}$
$HO_2 + O_3 \rightarrow 2O_2 + OH$	$k_{14} = 2.0 \times 10^{-15}$
Removal of HO$_2$ and OH	
$HO_2 + HO_2 \rightarrow H_2O_2 + O_2$	$k_{15} = 2.5 \times 10^{-12}$
$OH + NO_2(+M) \rightarrow HNO_3$	$k_{16} = 1.1 \times 10^{-11}$

$^*[M] = 2.46 \times 10^{19}$ molecule cm^{-3} (1 atm pressure, 300 K).

C1.9 Answers to problems

Atmospheric chemistry

C1.1. *Photolysis of HONO–CH$_3$CHO–Air mixtures.*
   The following mechanism can be deduced from the observed products:

Removal of HONO: $HONO + hv \rightarrow OH + NO$ (1)

$OH + HONO \rightarrow H_2O + NO_2.$ (2)(forms water)

Removal of CH$_3$CHO: $OH + CH_3CHO \rightarrow H_2O + CH_3CO.$ (3)(forms water)

Formation of NO is given by the difference between the rates of reactions 1 and 4:

$RO_2 + NO \rightarrow RO + NO_2$ (4)

$(R = CH_3COO_2).$

Formation of NO$_2$ results from reactions 2 and 4. Removal of NO$_2$ by reaction 5:

$OH + NO_2 \rightarrow HONO_2$ (5)

to form nitric acid and by the reaction sequence 6, 7, leading to formation of PAN:

$$CH_3CO + O_2 \rightarrow CH_3COO_2 \qquad (6)$$

$$CH_3COO_2 + NO_2 \rightleftharpoons PAN \qquad (7) \text{ (reaction is reversible)}$$

Formation of HCHO and CO_2 results from the following sequence in which OH is regenerated:

$$CH_3COO_2 + NO \rightarrow CH_3 + \underline{CO_2} + NO_2 \qquad (4)$$

$$CH_3 + O_2 \rightarrow CH_3O_2 \qquad (8)$$

$$CH_3O_2 + NO \rightarrow CH_3O + NO_2 \qquad (9) \ NO \rightarrow NO_2$$

$$CH_3O + O_2 \rightarrow \underline{HCHO} + HO_2 \qquad (10)$$

$$HO_2 + NO \rightarrow NO_2 + OH \qquad (11) \ NO \rightarrow NO_2$$

CO is formed by oxidation of HCHO:

$$OH + HCHO \rightarrow H_2O + HCO \qquad (12) \text{ forms water}$$

$$HCO + O_2 \rightarrow HO_2 + CO \qquad (13)$$

CH_3ONO_2 formation: $CH_3O + NO_2 \rightarrow CH_3ONO_2$ (14) competes with (10)

CH_3ONO formation: $CH_3O + NO \rightarrow CH_3ONO$ (15) competes with (10)

Dark behaviour is due to the reversibility of reaction 7 followed by reactions 4 + 8 to 15 removing NO (3 molecules).

Relative rates of OH + HONO and OH + CH$_3$CHO

$$-\frac{d(HONO)}{dt} = k_1[HONO] + k_2[HONO][OH]$$

$$\therefore \ -\int_{t_0}^{t} d\ln[HONO] \, dt = \int_{t_0}^{t} k_1 \, dt + k_2 \int_{t_0}^{t} [OH] \, dt$$

$$\ln \frac{[HONO]_0}{[HONO]_t} = k_1(t - t_0) + k_2 \int_{t_0}^{t} [OH] \, dt$$

$$-\frac{d[CH_3CHO]}{dt} = k_3[CH_3CHO][OH]$$

$$\therefore \ \ln \frac{[CH_3CHO]_0}{[CH_3CHO]_t} = k_3 \int_{t_0}^{t} [OH] \, dt.$$

Since the OH term is common

$$\frac{1}{(t-t_0)}\ln\frac{[\text{HONO}]_0}{[\text{HONO}]_t}=k_1+\frac{k_2}{k_3}\cdot\frac{1}{(t-t_0)}\ln\frac{[\text{CH}_3\text{CHO}]_0}{[\text{CH}_3\text{CHO}]_t}. \tag{i}$$

From Table C1.4

i	[HONO] $\times 10^{-14}$	$\ln\dfrac{C_0}{C}$	$\left\{\dfrac{\ln(C_0/C)}{(t-t_0)}\right\}$ min^{-1}	[CH$_3$CHO] $\times 10^{-14}$	$\ln\dfrac{C_0}{C}$	$\left\{\dfrac{\ln(C_0/C)}{(t-t_0)}\right\}$ min^{-1}
0	1.000			1.000		
3	0.833	0.1827	0.0609	0.881	0.1267	0.0422
6	0.700	0.3567	0.0595	0.793	0.2319	0.0386
9	0.591	0.5259	0.0584	0.722	0.3257	0.0362
12	0.501	0.6911	0.0576	0.664	0.4095	0.0341
15	0.427	0.8510	0.0567	0.616	0.4845	0.0323

Plot as equation i. Slope $=k_2/k_3$ and intercept $=k_1$ (expected values are: $k_2/k_3=0.66\times 10^{-11}/1.6\times 10^{-11}=0.41$; $k_1=7.5\times 10^{-4}\,\text{s}^{-1}$ or $0.045\,\text{min}^{-1}$)

Ozone during photolysis

Ozone is controlled by the reactions

$$\text{NO}_2+h\nu\rightarrow\text{NO}+\text{O}\qquad J_{\text{NO}_2}=2.5\times 10^{-3}\,\text{s}^{-1}$$

$$\text{O}+\text{O}_2(+\text{M})\rightarrow\text{O}_3\qquad \text{(fast)}$$

$$\text{NO}+\text{O}_3\rightarrow\text{NO}_2+\text{O}_2\qquad k=1.8\times 10^{-14}\,\text{cm}^3\,\text{molecule}^{-1}\,\text{s}^{-1}.$$

Photostationary state exists i.e.

$$[\text{O}_3]=\frac{J_{\text{NO}_2}[\text{NO}_2]}{k_{\text{O}_3+\text{NO}}[\text{NO}]}$$

$$\therefore\ [\text{O}_3]=1.39\times 10^{11}\frac{[\text{NO}_2]}{[\text{NO}]}.$$

The ratio $[\text{NO}_2]/[\text{NO}]$ can be estimated from Fig. C1.7

Initially $[\text{NO}_2]/[\text{NO}]=0.65$ $[\text{O}_3]=0.9\times 10^{11}$ 3.62 ppb

Finally $[\text{NO}_2]/[\text{NO}]=1.62$ $[\text{O}_3]=2.25\times 10^{11}$ 9.03 ppb

Dark $J_{\text{NO}_2}=0$ $[\text{O}_3]\rightarrow 0$ since NO in excess.

Therefore the concentration of O_3 is expected to rise initially to a local steady state of ~ 3.6 ppb which then increases steadily to 9 ppb. On cessation of photolysis $[\text{O}_3]$ will fall quite rapidly to zero.

C1.2. *Competitive rate method for determination of relative OH rate constants.*

(*i*) *Requirements*

1. Precise analytical method for relative changes in concentrations of reference and substrate, free from systematic error.
2. Substrate must be stable with respect to wall loss on the time scale of the experiments.
3. Chemical system must contain only one significant attacking reagent for both substrate and reference compound.

(*ii*) *Advantages*

1. Relatively simple technology (for ambient *P* and *T* measurements).
2. Provides data relevant to atmospheric conditions.
3. No absolute calibration of analytical measurements required.
4. Reproducibility is high and can be checked.
5. Impurities are not a great problem.

(*iii*) *Disadvantages*

1. Possibility of systematic error due to unknown chemistry.
2. Requirement of accurately known reference *k*.
3. Limitations imposed by radical sources and magnitude of radical concentration (e.g. for OH the minimum *k* is $\sim 10^{-13} \, cm^3 \, molecule^{-1} \, s^{-1}$).
4. Limited pressure and temperature range can be accessed.
5. Subject to wall reactions.

(*iv*) *Accuracy depends on*:   accuracy of reference reaction,
absence of systematic errors in chemistry,
magnitude of *k* relative to other removal mechanisms.
   Precision depends on: reproducibility of analytical measurement,
sampling reproducibility/artefacts,
radical concentration in steady state, decay rate of substrate.

(*v*) *Peroxyacetyl reactions*

Use thermal decomposition of $CH_3COO_2NO_2$ (PAN) in N_2 or air at 1 atm pressure. The rate constants can be measured relative to either K^* ($k_{ref} = k_r$) and rate of PAN decay (requires absolute measurement of concentration), or from relative decay rate of 2 substrates in the presence of PAN; the latter assumes CH_3CO_3 only attacking reagent. (Equations are similar to those given in text for HO_2 rate constants using thermal decomposition of HO_2NO_2.)

C1.3. *Mechanism of Cl_2 photosensitised decomposition of ozone.*

The following mechanism can explain the observed kinetics products:

Photodissociation of Cl_2: $Cl_2 + h\nu$ ($\lambda = 350$ nm) $\rightarrow Cl + Cl$

$$Cl \text{ atom source rate} = k_a[Cl_2]$$

Removal of O_3 and formation of ClO:

$$Cl + O_3 \rightarrow ClO + O_2. \tag{1}$$

Square root dependence of ClO and lack of loss of Cl_2 overall indicates:

$$ClO + ClO \rightarrow Cl_2 + O_2 \tag{2}$$

$\phi > 2$ indicates Cl is regenerated e.g.,

$$ClO + ClO \rightarrow ClOO + Cl \tag{3}$$

(n.b. $ClO + O_3 \rightarrow Cl + 2O_2$ would give a non-zero-order kinetic decay of O_3). $ClOO$ is unstable and in equilibrium with Cl

$$ClOO \rightleftharpoons Cl + O_2. \tag{4}$$

The above reactions provide a basic mechanism consistent with the kinetics.

Formation of $OClO$

$$ClO + ClO \rightleftharpoons OClO + Cl \tag{5, -5}$$

[minor route for $ClO + ClO$ reaction at low temperature. Maximum and $[O_3]$ dependence of $[OClO]_{max}$ can be explained by $OClO$ loss by (-5)].

At high Cl_2 and O_3 concentrations higher oxides of chlorine may be formed:

$$OClO + O_3 \rightarrow ClO_3 + O_2 \tag{6}$$

$$2ClO_3 \rightarrow Cl_2O_6 \text{ etc.} \tag{7}$$

Electronically excited O_2 can be produced in reactions 1 ($\Delta H^\ominus = -162$ kJ mol^{-1}) and 2 ($\Delta H^\ominus = -204$ kJ mol^{-1}) but not in reaction 4 ($\Delta H^\ominus = +32$ kJ mol^{-1}) when $O_2(^3\Sigma)$ must be formed. ($\Delta H_f^\ominus[O_2(^1\Delta)] = 94.3$ kJ mol^{-1}, $\Delta H_f^\ominus[O_2(^1\Sigma)] = 156.9$ kJ mol^{-1}).

Time dependence of $[Cl]$—increases with time to maintain constant rate until reactions other than reaction 1 remove Cl.

Quantum Yield Expression—We consider the initial photolysis and reactions $1 \rightarrow 4$ only. For steady state of $[Cl]$, $[ClOO]$ and $[ClO]$ and if α is the fraction of ClO radicals which regenerate Cl we have:

$$Cl_2 \rightarrow 2Cl \xrightarrow{-2O_3} 2ClO \longrightarrow 2\alpha Cl \xrightarrow{-2\alpha O_3} 2\alpha ClO \longrightarrow \alpha(2\alpha Cl) \xrightarrow{-2\alpha^2 O_3} \text{etc.}$$
$$\qquad\qquad\downarrow \qquad\qquad\qquad\qquad\qquad \downarrow$$
$$\qquad (1-\alpha)Cl_2 \qquad\qquad\quad 2\alpha(1-\alpha)Cl_2$$

$\therefore$ Total amount of O_3 removed per photon absorbed $= 2(1 + \alpha + \alpha^2 \dots)$. This is a geometric progression the sum of which, for $\alpha < 1$, is given by

$2(1 - \alpha^n)/(1 - \alpha)$.

If $n = \infty$, $\alpha^n = 0$

$$\therefore \phi = \frac{2}{1 - \alpha}.$$

If we neglect the OClO forming route:

$$\alpha = \frac{k_3}{k_2 + k_3}; \quad \frac{k_3}{k_2} = (\alpha^{-1} - 1)^{-1}.$$

From the data in Table C1.5

T/k	α	k_3/k_2
278.2	0.342	0.520
287.2	0.375	0.600
297.4	0.432	0.761
306.9	0.467	0.875
318.3	0.510	1.041

Atom route increases relative to Cl_2 formation route as T increases. (Could plot Arrhenius plot and get $E_3 - E_2$ and A_3/A_2.)

Origins of O_2^*   See above, the involvement of $O_2(^1\Delta)$ or $O_2(^1\Sigma)$ can explain the effect of N_2, O_2 on ϕ.

Higher ϕ in N_2   In O_2, $O_2^* + O_3 \rightarrow 2O_2 + O(^2P)$

$O(^3P) + O_2 \rightarrow O_3$.   no ozone loss

In N_2, $O(^3P) + O_3 \rightarrow 2O_2$

$+ Cl_2 \rightarrow Cl + ClO$ leads to more ozone loss

$+ ClO \rightarrow Cl + O_2$.

Radiation at 250 nm   No Cl_2 photolysis; O_3 photolysis $\rightarrow O(^1D) + O_2(^1\Delta)$. In O_2 there is little decay due to $O(^1D)$ quenching to $O(^3P)$ and scavenging of $O(^3P)$ by O_2 form O_3 again. In N_2 removal of O_3 as above.

at 600 nm   Very weak Cl_2 photolysis—very slow O_3 decay. O_3 absorption produces $O(^3P)$ only. In O_2 there is no decay.

Stratospheric ozone—effect of Cl and ClO.

In the stratosphere the inefficient Cl atom recycling by the slow reaction 3:

$ClO + ClO \rightarrow ClOO + Cl$ (3)

is replaced by the rapid Cl recycling reactions:

$O + ClO \rightarrow Cl + O_2$

$NO + ClO \rightarrow Cl + NO_2$.

The former leads to catalytic removal of ozone by fast cycles, with a long kinetic chain length. The reaction with NO is a null reaction since O_3 is regenerated by photolysis of NO_2 to give $O + NO$, followed by $O(^3P)$ reaction with O_2.

C1.4. Tropospheric photochemistry

(a) Steady state concentration of OH

Equating sources and sinks of odd hydrogen radicals we obtain:

$$2J_5[HCHO] + 2J_1\left\{\frac{k_2[H_2O]}{k_2[H_2O] + k_3[M]}\right\}[O_3] = k_{16}[OH][NO_2] + 2k_{15}[HO_2]^2.$$

Equating interconversion terms we get:

$$(k_8[CO] + k_9[CH_4])[OH] = (k_{13}[NO] + k_{14}[O_3])[HO_2]$$

which gives

$$[HO_2]^2 = [OH]^2\left\{\frac{k_8[CO] + k_9[CH_4]}{k_{13}[NO] + k_{14}[O_3]}\right\}^2.$$

The expression for OH is given by the roots of the quadratic equation:

$$a(OH)^2 + b(OH) - c = 0$$

where $a = 2k_{15}\left\{\dfrac{k_8[CO] + k_9[CH_4]}{k_{13}[NO] + k_{14}[O_3]}\right\}^2$

$b = k_{16}[NO_2]$

$c = 2J_5[HCHO] + 2J_1\left\{\dfrac{k_2[H_2O]}{k_2[H_2O] + k_3[M]}\right\}[O_3].$

i.e. $[OH] = \{-b \pm (b^2 + 4ac)^{1/2}\}/2a$, giving $[OH] = 4.4 \times 10^6$ molecule cm^{-3}

at low NO_x this approximates to $[OH] = \left(\dfrac{c}{a}\right)^{1/2}$ giving 4.6×10^6 molecule cm^{-3}

$\therefore$ OH will increase with the square root of the production term, c, which is a linear function of light intensity, $[H_2O]$, $[O_3]$. Hence the effects of atmospheric variables follow.

OH will decrease with the square root of the removal term, a, which is a complex function of temperature (via k_{15}), CO (increases a), NO (decreases a).

(b) Laboratory simulation

This is a complex coupled system involving O_x, HO_x, NO_x and CH_xO_y. At the concentrations present in the atmosphere the interconversion reactions are too slow to conduct in an enclosed system without surface interference.

Certain aspects of the system, suitably selected, might be simulated, e.g.

O_x, HO_x system might be simulated in O_3–H_2O photolysis at $\lambda < 310$ nm. Could obtain information on $HO_2 + HO_2$, $OH + HO_2$ and $HO_2 + O_3$.

O_x, NO_x system might be simulated to check photostationary states etc, and rates of $NO + O_3$, $NO_2 + O_3$ etc.

O_x, CH_4 system might be investigated as O_3–CH_4–O_2 photolysis at $\lambda < 310$ nm.

Chapter C2
Hydrocarbon Oxidation
Mechanisms in Combustion Systems

R. A. COX

C2.1 Introduction

The complete oxidation of hydrocarbons to CO_2 and H_2O represents one of the highest exothermicities in the field of chemistry:

$$> CH_2(l) + 1.5 O_2(g) \rightarrow CO_2(g) + H_2O(l) + 653 \, kJ \, mol^{-1}.$$

The convenience of hydrocarbons as an energy source, and the fact that partial oxidation of hydrocarbons leads to a variety of molecules used as building blocks for a huge range of useful chemicals, means that modern society probably depends more on the combustion of hydrocarbons than on any other chemical reaction. The various applications have been described by Cullis[1] in terms of a typical temperature–pressure ignition diagram for a hydrocarbon–O_2 (or air) mixture as shown in Fig. C2.1. In the present discussion we describe what is known about the chemical mechanism of hydrocarbon oxidation as we follow the locus of a line of increasing temperature from the slow combustion region through the cool flame peninsula to high temperature ignition.

C2.2 Low temperature oxidation

(i) Slow reaction

The principal features of the mechanism of the slow oxidation of hydrocarbons are best appreciated by turning first to another system, the slow combustion of CH_3CHO, which occurs at temperatures between 400–450 K.[2] The major product is peracetic acid formed with the overall stoichiometry:

$$CH_3CHO + O_2 \rightarrow CH_3CO_3H \quad \Delta H = -102.6 \, kJ \, mol^{-1}. \tag{1}$$

The reaction is difficult to study experimentally because of heterogeneous effects but most of the observations can be explained in terms of peracetic acid formation via a gas phase chain reaction involving the following reactions

$$CH_3CHO + O_2 \rightarrow CH_3CO + HO_2 \qquad \text{Initiation} \tag{2}$$

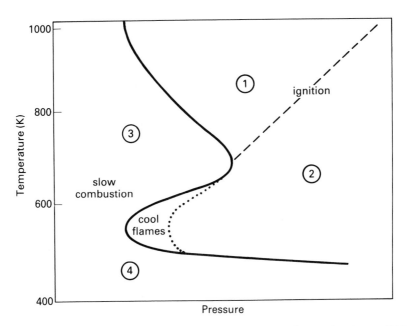

Fig. C2.1. Schematic ignition diagram for a hydrocarbon–oxygen mixture, showing applications of hydrocarbon oxidation. (1) Conversion by rapid combustion of chemical energy, e.g. Turbojet, Furnace. (2) Conversion, by ignition, of chemical energy to heat and mechanical energy, e.g. internal combustion engine. (3) Intermediate temperature oxidation to useful chemicals, e.g. O-heterocyclics. (4) Controlled conversion to useful chemicals, e.g. alcohols, carbonyls, peroxides. Adapted from reference 1.

$$CH_3CO + O_2 \rightarrow CH_3COO_2 \qquad\qquad \text{Propagation} \quad (3)$$

$$CH_3COO_2 + CH_3CHO \rightarrow CH_3CO_3H + CH_3CO \qquad\qquad (4)$$

$$CH_3COO_2 + HO_2 \rightarrow \text{products (homogeneous)} \qquad\qquad (5)$$

$$CH_3COO_2 \rightarrow \text{wall} \quad \text{(heterogeneous)} \qquad\qquad (6)$$

Termination

At the upper end of the temperature range, the rate of acetaldehyde oxidation shows autocatalytic behaviour, i.e. acceleration of the reaction rate, reaching a maximum value at ~ 30–50% reaction. This is due to the instability of the peracid which decomposes to produce CH_3 and OH which, in turn, react rapidly to regenerate acetyl radicals:

$$CH_3CO_3H \rightarrow CH_3 + CO_2 + OH \qquad\qquad (7)$$

$$OH + CH_3CHO \rightarrow H_2O + CH_3CO \qquad\qquad (8)$$

$$CH_3 + O_2 + M \rightarrow CH_3O_2 + M \qquad\qquad (9)$$

$$CH_3O_2 + CH_3CHO \rightarrow CH_3OOH + CH_3CO. \qquad\qquad (10)$$

The process of formation and decomposition of the peracid leads to an increase in the number of radical chain centres and constitutes a branching sequence. Because branching involves a stable product that builds up, it is termed *degenerate branching*. The theory of degenerately branched chain reactions, of which hydrocarbon oxidation at low temperatures is the best known example, was developed in full by Semenov.[3]

Alkanes do not oxidise readily at low temperatures around 400 K. The origin of the kinetic differences between simple alkanes and aldehydes arises in the rate determining propagation step:

$$RO_2 + RH \rightarrow RO_2H + R. \tag{11}$$

The C—H bond dissociation energy, $D(C—H)$, in alkanes is in the range 384–418 kJ mol^{-1}, leading to much higher activation energies for reaction 11 for alkanes (~ 80 kJ mol^{-1}), than are found for aldehydes (40–60 kJ mol^{-1}) with their less stable C(O)—H bonds. Furthermore autocatalysis requires higher temperatures than for aldehydes, since the degenerate branching agents, the alkylhydroperoxides are more thermally stable than the peracids.

The formation of peroxy species in the reaction with O_2 is exothermic (two new bonds are formed) and, in situations of rapid or accelerating reaction rate, self-heating in the reacting gases occurs when heat loss to the container cannot compensate for the chemical heat release. This feature can lead quickly to an unstable regime when self-ignition occurs. Self ignition of hydrocarbons and other fuels at low temperatures and pressures gives rise to the well-known phenomenon of cool flames which will now be discussed.

(ii) Cool flames

Cool flames, characterised by blue luminosity occurring in the low temperature oxidation of hydrocarbons, have the following distinctive properties:
(a) they are preceded by a relatively long induction period; (b) only 5–10% of the fuel is consumed during the period up to and during the appearance of the flame; (c) there may be multiple occurrences of the cool flame in the same reacting mixture; (d) there are sharp temperature excursions but only of 100–200 K; (e) the reaction rate after passage of the cool flame may be slower than just before.

Many discussions of the chemistry of cool flames have been presented and a recent paper by Benson[4] gives an overall view. Their curious features can be understood in terms of the degenerately branched chain reaction involving hydroperoxides, which we will now consider in terms of oxidation of a general hydrocarbon RH, which is assumed to occur by a chain reaction involving the radicals OH, R (alkyl radical) and RO_2 (peroxy radical):

$$OH + RH \longrightarrow R \longrightarrow RO_2 \longrightarrow products + OH + heat$$

fuel $\longrightarrow$ ROOH + OH + heat

$\longrightarrow$ RO + OH

branching $\longleftarrow$

The overall scheme involves the conversion of hydrocarbon to products by a chain reaction of long kinetic chain length, initiated by a slow generation of radical centres involving reaction of fuel molecules with O_2

$$RH + O_2 \rightarrow R + HO_2. \tag{12}$$

A small fraction of the exothermic propagation steps, the detailed nature of which will be considered later, leads to formation of hydroperoxide branching agents which accumulate during the induction period. The decomposition of the branching agent leads to autocatalysis, acceleration of the reaction rate and consequent self-heating of the gas. The cool flame occurs when a critical concentration of branching agent has accumulated. During cool flames, the temperature rises by as much as 100 K but only small amounts of fuel are consumed before the flame is quenched. The quenching is a consequence of a region of negative temperature dependence of the reaction rate at the higher temperature. The existence of this phenomenon has been demonstrated repeatedly in the slow oxidation of many simple alkanes and other hydrocarbons and organic species. Data for propane[5] are shown in Fig. C2.2 where the negative temperature coefficient is observed between 600 and 650 K. The most satisfactory explanation lies in the change in chemical mechanism of the $R + O_2$ reaction. At temperatures above about 680 K, the addition of O_2 to R becomes reversible and R and RO_2 are equilibrated; the alternative abstraction reaction of O_2 to form the conjugate alkene becomes important, and competes with the main propagation chain

$$R + O_2 \rightleftharpoons RO_2 \tag{13, -13}$$

$$R + O_2 \rightarrow \text{olefin} + HO_2. \tag{14}$$

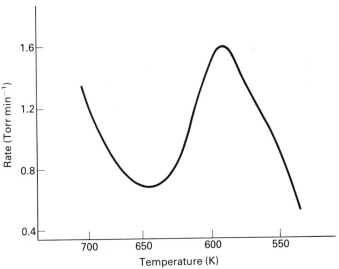

Fig. C2.2. Variation with temperature of the maximum rate of oxidation of propene; total pressure = 160 Torr, with a [Fuel]:[O_2] ratio of 1:2. Adapted from reference 5.

The HO$_2$ radical is much less reactive than OH and cannot carry the propagation chain so rapidly by the reaction

$$HO_2 + RH \rightarrow H_2O_2 + R \tag{15}$$

in competition with termination

$$HO_2 + HO_2 \rightarrow H_2O_2 + O_2. \tag{16}$$

The overall reaction forming H$_2$O$_2$ + alkene is thermoneutral, so the self-heating effect disappears. The chemical route to the hydroperoxide branching agent is cut off and replaced by the formation of H$_2$O$_2$ which is relatively stable towards fission into radicals below 800 K, and consequently cannot act as a branching agent at the lower temperatures. The net result is that the hydroperoxide branching agent, which reaches a critical concentration just prior to the cool flame, is destroyed more rapidly than it is created during the brief time of temperature elevation, and the cool flame is quenched when the hydroperoxide is exhausted. A second cool flame can only propagate when the branching agent has built up again to the critical concentration. This behaviour of peroxide species is illustrated in Fig. C2.3, which shows experimental measurements of the oxidation of isobutane in the cool flame cool regime, as reported by Pollard in his comprehensive review of hydrocarbon oxidation.[6]

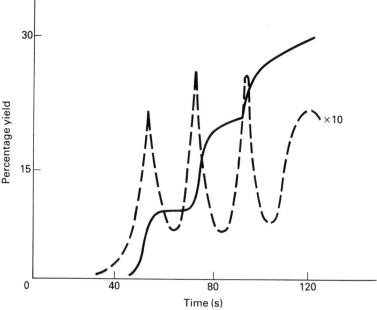

Fig. C2.3. Experimentally determined yields of t-butyl hydroperoxide (------; × 10) and hydrogen peroxide (——) in the cool flame oxidation of isobutane. Initial temperature = 583 K; [i-C$_4$H$_{10}$] : [O$_2$] = 1 : 2; initial isobutane concentration = 1.44 × 10^{-3} mol cm^{-3}. Adapted from reference 6.

(iii) Chemical nature of propagating steps in hydrocarbon oxidation in the cool flame regime

During the last 15 years there has been much discussion concerning the nature of the principal chain-propagating steps in the cool flame oxidation of hydrocarbons. The discussion originally focussed on the seemingly contradictory results when high and low molecular weight hydrocarbons are used as fuels. Specifically it is found that a low molecular weight ($<C_4$) alkane yields mainly its conjugate alkene as the initial product of low temperature oxidation, and oxygenated products formed in the cool flame were believed to result from oxidation of the alkene. Higher alkanes ($\geqslant C_4$) generally form a variety of oxygenated species, including O-heterocyclics as well as alkenes with fewer C atoms, as primary products. The 'alkene' theory proposed by Knox[7] involves reaction 14 as the propagation step at all temperatures with chain propagation via the unspecified overall reaction

$$HO_2 + \text{olefin} \rightarrow \text{carbonyl products} + OH. \tag{17}$$

Heterogeneous reactions were also suggested to explain some of the surface sensitive features of low temperature oxidation of C_2–C_4 alkanes. A better description of chain propagation appears to be offered by the alkylperoxy radical isomerisation theory proposed by Fish,[8] which was developed mainly as a result of studies of the oxidation of alkanes with carbon number >4. The mechanism involves isomerisation of the peroxy radical, formed by addition of O_2 to R, by internal H atom abstraction:

followed by cleavage of the ROOH radical in a variety of different ways to form OH and oxygenated products, e.g. ring closure

2,4 disubstituted oxetan

or β scission

At low temperatures, the oxidation of the hydroperoxy alkyl radical may occur in competition with the unimolecular cleavage reaction, as has been shown from the formation of certain characteristic products:

Note that hydroperoxides can be formed via this route, providing branching agents.

The rate constants for some of the unimolecular isomerisation and decomposition reactions have been estimated by thermochemical kinetic techniques.[8,9] The rate constants are highly dependent on structure; the ring size in the transition state for internal H-atom abstraction and the nature of the H-atom abstracted (primary, secondary or tertiary) influence both the Arrhenius pre-exponential factor and the activation energy. For a 1,5-transfer of a secondary H atom, $A \simeq 1 \times 10^{12}\,\text{s}^{-1}$ and $E \simeq 100\,\text{kJ}\,\text{mol}^{-1}$ giving a lifetime for RO_2 with respect to isomerisation of $\sim 10^{-5}\,\text{s}$ at 750 K. However the isomerisation of RO_2 is reversible since the reaction is endothermic by $30\,\text{kJ}\,\text{mol}^{-1}$ and hence the reverse reaction will have a lower activation barrier. Thus RO_2 and ROOH will be equilibrated and the overall propagation rate will depend on the rates of decomposition of ROOH to give OH and other competitive processes involving RO_2 and ROOH.

The intramolecular processes are generally slower when transition states with small ring size and mainly primary H atoms are involved, as is the case for ethane, propane and butane. As a result the unimolecular propagation steps are slower for these alkanes and alternative radical reactions occur which may terminate chains or form unreactive radicals such as HO_2. In addition propagation via radical + radical reactions, e.g. reaction 18:

$$RO_2 + RO_2 \rightarrow 2RO + O_2 \tag{18}$$

is now believed to be important in the oxidation of the lower alkanes.

The mechanism therefore depends on the nature of the alkyl peroxy radical, and can be summarised by the scheme which is shown below. The predominant reaction path will depend on the relative values of the equilibrium constants K_{13}, K_{19} and K_{20} and the rate constants $k_{14}, k_{18}, k_{11}, k_{21}$ and k_{22}. Generally it can be said that k_{11} is slow for all R groups unless labile H's (e.g. aldehydes) are present. Reaction 18 is only rapid if R is small and unbranched, whilst very limited measurements of k_{14}, k_{13} and K_{13} are available for only a few radicals. K_{19}, K_{20}, k_{21} and k_{22} can only be estimated, since no experimental measurements are available, so all that can be said at the present time is that this is a plausible picture, but much

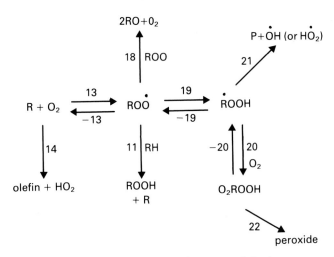

work is needed to establish unequivocally the rates of the important elementary steps and to put the model on a quantitative basis.

C2.3 Oxidation mechanisms at higher temperatures

At temperatures above ~ 700 K the equilibrium $(13, -13)$ shifts so that $[R] \gg [RO_2]$ and alternative reactions of the alkyl radical occur. For low molecular weight alkyl radicals reaction 14 with O_2 is favoured. At temperatures above 750 K hydrogen peroxide forms in the subsequent reactions of HO_2 produced in reaction 12, i.e.

$$HO_2 + RH \rightarrow H_2O_2 + R \tag{15}$$

$$HO_2 + HO_2 \rightarrow H_2O_2 + O_2. \tag{16}$$

Hydrogen peroxide can decompose at these higher temperatures:

$$H_2O_2 + M \rightarrow 2OH + M \tag{23}$$

so that, when it is formed in a propagation step, e.g. reaction 15, H_2O_2 constitutes a source of degenerate chain branching. This proposed mechanism cannot be fully tested since the rates and temperature dependence of H-abstraction reactions involving HO_2 are not well known. There is only indirect evidence that chain branching through H_2O_2 is responsible for single stage ignition at higher temperatures.

For larger alkyl radicals, decomposition of the radicals becomes important at temperatures around 770 K

$$R \rightarrow olefin + R' \tag{24}$$

$$R \rightarrow olefin + H \tag{25}$$

The significance of this process is that it causes a reduction in carbon number so that large radicals, capable of fast intramolecular processes, are replaced by simple alkyl radicals, CH_3 and C_2H_5, which, as we have seen, are oxidised by a different mechanism. For example the 2-methyl-butyl radicals derived from isopentane can decompose as follows

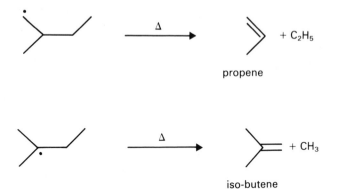

propene

iso-butene

At even higher temperatures, above ~ 950 K, when ignition is generally under-way in a closed system, further changes in mechanism occur. The equilibrium 26 shifts in favour of the HO_2 dissociation products:

$$HO_2 + M \rightleftharpoons H + O_2 + M. \tag{26}$$

and the chain branching step (27)

$$H + O_2 \rightarrow OH + O \tag{27}$$

dominates the rate of combustion.

At these temperatures even the simple alkyl radicals decompose to yield H atoms, e.g.

$$C_2H_5 + M \rightarrow C_2H_4 + H + M. \tag{29}$$

An exception is the methyl radical which, at low temperatures, can only be oxidised via CH_3O_2 or via recombination to give C_2H_6. Above 1000 K, CH_3 can be oxidised in the chain branching step:

$$CH_3 + O_2 \rightarrow HCHO + H + O. \tag{30}$$

C2.4 Major problem areas remaining

It is clear from the above discussion that there is a considerable need for the establishment of a better kinetic data base for those reactions which have emerged as key steps from the large number of investigations of hydrocarbon combustion. Of particular importance are the reactions of HO_2 and RO_2 radicals with alkanes, alkenes and aldehydes, and the reactions of alkyl radicals with O_2. This latter reaction is of particular interest as very recent work seems to point to a common intermediate to both addition and HO_2 formation,[10] implying a common reaction intermediate for the overall $C_nH_{2n+1}+O_2$ system:

$$O_2+C_nH_{2n+1}\leftrightarrow C_nH_{2n}OOH\leftrightarrow C_nH_{2n}+H\dot{O}_2.$$

Absolute rate constants for alkyl radical decomposition are also required. With regard to mechanisms, the outstanding problems lie in the oxidation mechanism of olefins and other unsaturated compounds, and also combustion of aromatics for which the key elements in the mechanism have not been established at the present time.

C2.5 Autoignition of hydrocarbons at high pressure

(i) Modelling of autoignition

The phenomena of 'knock' in petrol engines and compression ignition of diesel fuels involve autoignition of hydrocarbon fuels at high pressure, in which oxidation occurs spontaneously rather than by spark ignition. Models describing this process have been developed, based on knowledge of the chemistry of hydrocarbon oxidation, coupled with consideration of the heat release and loss processes.[11,12]

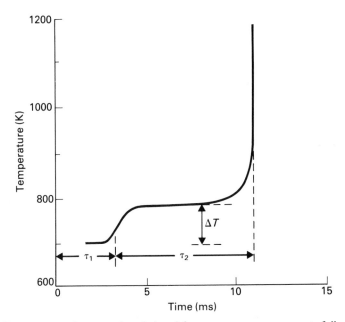

Fig. C2.4. Temperature–time record as deduced from pressure measurements following rapid compression of octane–air mixture in a rapid compression machine. Adapted from Hirst S. L. and Kirsch L. J. (1980) *Combustion Modelling in Reciprocating Engines*, (Ed. by J. N. Muttari and C. A. Amann) Plenum, New York.

The characteristics of autoignition at high pressure have been investigated in a rapid compression machine. A typical example of the pressure–time recording following compression heating to 700 K is shown in Fig. C2.4. At this temperature, two-stage ignition occurs with a clearly defined cool flame pulse followed by a period of very slow rise in temperature before the onset of rapid and complete ignition. The ignition delay times, τ_1 and τ_2, for cool flame and ignition respectively, have been found to show characteristic dependences on the temperature, pressure and composition of the mixtures.[12] Any model of autoignition must be capable of predicting these quantities over a range of conditions.

Due to the high complexity of the hydrocarbon oxidation mechanisms, simplification of the chemistry is required for model formulation. In the 'Shell' model,[11,12] a generalised kinetic scheme, which involves chemical entities representing a variety of individual species undergoing a set of generalised reactions is used. This approach can be justified since different types of fuel molecules exhibit a broadly similar pattern of behaviour and the kinetic behaviour of the organic free radicals involved in the mechanism also shows generality, radicals of the same type exhibiting similar kinetics.

The Shell model contains the following reactions, believed to be the minimum required to describe two-stage autoignition:[12]

$RH + O_2 \rightarrow 2R$	Initiation
$R \rightarrow R + products + heat$	Propagation cycle
$R \rightarrow R + B$	Propagation forming branching agent, B.
$R \rightarrow R + Q$	Propagation forming Q
$R + Q \rightarrow R + B$	Propagation converting Q to B
$B \rightarrow 2R$	Branching
$R \rightarrow products$	Linear termination
$2R \rightarrow products$	Quadratic termination.

It is assumed that all propagating radicals are coupled together according to the steady state hypothesis and rate equations are written in terms of the total radical concentration R. Both linear and quadratic termination reactions involving R were required to describe the first stage (cool flame) ignition. The second stage, 'hot ignition', is caused by the formation of the intermediate, Q, which provides a secondary source of branching agent at the higher temperature following the cool flame. The build up in concentration of Q allows the system to 'break out' of the temperature stabilised region where the reaction rate has a negative temperature coefficient.

Consumption of fuel and O_2 and chemical heat release are related to the propagation rate and are based on an assumed overall reaction stoichiometry; heat loss from the gas is treated in a simple Newtonian convective model, i.e. heat loss $= \phi V(T - T_{wall})$, where T is the gas temperature and T_{wall} that of the vessel wall. With this formulation the time dependence of RH, O_2, R, B, Q and T can be described by five differential equations plus a linear algebraic equation relating fuel and O_2 consumption. The model is capable of simulating the temperature/time profile of a two stage ignition as well as the variation of induction periods with temperature. Furthermore, by expressing the generalised rate coefficients for the reactions in the parameterised form:

$$k_i = A_i \exp(-E_i/RT) \cdot [O_2]^{x_i}[RH]^{y_i}$$

the dependence of induction periods on the concentrations of fuel and O_2 and on the total pressure can be predicted. This parameterisation is important for incor-

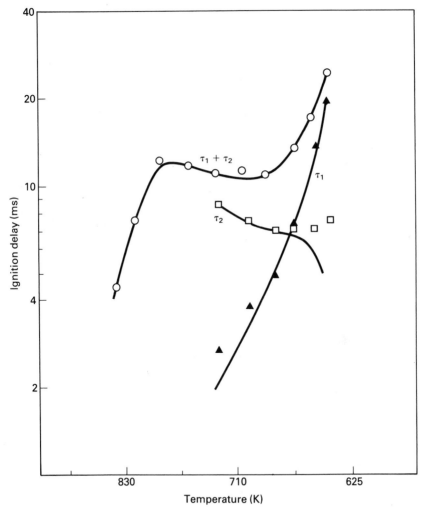

Fig. C2.5. Temperature dependence of the ignition delays for cool flame, τ_1, and for hot ignition, τ_2, in a 90% iso-octane, 10% n-heptane primary reference fuel, 0.9 stoichiometric in air. Points show the experimental data, lines are computed from the 'Shell' model. Adapted from reference 12.

porating the model into an internal combustion engine cycle, since the gas temperature and the concentrations of fuel and oxygen vary with time and conditions of operation.

The parameters in the model were adjusted by a combination of intuition and experimental constraint to fit the observed results in the rapid compression machine over a range of composition, temperature and fuel. Some examples of the fits for ignition delay time vs temperature are shown in Fig. C2.5. The values taken by the parameters, particularly the indices x_i and y_i, provide important pointers to the detailed nature of the chemical mechanism. For a detailed description of this model involving generalised reactions, the student is referred to the papers of Halstead, Kirsch and co-workers.[11,12] It is of interest to note that certain parameters in the model were required to take on values which pointed to the importance of O_2 in determining the competition between linear propagation and branching and of alkyl radical decomposition in linear termination.

(ii) Relationships between structure and autoignition properties

Further discussion of the chemical aspects of autoignition modelling has been recently provided by Cox and Cole (1985).[13] They refined and extended the Shell model using 15 generalised reactions involving five types of radical species, and used rate parameters which were, as far as possible, compatible with the expanding body of kinetic information on the type of elementary processes involved. The model was capable of predicting experimental ignition delay times in the range 700–850 K for hydrocarbon–air mixtures at high pressure.

The mechanism adopted is shown schematically in Fig. C2.6, and its relationship to the basic Shell model may be appreciated. The features of the alkylperoxy radical isomerisation theory (section C2.2), as well as HO_2 and H_2O_2 chemistry, are incorporated in some detail. Where required for best agreement with experimental results, certain rate parameters were altered from those assigned on the basis of representative literature values, which have been obtained mainly from studies of hydrocarbon oxidation at ~ 750 K in H_2–O_2 mixtures (Baldwin et al.[14]). It was not necessary, however, to have recourse to unreasonable changes, guesswork or parameterisation.

It was found that the cool flame induction period at temperatures around 750 K was very sensitive to the fraction of propagation steps involving ROOH which formed branching agents. Since the relative rates of the various intramolecular processes involving alkylperoxy radicals and their hydroperoxide isomers, $\dot{R}OOH$, are known to be highly dependent on chemical structure, this points to a possible rationalisation of fuel performance with respect to autoignition under engine test conditions, at least for alkane fuels.

In subsequent work, models were formulated as above for the primary refer-

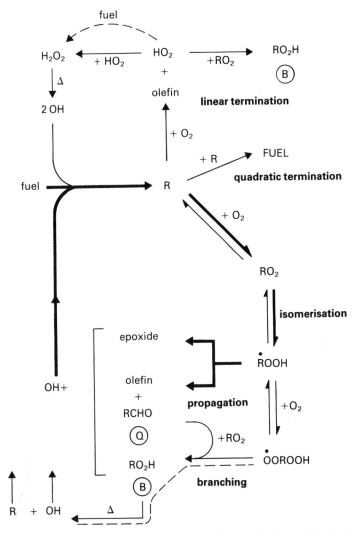

Fig. C2.6. Schematic diagram of the chemical mechanism for autoignition of a C_8-hydrocarbon–air mixture. Adapted from reference 13.

ence fuels iso-octane (100 RON) and n-heptane (0 RON).* Figure C2.7 illustrates the major pathways in the oxidative breakdown of iso-octane at intermediate temperatures, where isomerisation of alkylperoxy radicals is important. 50% of the OH radical attack on the alkane occurs at the primary H-atom sites, yielding a terminal peroxy radical, A. The rates of the subsequent isomerisations involving H-transfer and of the cleavage reactions leading to OH and oxygenated products can be estimated. For radical A, the 1,6 H-transfer dominates with only 5% of all

*Research Octane Number (RON) is used to classify motor fuels according to their propensity to 'knock', as determined in a standard test engine.

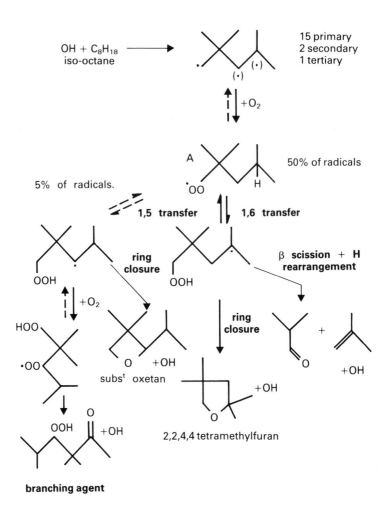

Fig. C2.7. Schematic diagram of the major pathways in the mechanism of the oxidative breakdown of iso-octane at temperatures in the region 700–800 K.

radicals from iso-octane undergoing 1,5-transfer of an H atom at 753 K. Furthermore, the work of Baldwin *et al.*[14] has shown that oxidation of the hydroperoxyalkyl radicals (which in this model is assumed to provide the main route to branching agent) is only competitive with cleavage or ring closure for the radicals formed by 1,5 H-transfer. There is, therefore, a restriction in the fraction of the radicals that can form branching agents in this way and the restriction is a function of radical structure. Examination of the analogous pathways for oxidation of n-heptane at 753 K, shows that up to 26% of the radicals formed by attack of OH on n-heptane can undergo 1,5-transfer of an H atom, and potentially form branching agents. Thus, in accordance with experience, we expect shorter cool flame induction periods and a higher propensity to autoignition in n-heptane, compared with iso-octane, on this basis.

C2.6 Suggestions for further reading

Hydrocarbon oxidation and combustion

Minkoff G. J. and Tipper C. F. H. (1962) *Chemistry of Combustion Reactions*, Butterworths, London.

Walker R. W. (1975) Specialist Periodical Report. A Critical Survey of Rate Constants for Reactions in Gas Phase Hydrocarbon Oxidation. In *Reaction Kinetics*, Vol 1 and Vol 2, Chem. Soc., London.

Bamford C. H. and Tipper C. F. H. (Eds) (1977) *Comprehensive Chemical Kinetics*, Vol 17, *Gas Phase Combustion*, Elsevier Scientific, Amsterdam.

Cool flame chemistry

Benson S. W. (1981) The Kinetics and Thermochemistry of Chemical Oxidation with Application to Combustion and Flames, *Prog. Energy Combust. Sci.*, **7**, 125.

Modelling hydrocarbon combustion

Westbrook C. K. and Dryer F. L. (1984) Chemical Kinetic Modelling of Hydrocarbon Combustion, *Prog. Energy. Combust. Sci.*, **10**, 1.

Halstead M. P., Kirsch L. J., Prothero A. and Quinn C. P. (1975) A Mathematical Model for Hydrocarbon Autoignition at High Pressure, *Proc. R. Soc. London, Ser. A*, **345**, 515.

C2.7 References

1 Cullis C. F. (1967) *Chem. Britain*, **3**, 370.
2 Dixon D. J. and Skirrow G. (1977) Gas Phase Combustion. In *Comprehensive Chemical Kinetics*, Vol 17 (Ed. by C. H. Bamford and C. F. H. Tipper), p. 369, Elsevier Scientific, Amsterdam.
3 Semenov N. N. (1958) *Some Problems of Chemical Kinetics and Reactivity*, Pergamon, London.
4 Benson S. W. (1981) *Prog. Energy Combust. Sci.*, **7**, 125.
5 Seakins N. and Hinshelwood C. N. (1963) *Proc. R. Soc. London, Ser. A*, **276**, 324.
6 Pollard R. T. (1977) Gas Phase Combustion. In *Comprehensive Chemical Kinetics*, Vol. 17, (Ed. by C. H. Bamford and C. F. H. Tipper), p. 249. Elsevier Scientific, Amsterdam.
7 Knox J. H. (1965) *Combust. Flame*, **9**, 297.
8 Fish A. (1968) *ACS Adv. Chem. Ser.*, **76**, 69.
9 Baldwin R. R., Hisam M. W. M. and Walker R. W. (1982) *J. Chem. Soc., Faraday Trans. 1*, **78**, 1615.
10 Slagle I. R. and Gutman D. (1985) *J. Am. Chem. Soc.*, **107**, 5342.
11 Halstead M. P., Kirsch L. J., Prothero A. and Quinn C. P. (1975) *Proc. R. Soc. London, Ser. A*, **346**, 515.
12 Halstead M. P., Kirsch L. J. and Quinn C. P. (1977) *Combust. Flame*, **30**, 45.
13 Cox R. A. and Cole J. A. (1985) *Combust. Flame*, **60**, 109.
14 Baldwin R. R., Bennett J. P. and Walker R. W. (1980) *J. Chem. Soc., Faraday Trans. 1*, **76**, 1075.

C2.8 Problems

Hydrocarbon oxidation mechanisms in combustion systems

C2.1. The following scheme describes the reactions occurring in the oxidation of alkanes in the pre-cool flame region at 600–800 K

$$C_nH_{2n+1} + O_2 \underset{-2}{\overset{2}{\rightleftharpoons}} C_nH_{2n+1}OO$$

$$\xrightarrow[-3]{3} \alpha\text{-}C_nH_{2n}OOH \xrightarrow{5} \begin{array}{l} C_nH_{2n} \\ \text{conjugate alkene} \\ + HO_2 \end{array}$$

$$\xrightarrow[-4]{4} \beta\text{-}C_nH_{2n}OOH \xrightarrow{6a} \text{carbonyl} + \text{alkene} + OH$$

$$\xrightarrow{6b} \text{heterocyclic} + OH$$

$$\downarrow 1$$

$$C_nH_{2n} + HO_2$$

α-isomerisation (reaction 3) involves internal H-abstraction from C atom adjacent to a peroxy group.

β-isomerisation (reaction 4) involves internal H-abstraction from C atom remote from a peroxy group.

Assuming $k_5 \gg k_3$ and k_{-3}, $(k_{6a} + k_{6b}) \gg k_4$ and k_{-4}, and $k_2 \gg (k_3 + k_4)$, derive an expression giving the relative rates of the product formation via the 'low temperature' mechanism and 'high temperature' mechanism for hydrocarbon oxidation (n.b. assume $C_nH_{2n+1}OO$ is in a steady-state).

Discuss the dependence of the predominant mechanism on the molecular weight and structure of the parent alkane from which the alkyl radical is derived, with particular reference to propane and n-pentane for which rate data are given in the Table.

Reaction	Propane			n-Pentane		
	A	E/R	k_{750}	A	E/R	k_{750}
1	5.6(−12)	2 500	2.0(−13)	5.6(−12)	2 500	2.0(−13)
2	2.0(−12)	0	2.0(−12)	2.0(−12)	0	2.0(−12)
−2	2.0(14)	14 500	8.0(5)	2.0(14)	14 500	8.0(5)
3	4.8(13)	17 400	3.8(3)	1.6(13)	15 000	3.1(4)
−3	0.4(13)	13 600	5.3(4)	0.12(13)	11 200	3.9(5)
4	3.0(12)	14 100	2.1(4)	2.0(12)	12 000	2.2(5)
−4	0.25(12)	10 300	2.7(5)	0.17(12)	8 200	3.0(6)
5	2.0(13)	10 000	3.2(7)	2.0(13)	10 000	3.2(7)
6a	1.6(14)	14 900	3.7(5)	6.0(13)	13 200	1.3(6)
6b	1.1(11)	7 942	2.5(6)	1.1(11)	7 942	2.5(6)

Units A, k: s^{-1} or cm^3 molecule^{-1} s^{-1}; E/R: K.
The values used for reactions 3 and 4 may be rationalised as follows:

C_3H_6: reaction 3 is a 1,4 primary* shift: $A = [(8 \times 10^{12}) \times 6]s^{-1}$
$E = 145 \, kJ \, mol^{-1}$

reaction 4 is a 1,5 primary* shift: $A = [(1 \times 10^{12}) \times 3]s^{-1}$
$E = 117 \, kJ \, mol^{-1}$

C_5H_{12}: reaction 3 is a 1,4 secondary* shift: $A = [(8 \times 10^{12}) \times 2]s^{-1}$
$E = 125 \, kJ \, mol^{-1}$

reaction 4 is a 1,5 secondary* shift: $A = [(1 \times 10^{12}) \times 2]s^{-1}$
$E = 100 \, kJ \, mol^{-1}$

Reverse reaction, -3 and -4, assume

$\Delta S = 21 \, J \, mol^{-1} \, K^{-1}$. $A_{-n} = A_n/12.2$

$$\Delta H = 31 \, kJ \, mol^{-1} \cdot \left(\frac{E}{R}\right)_{-n} = \left(\frac{E}{R}\right)_n - 3800 \, K$$

for $n = 3, 4$.

C2.2. The following five generalised reactions involving radical R and branching agent B describe the pre-cool flame chemistry in the spontaneous non-isothermal oxidation of hydrocarbons in air:

$RH + O_2 \xrightarrow{k_1} 2R$ initiation (rate $v_q = k_i[RH][O_2]$)

$R \xrightarrow{k_p} products + R$ propagation

$R \xrightarrow{fk_p} B + R$ propagation forming branching agent

$B \xrightarrow{k_B} 2R$ branching

$R \xrightarrow{k_1} products$ linear termination.

Write down the differential equations describing the time dependence of R and B and derive an expression for the local steady state concentration of radicals in terms of B and for its variation with time in terms of the net branching factor, ϕ, defined as $\phi = 2fk_p - k_1$ ($\phi > 0$).

Deduce an expression for the induction period to the cool flame, as defined by the time taken for the local radical concentration to rise from its initial value (no branching agent present) to a value R_{crit}. This value corresponds to the point in time when heat release from the chemistry just offsets heat loss to the surroundings.

Beyond this point the radical concentration continues to grow with an accompanying rise in temperature. The rise in radical concentration during the cool flame leads to additional termination due to:

$R + R \xrightarrow{k_2} products$ quadratic termination.

During the cool flame the concentrations of radicals maximise, and then fall to

---

*Numbers refer to the positions on the molecule chain relative to the terminal O atom. Primary and secondary refer to the type of H-atom abstracted in the isomerisation.

a new value of R_{crit}, whilst the temperature rises then stabilises at a new, higher value. The dependence of the local radical concentration $R(T)$ on temperature may be obtained by assuming that both B and R are in local steady state. Solve the simultaneous steady state equations for B and R to obtain an expression for $R(T)$ in terms of the net branching factor ϕ. How does ϕ change during the cool flame?

C2.9 Answers to problems

Hydrocarbon oxidation mechanisms in combustion systems

C2.1. *Low and high temperature oxidation mechanisms*:
The low temperature mechanism involves the formation of products through the alkyl-peroxy radical isomerisation mechanism for which the isomerisation involving internal H-abstraction (α or β to peroxy group) is rate determining, i.e.

$$\text{Rate (low } T) = (k_3 + k_4)[RO_2]. \qquad (R = C_nH_{2n+1}).$$

The high temperature mechanism involves the formation of products through the direct reaction of alkyl radicals with O_2 to form the conjugate alkene which is rate determining

$$\text{Rate (high } T) = k_1[R][O_2].$$

Writing steady state for $[RO_2]$

$$\frac{d[RO_2]}{dt} = k_2[R][O_2] - k_{-2}[RO_2] - k_3[RO_2] - k_4[RO_2] = 0$$

$$\therefore \frac{[RO_2]}{[R]} = \frac{k_2[O_2]}{k_{-2} + k_3 + k_4}$$

$$\therefore \frac{\text{Rate (low)}}{\text{Rate (high)}} = \frac{(k_3 + k_4)[RO_2]}{k_1[R][O_2]} = \frac{(k_3 + k_4)k_2}{k_1(k_{-2} + k_3 + k_4)}$$

$$= \frac{k_2}{k_1}\left\{\frac{1}{\dfrac{k_{-2}}{(k_3 + k_4)} + 1}\right\}.$$

For propane at 750 K $\quad \dfrac{k_{-2}}{k_3 + k_4} = 32 \quad \therefore \dfrac{R_L}{R_H} = 0.0303 \quad \dfrac{k_2}{k_1} = 0.30.$

For pentane $\quad \dfrac{k_{-2}}{k_3 + k_4} = 3.2 \quad \therefore \dfrac{R_L}{R_H} = 0.238 \quad \dfrac{k_2}{k_1} = 2.38.$

$\therefore$ A low temperature mechanism is much more important for pentane at 750 K than for propane.

The predominant mechanism depends on the following processes:

k_1 —structure independent (except for radicals with no conjugate olefin)
k_2 —structure independent ⎫
k_{-2}—structure independent ⎬ to first approximation

$\left.\begin{array}{c} k_3 \\ k_4 \end{array}\right\}$—highly structure dependent fast isomerisation favoured by:

—large ring size in transition state

—weaker C—H bond (i.e. $3° > 2° > 1°$).

For propane

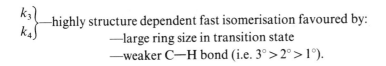

| small ring size 1⁰H abstraction | not possible | α 2⁰H abstraction | larger ring β 1⁰H abstraction |

increase $k_3 + k_4$

For n-pentane

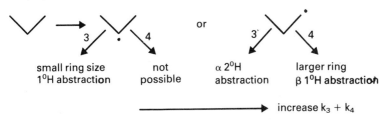

α,β 2⁰H abstraction   α β,γ 2⁰H abstraction   α, β, γ 2⁰ H   abstraction
γ 1⁰ H abstraction   δ 1⁰ H abstraction   δ 1⁰ H   abstraction

increase $k_3 + k_4$

Clearly the isomerisation rate for longer chain HC molecules will increase.

C2.2. *Mathematical model of cool flame behaviour*
(after Halstead, Kirsch, Prothero and Quinn.[11])

Differential equations for radical R and branching agent B:

$$\frac{dR}{dt} = v_q + 2k_B B - k_1 R \qquad (i)$$

$$\frac{dB}{dT} = f k_p R - k_B B \qquad (ii)$$

Local steady state of R:

from (i) if $\dfrac{dR}{dt} = 0 \qquad v_q + 2k_B B - k_1 R = 0 \qquad$ (iii)

$$\therefore R = \frac{v_q + 2k_B B}{k_1}$$

Variation of R with time:

R changes with time as B changes
differentiate (iii) with respect to t, assume v_q constant

$$2k_B \frac{dB}{dt} - k_1 \frac{dR}{dt} = 0 \qquad (iv)$$

(iii) also gives $k_B B = \dfrac{1}{2}\{k_1 R - v_q\}$.

Substitute in (ii) $\dfrac{dB}{dt} = fk_p R + \dfrac{v_q}{2} - \dfrac{k_1 R}{2}$

$\therefore \quad \dfrac{dB}{dt} = \phi\,\dfrac{R}{2} + \dfrac{v_q}{2} \qquad \phi = 2fk_p - k_1$.

branching factor

Substitute for $\dfrac{dB}{dt}$ in (iv): $2k_B\left\{\dfrac{\phi R}{2} + \dfrac{v_q}{2}\right\} - k_1\,\dfrac{dR}{dt} = 0$

$\therefore \quad \dfrac{dR}{dt} = \dfrac{k_B}{k_1}\{\phi R + v_q\}.$ \hfill (v)

Induction period

Defined by time $\tau = t_{crit} - t_0$ for radical concentration to change from R_0 to R_{crit}.

Obtained by integrating equation v between these boundary conditions.

Separate variables:

$$\int_{R_0}^{R_c} \dfrac{dR}{\left(1 + \dfrac{\phi . R}{v_q}\right)} = \int_{t_0}^{t_c} \dfrac{k_B v_q}{k_1}\,dt$$

$$\therefore \quad \left\{\dfrac{v_q}{\phi}\,\ln\left(1 + \dfrac{\phi . R}{v_q}\right)\right\}_{R_0}^{R_c} = \dfrac{k_B v_q}{k_1}(t_c - t_0)$$

$$\therefore \qquad\qquad \tau = \dfrac{k_1}{k_B \phi}\cdot\left[\ln\left(1 + \dfrac{\phi}{v_q}\cdot R_c\right) - \ln\left(1 + \dfrac{\phi}{v_q} R_0\right)\right]$$

Since $R_0 = \dfrac{v_q}{k_1}$, $\tau = \dfrac{k_1}{k_B \phi}\left\{\ln\dfrac{[1 + \phi/(v_q R_c)]}{(1 + \phi/k_1)}\right\}$

When quadratic termination occurs there is an additional $-2k_2 R^2$ term in (i); (ii) remains the same; (iii) contains the additional $-2k_2 R^2$ term. If B is in local steady state $dB/dt = 0$ in (ii) i.e. $B(T) = fk_p R/k_B$. B will change as temperature changes in the cool flame, as will R the local steady state of radicals. Substitute for B in modified equation iii:

$$v_q + 2fk_p R - k_1 R - 2k_2 R^2 = 0$$

i.e. $2k_2 R^2 - \phi R - v_q = 0$ $\phi =$ net branching factor as before

$$R(T) = \dfrac{\phi \pm (\phi^2 + 8k_2 v_q)^{1/2}}{4k_2} \hfill (vi)$$

Cool flame is quenched by fall in radical concentration leading to negative temperature dependence of reaction rate.

The negative T dependence of R results in the value of ϕ which changes from positive to negative during the cool flame temperature rise. Except at the temperature where $\phi = 0$ (T_x) we can assume that

$$\phi^2 \gg 8k_2 v_q$$

rearranging (vi) $R(T) = \dfrac{1}{4k_2} \left\{ \phi \pm \phi \left(1 + \dfrac{8k_2 v_q}{\phi^2} \right)^{1/2} \right\}.$

Taking $+$ve root for $T < T_x$ i.e. $\phi(T) > 0$
$\quad\quad\quad\;\; -$ve root for $T > T_x$ i.e. $\phi(T) < 0$
(n.b. $R(T) \geqslant 0$)

$T < T_x$: $R(T) = \dfrac{\phi}{2k_2}$ decreasing with T

$T > T_x$: $R(T) = \dfrac{1}{4k_2} \left\{ \phi - \phi \left(1 + \dfrac{1}{2} \cdot \dfrac{8k_2 v_q}{\phi^2} + \dfrac{1}{2}\left(-\dfrac{1}{2}\right) \cdot \dfrac{1}{2!} \left\{\dfrac{8k_2 v_q}{\phi^2}\right\}^2 \cdots \right) \right\}$

Binomial expansion

$$= 0 - \dfrac{v_q}{\phi} + \dfrac{k_2}{2}\dfrac{v_q^2}{\phi^3}$$

$\approx -\dfrac{v_q}{\phi}$ passes through minimum since v_q
$\quad\quad\quad$ increases rapidly with T.

Chapter C3
Combustion Probes

M. J. PILLING

C3.1 Introduction

Combustion is associated with a variety of phenomena including cool, diffusion and laminar flames and detonations. The problems which stand in the way of a detailed understanding are immense and embrace not only complex chemical kinetics but also a wide range of transport and hydrodynamic properties.[1] Light emission provided one of the earliest means of probing flames and a thorough discussion of many of the classical spectroscopic techniques may be found in references 2 and 3. Whilst some conclusions could be drawn, from such measurements, on chemical processes occurring in flames, most of the information obtained was concerned with the physical characteristics and, particularly, the temperature. Line reversal and measurements of rotational and vibrational temperatures of electronically excited radicals, e.g. $OH(A\,^2\Sigma^+)$, $C_2(A\,^3\Pi)$ and $CH(A\,^2\Delta)$ were widely employed and demonstrated that the required equilibria were not always established. For example estimates of the temperature of the inner cone of a premixed acetylene/oxygen flame varied by over 3000 K, depending on the technique used.[3]

Recent years have seen a resurgence of interest in flame diagnostics, consequent on the development of a wide range of laser techniques for probing flames. A variety of properties may now be examined, ranging from temperature and species concentration to pressure and density gradients, sizes of particles and hydrodynamic flow patterns.[4] The great advantage of many of these techniques is that they are non-intrusive, so that the flame is not disturbed, in any significant way, as it is probed. The aim of this chapter is to provide a brief survey of some aspects of this rapidly expanding field, emphasising those areas of primary interest to the kineticist, namely the determination of species concentrations and temperature. We shall concentrate on aspects which are aimed primarily at determining flame characteristics, and ignore the large and important field which exploits the use of flames as a means of obtaining, for example, spectroscopic data on free radicals.

C3.2 Laser induced fluorescence (LIF)

The technique of laser induced fluorescence was discussed at some length in Chapter B1, where its sensitivity and its specificity were stressed. Its advantages in flames are its ability to probe radicals in their ground states (unlike thermally excited emission spectroscopy or chemiluminescence) and its potential for spatial resolution. LIF finds increasing application as a flame diagnostic and is used particularly to probe OH, CH and CN, although other radicals have also been studied.[5]

Conditions in flames are hardly ideal for LIF. High pressures lead to substantial quenching of the excited state and to a consequent loss of sensitivity. Differential quenching rates also lead to problems of interpretation, it is not always straightforward to relate *relative* concentrations of rovibronic states to LIF intensities. For example, measurements on OH have demonstrated that the ratio of rotational to electronic collisional relaxation decreases with rotational quantum number, N, a consequence of the larger energy spacing at high N.[6] Finally, background emission, Rayleigh scattering and Mie scattering from particles all present problems.

It is instructive to examine a simple, steady-state model of the rates of excitation and de-excitation in a two level system:

$$X_1 + h\nu \rightarrow X_2^* \qquad \text{absorption}, k_1 = B_{12}\rho \qquad (1)$$

$$X_2^* \rightarrow X_1 \qquad \text{spontaneous emission}, k_2 = A_{21} \qquad (2)$$

$$X_2^* + Q \rightarrow X_1 + Q \qquad \text{quenching}, k_3 = k_q[Q] \qquad (3)$$

where X_1 and X_2^* are the ground and excited states, Q is a quenching molecule, B_{12} and A_{21} are the Einstein B and A coefficients and ρ is the radiation energy density. Applying the steady-state approximation to X_2^*,

$$[X_2^*] = B_{12}\rho[X_1]/(k_q[Q] + A_{21}) \qquad (4)$$

and the fluorescence intensity, I_f, is:

$$I_f = \left\{ \frac{h\nu}{4\pi} B_{12}\rho\Omega V \right\} \left\{ \frac{A_{21}}{k_q[Q] + A_{21}} \right\} \cdot [X_1] \qquad (5)$$

where V is the focal volume and Ω the solid angle subtended by the collection optics. This treatment is modified if the decay-time of X_2^* is comparable to or longer than the laser pulse width, but the overall conclusions are unchanged. The second bracketed term in equation 5 demonstrates the reduction in the fluorescence yield by quenching and this factor is typically only $10^{-2} - 10^{-3}$. Equation 5 also demonstrates how absolute concentrations of the ground state species, X_1, may only be evaluated if $k_q[Q]$ is known. The treatment is further complicated in multilevel systems when collisional transfer between the component states of X_1 and of X_2^* must also be considered.

At high laser intensities, stimulated emission from X_2^* must also be included:

$$X_2^* + h\nu \rightarrow X_1 + 2h\nu \quad \text{stimulated emission, } k_6 = B_{21}\rho. \tag{6}$$

If the laser intensity is sufficiently high, saturation is achieved (section B1.2) and $[X_1]$ and $[X_2^*]$ become comparable: we can no longer assume, as we have implicitly done so far, that $[X_1] \gg [X_2^*]$ and we must now account for the depletion of X_1 by setting $[X_1] + [X_2^*] = [X_1']$, where $[X_1']$ is the total population of X_1 in the flame prior to laser perturbation, i.e. $[X_1']$ is the quantity we require. We now find that

$$[X_2^*] = B_{12}\rho[X_1']/\{k_q[Q] + A_{21} + (B_{12} + B_{21})\rho\}$$

and when $(B_{12} + B_{21})\rho \gg k_q[Q] + A_{21}$, saturation is achieved and

$$[X_2^*] = [X_1']\{B_{12}/(B_{12} + B_{21})\}$$
$$= [X_1'](1 + g_1/g_2)^{-1}$$

where g_1 and g_2 are the degeneracies of levels 1 and 2. Under saturation conditions, absolute concentrations may be calculated directly from fluorescence intensities with a knowledge of only the geometrical factors, Ω and V, and of A_{21}.[5] Saturation is more difficult to achieve in a flame than in a low pressure LIF experiment because of the large $k_q[Q]$ term. Figure C3.1 shows a plot of fluorescence intensity vs laser intensity for C_2 in an acetylene/oxygen flame, demonstrating the approach to, but not the establishment of, saturation. Saturated laser induced fluorescence is more readily observed in low pressure flames. Kohse–Höinghaus et al.[8] determined concentrations of OH and CH, at levels of $\sim 10^{13}$ and 10^{15}cm^{-3}, with quoted uncertainties of $\lesssim 25\%$ in C_2H_2/O_2 flames at 10 Torr, whilst Salmon et al.[9] have measured [NH] in a 50 Torr $CH_4/N_2O/Ar$ flat flame, demonstrating a 96% degree of saturation. It should also be recalled that our treatment refers to a simple two level system and a more complex discussion is required for a multilevel radical system.

In addition to their use in identifying radicals and determining relative and, in some cases, absolute concentrations, LIF measurements have been widely used for measuring temperatures. As discussed above, care has to be taken when interpreting temperatures inferred from rotational distributions, especially at high rotational quantum numbers, because of differential quenching rates. The rotational spectrum of OH has been widely employed, and additives such as NO (rotational spectrum) or In (relative intensities of LIF excited from $^2P_{1/2,3/2}$) are also used. It is preferable to observe several rotational lines, although a premium is also placed on recording speed in an unstable system like a flame. A rapid scanning, frequency-doubled ring dye laser has been employed with OH, which sweeps 5cm^{-1} in $250 \mu s$, thus permitting rapid measurement on several adjacent rotational lines.[4] Temperatures are typically measured to a precision of $\pm 100 \text{K}$ by the LIF technique and similar uncertainties have been reported under conditions of satura-

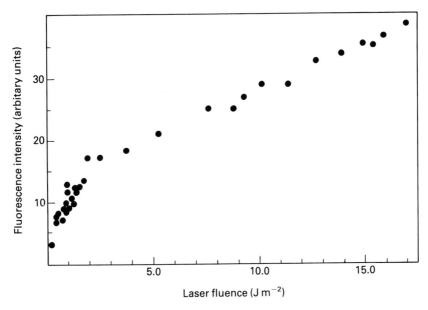

Fig. C3.1. Plot of the laser induced fluorescence signal for the C_2 Swan bands ($A\,^3\Pi_g$–$X\,^3\Pi_u$) as a function of laser intensity, demonstrating the approach to saturation. Adapted from reference 7 with permission.

tion.[8] Problem C3.1 illustrates the use of rotational spectra to determine temperatures.

A discussion of the use of atomic transitions may be found in reference 10. A stoichiometric methane/air flame was seeded with indium which was excited on one of two transitions with an excimer pumped dye laser. At a laser wavelength of 410 nm, the ground state was excited via the ($6s\,^2S_{1/2}$–$5p\,^2P_{1/2}$) transition, whilst the upper spin-orbit component was excited at 451 nm ($6s\,^2S_{1/2}$–$5p\,^2P_{3/2}$). Both excitations were monitored via fluorescence at 451 nm and temperatures deduced from the ratio of the fluorescence intensities for the two excitation wavelengths. The quoted uncertainty in the measurements was $\pm\,200$ K at a temperature in the centre of the flame of ~ 2150 K. The main advantage of the technique is its good spatial resolution and its potential for temporal resolution. The latter could be achieved by exciting the fluorescence using two dye lasers, operating at 410 and 451 nm, with one delayed relative to the other by at least 10 ns, together with two gated detectors.

Finally, LIF has been used to probe the two-dimensional image of [OH] in a flame. The technique is illustrated in Fig. C3.2.[11] The fluorescence was excited by a Nd:YAG pumped dye laser, focussed by a cylindrical lens to a sheet ~ 0.5 mm $\times$ 35 mm; the fluorescence was imaged onto an optical multichannel analyser thus providing a two dimensional image of [OH] in a plane defined by the intersection of the laser beam and the flame. The 1 μs gate and the possibility of

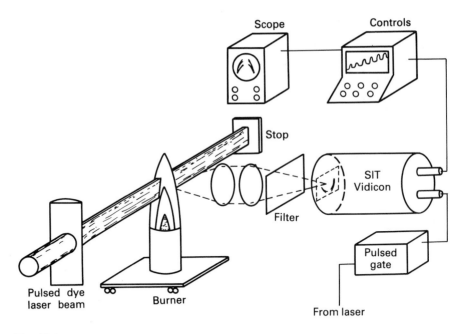

Fig. C3.2. LIF technique for forming a two-dimensional image of [OH] in a flame. The LIF is excited in the plane where the laser intersects the flame and is imaged on to a vidicon tube. Reproduced from reference 11 with permission.

obtaining single shot images mean that the technique has great potential in studying, for example, turbulent combustion. Dyer and Crosley[11] realised this potential by Q switching the Nd:YAG laser twice in the same flashlamp pulse and directing the successive images onto separate halves of a vidicon using an oscillating mirror. A time separation between the two pulses of $\sim 100\,\mu$s was achieved.

C3.3 Coherent anti-Stokes Raman spectroscopy

Rayleigh scattering has been employed in flames to study number densities and temperatures,[4] but runs into difficulties if the concentration of particulates is high. Spontaneous Raman spectroscopy (SRS) has also been used, particularly with pulsed lasers to overcome the luminous background. SRS is not so sensitive as LIF and has only been employed on majority species such as N_2, although the signal level is greatly enhanced with spontaneous resonant Raman spectroscopy. Temperature measurements are made using, for example, the Stokes/anti-Stokes intensities or the rotational profiles.

The Raman technique most widely employed in flames, however, is that of coherent anti-Stokes Raman spectroscopy or CARS.[12] The flame is irradiated with two laser beams, frequencies ω_1 and ω_2, which interact through the third order non-linear susceptibility of the sample and generate a coherent beam at frequency ω_3, where $\omega_3 = 2\omega_1 - \omega_2$ (Fig. C3.3); ω_3, the CARS signal, becomes very

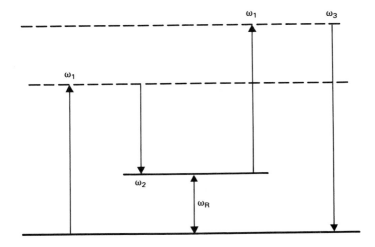

Fig. C3.3. Generation of a CARS (coherent anti-Stokes Raman spectroscopy) signal. ω_1 and ω_2 are the laser frequencies, ω_R a Raman active transition in the sample and ω_3 the CARS signal.

intense when $\omega_1 - \omega_2$ is equal to a Raman active transition ω_R. A detailed discussion of the theory of CARS spectroscopy may be found in reference 12. The great advantage of CARS for flames is that the emission at ω_3 is *coherent* and its consequent small divergence allows easy discrimination of the signal against the background luminosity of the flame.

A schematic experimental arrangement is shown in Fig. C3.4. A frequency-doubled, Q-switched Nd:YAG laser is generally employed to generate ω_1; the YAG laser also pumps a dye which provides ω_2 and the two pulses are combined spatially and temporally in the flame.[5,13] The coherent ω_3 signal is separated by prisms and detected. By employing a broadband ω_2 pulse, and detecting with a monochromator and OMA, a complete CARS spectrum may be generated in a single shot, thus providing, once again, a powerful means for studying turbulent flames. An interesting application, in a highly non-ideal system, is in the study of an internal combustion engine.[14] The technique is limited to majority species, but has been widely employed to measure temperatures by matching calculated and experimental spectra, which are generally only partially rotationally resolved (see Fig. C3.5).

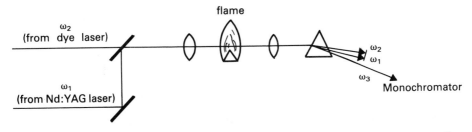

Fig. C3.4. Schematic diagram of the experimental generation of a CARS signal in a flame. Details of experimental systems may be found in reference 5.

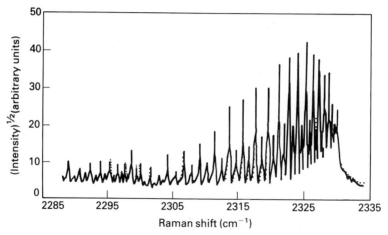

Fig. C3.5. Experimental CARS spectrum of the nitrogen Q branches for $v = 0$ and $v = 1$ in a sooting laminar ethylene/air diffusion flame. The broken curve, just visible behind the full experimental trace, is a calculated curve for $T = 1370$ K. Reproduced from reference 15 with permission.

C3.4 Miscellaneous spectroscopic techniques

(i) Laser absorption spectroscopy

Laser absorption spectroscopy is not so sensitive as LIF and cannot provide such spatially precise information. It is a 'line of sight' technique, so that the signal only gives a measure of the concentration integrated along the light path and, therefore, through the flame, where considerable variations may exist. Its major advantages over LIF is its independence of quenching processes and of the requirement of a fluorescing excited state. IR absorption, using diode lasers, is a promising technique. The vibration/rotation spectra of flame components are well resolved and the profile of a single (v, J) state may be scanned, and the concentration calculated straightforwardly from the integrated absorption intensity of the line. In this way, no assumptions need be made about the lineshape, which may be a complex function of pressure and temperature because of collisional broadening. The potential sensitivity of the technique is illustrated by measurements on CO $(v = 1–4)$ in a discharge-flow system where concentrations down to 10^{12} molecule cm^{-3} were detected.[16]

(ii) Optoacoustic spectroscopy

If sodium, seeded in a flame, is excited to the 3^2P level by a pulsed dye laser, electronic quenching rapidly converts some of the energy to translation, generating a sound wave which may be detected by a microphone. The time of arrival of the sound wave at the microphone may be used to probe the local density and temperature with good spatial resolution.[5] Alternatively, the flame may be irradiated with a cw laser, such as an argon ion laser, which is chopped at a few kHz; a sound wave, at the modulation frequency, is generated in the flame and may again be detected by a microphone. The principle of such a device, which is the basis of the spectrophone, predates lasers. The great sensitivity of the technique and its freedom from interference from scattered light has led to its application in the study of light absorption by soot particles.[17]

(iii) Optogalvanic spectroscopy

Electrical properties of flames have always been an important area, either as a diagnostic or as a means of studying reactions of ions at high temperatures. The use of lasers to probe the electrical properties has produced a technique of quite remarkable sensitivity, since the collisional ionisation rate can be enhanced by orders of magnitude by saturating an optical transition:

$$A + hc\omega_1 \rightarrow A_1^*$$

$$A_1^* + M \rightarrow A^+ + e^- + M.$$

Smyth et al. have exploited optogalvanic spectroscopy to probe highly excited states of sodium.[18] The technique is based on the fact that the efficiency of collisional ionisation increases with the energy of the excited atom. Highly excited states can, therefore, be detected by observing the increase in optogalvanic signal on irradiating the system with a further tunable laser:

$$A_1^* + hc\omega_2 \rightarrow A_2^*$$

$$A_2^* + M \rightarrow A^+ + e^- + M.$$

Figure C3.6 shows a spectrum obtained by irradiating an H_2/air flame, seeded with sodium, with two counterpropagating beams, frequencies ω_1 and ω_2, from unfocussed N_2 pumped dye lasers. ω_1 saturates the $Na(3^2P-3^2S)$ transition and the subsequent excitation to the nd,ns levels by ω_2 is detected by measuring the current changes in the flame resulting from subsequent collisional ionisation. Sodium concentrations as low as 10^9 atom cm^{-3} were employed, with laser pulse energies of 35–70 μJ (ω_1) and 1–300 μJ (ω_2).[18]

(iv) Multiphoton ionisation

Resonance enhanced multiple photon ionisation (REMPI—see section B1.3) provides a sensitive means of detecting non-fluorescing radicals. It is, however, more intrusive than most laser techniques, since it requires an ionisation probe to be inserted into the flame. Early applications were addressed at probing H and O atoms using $2 + 1$ REMPI.[19] The transitions involved for $O(^3P)$ are shown in Fig. C3.7. Although atoms are easy to detect by resonance fluorescence or resonance

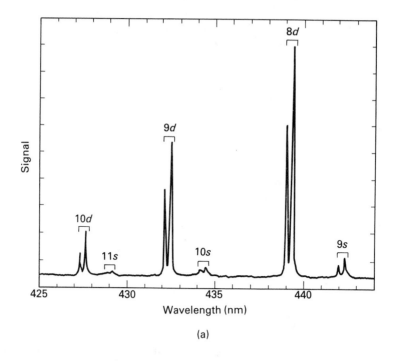

(a)

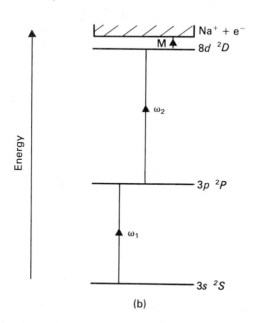

(b)

Fig. C3.6. (a) Optogalvanic spectrum for stepwise excitation of sodium ($nd,ns \leftarrow 3p \leftarrow 3s$) in a hydrogen/air flame. The splitting of each transition is equal to the ($3p\ ^2P_{3/2} - 3p\ ^2P_{1/2}$) separation. Reproduced, with permission, from reference 18. (b) Energy level diagram illustrating the optical and collisional processes involved.

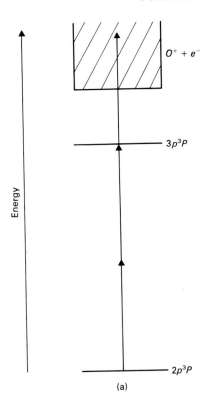

Energy

$O^+ + e^-$

$3p\,^3P$

$2p\,^3P$

(a)

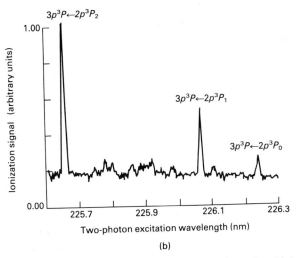

$3p\,^3P\leftarrow2p\,^3P_2$

1.00

Ionization signal (arbitrary units)

$3p\,^3P\leftarrow2p\,^3P_1$

$3p\,^3P\leftarrow2p\,^3P_0$

0.00

225.7   225.9   226.1   226.3

Two-photon excitation wavelength (nm)

(b)

Fig. C3.7. (a) Energy level diagram showing $2+1$ resonance enhanced multiple photon ionisation for $O(2\,^3P)$. The $O(3\,^3P)$ states were formed by 2 photon absorption of 226 nm radiation, produced by frequency doubling 573 nm pulses from a rhodamine 6G dye laser, pumped by an Nd:YAG laser. The doubled dye laser output was then mixed with the 1.06 μm Nd:YAG fundamental to produce the final pumping radiation at frequency ω_1. (b) Trace of the signal obtained. (Reproduced, with thanks, from a spectrum supplied by J. E. M. Goldsmith.)

absorption in the studies of elementary reactions, these techniques cannot be used with flames and REMPI represents the first sensitive technique for detecting important atomic species such as H and O. A variety of radicals have recently been detected in flames, for example, Tjossem and Cool[20] report detection of C_2O via $3+1$ REMPI in a $CH_4/O_2/Ar$ flame. Importantly, given the multiple photon nature of the technique, they also demonstrated that the $2+1$ REMPI signal from NO, diluted in N_2, is linear in [NO] over dilutions of 0.2–100 ppm in 70–700 Torr N_2. Cool has also discussed the experimental requirements for quantitative concentration measurements elsewhere.[21]

C3.5 Laser perturbation

An interesting technique for studying kinetics in flames has been developed by Kirsch and Morley at Shell's Thornton Research Centre.[22] They set up a one or two stage ignition above a flat burner in premixed heptane/air and probed the OH concentration with a dye laser. A cool flame was established some distance above the burner and a second stage ignition was observed, under some conditions, at still greater distances. It became apparent that the LIF emanating from the dark, hot region above the cool flame was non-linear in laser intensity and that the bulk of the signal was produced by photolysis of H_2O_2 generated in the flame. Kirsch and Morley exploited this observation by using a second laser to perturb [OH] by photolysis of H_2O_2 and push it well above its 'natural' value. The dye laser was then used in a pump and probe experiment to monitor the decay of [OH] back to its pre-perturbation value. The decay showed at least three exponential components and, although only preliminary results have been obtained so far, the technique promises detailed kinetic information on the sequence of reactions, described in Chapter C2, which make up cool flame kinetics.

C3.6 Mass spectrometry

Mass spectrometry has two major attractions for the study of flame species, namely its sensitivity and its universality—it can be used to detect both molecules and radicals, regardless of their ability to absorb or emit light. Its major disadvantage is its invasiveness and the consequent doubts about the extent to which the probe perturbs and, indeed, correctly samples the flame.

Figure C3.8 shows the experimental system employed at British Gas[23] where the flame is sampled using a quartz probe, which is part of a 3-stage molecular beam sampling inlet to a quadrupole mass spectrometer. A plot of the dependence of the concentration of a variety of species, as a function of distance from the burner for a methane/O_2 flame, is shown in Fig. C3.9. The technique has provided probably the most pertinent data so far obtained, from a purely chemical kinetic viewpoint, provided one can confidently assume that perturbations are minimal.

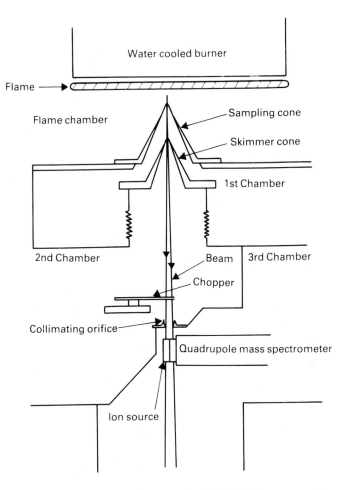

Fig. C3.8. Schematic diagram of the sampling system for the British Gas mass spectrometer flame probe. Reproduced, with permission, from reference 23.

The interesting aspect of the concentration profiles shown in Fig. C3.9 is the growth of the C_2 species, and especially C_2H_4, which can be directly linked to CH_3 via the reactions:

$$CH_3 + H \rightarrow CH_2 + H_2$$
$$CH_3 + CH_3 \rightarrow C_2H_4 + H_2$$
$$CH_3 + CH_2 \rightarrow C_2H_4 + H.$$

C3.7 Conclusions

The complexity of flames derives, at least in part, from the coupling between kinetics, temperature and hydrodynamics. Kineticists tend to concentrate, quite

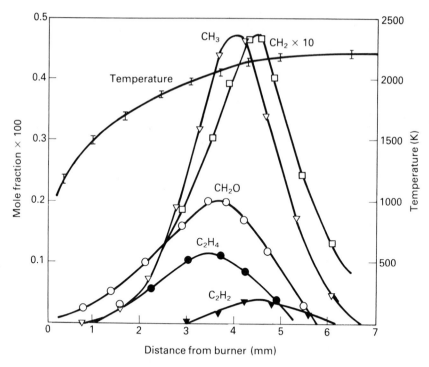

Fig. C3.9. Species concentration and temperature profiles in a stoichiometric CH_4/O_2 flame, at 20 Torr total pressure, as detected by mass spectrometry. Reproduced, with permission, from reference 24.

naturally, on modelling the chemistry and many detailed reaction schemes have been proposed. The new techniques described above still have to make a significant impact on the development and refinement of these models, many of which are based on shock tube data and macroscopic flame behaviour, such as burning velocity.

The kinetic schemes incorporated into hydrodynamic models of flames are more schematic and tend to rely on a global description, although restricted multistep schemes can now be included. The advances made in the mathematical description of laminar flames and even of turbulent flames[1] provide an increasingly sound basis on which detailed modelling can proceed. Jones and Whitelaw[24] recently gave an account of the current status of work in this area, demonstrating the imminence of numerical models which will be capable of accurate calculation of the concentrations of major species, but stressing the inter-dependence of such models on experimental measurements based on many of the techniques discussed in this chapter.

C3.8 Suggestions for further reading

Crosley D. R. (Ed.) (1980) *Laser Probes for Combustion Chemistry.* ACS Symposium Series, 134, American Chemical Society, Washington D.C.

Gaydon A. G. and Wolfhard H. G. (1979) *Flames*, 4th edn. Chapman and Hall, London.
Penner S. S., Wang C. P. and Bahadori M. Y. (1984) 20*th Symp. (Int.) on Combustion*, 1149.

C3.9 References

1 Williams F. A. (1985) *Combustion Theory*, 2nd edn. Benjamin/Cummings, Menlo Park.
2 Gaydon A. G. and Wolfhard H. G. (1979) *Flames*, 4th edn. Chapman and Hall, London.
3 Mavrodineanu R. and Boitaux H. (1965) *Flame Spectroscopy*, Wiley, New York.
4 Penner S. S., Wang C. P. and Bahadori M. Y. (1984) 20*th Symp. (Int.) on Combustion*, 1149.
5 McDonald J. R. (1980) *Laser Probes for Combustion Chemistry*, (Ed. by D. R. Crosley), ACS Symposium Series, **134**, 19. A. C. Eckbreth, *ibid.*, 271.
6 Crosley D. R. (1980) *Laser Probes for Combustion Chemistry*, (Ed. by D. R. Crosley), ACS Symposium Series, **134**, 1.
7 Baronavski A. P. and McDonald J. R. (1977) *Appl. Opt.*, **16**, 1897.
8 Kohse-Höinghaus K., Heidenreich R. and Just Th. (1984) 20*th Symp. (Int.) on Combustion*, 1177.
9 Salmon J. T., Lucht R. P., Sweeney D. W. and Laurendeau N. M. (1984) 20*th Symp. (Int.) on Combustion*, 1187.
10 Alden M., Grafström P., Lundberg H. and Svanberg S. (1983) *Opt. Letters*, **8**, 241.
11 Dyer M. J. and Crosley D. R. (1982) *Opt. Letters*, 7, 382; (1984), **9**, 217.
12 Druet S. and Taran J. P. (1979) *Chemical and Biochemical Applications of Lasers*, (Ed. by C. B. Moore), p. 187. Academic Press, New York.
13 Eckbreth A. C. and Hall R. J. (1979) *Combust. Flame*, **36**, 87.
14 Stenhouse I. A., Williams D. R., Cole D. R. and Swords M. D. (1979) *Appl. Opt.*, **18**, 3819.
15 Farrow R. L., Lucht R. P., Flower W. L. and Palmer R. E. (1984) 20*th Symp. (Int.) on Combustion*, 1307.
16 Dreier T. and Wolfrum J. (1981) 18*th Symp. (Int.) on Combust.*, 801.
17 Killinger D. K., Moore J. and Japar S. M. (1980) *Laser Probes for Combustion Chemistry*, (Ed. by D. R. Crosley), ACS Symposium Series, **134**, 457.
18 Smyth K. C., Schenck D. K. and Mallard W. G. (1980) *Laser Probes for Combustion Chemistry*, (Ed. by D. R. Crosley), ACS Symposium Series, **134**, 175.
19 Goldsmith J. E. M. (1983) *J. Chem. Phys.*, **78**, 1610.
20 Tjossem P. J. H. and Cool T. A. (1984) 20*th Symp. (Int.) on Combustion*, 1321.
21 Cool T. A. (1984) *Appl. Opt.*, **23**, 1559.
22 Kirsch L. J. and Morley C. J., private communication.
23 Harvey R. and Maccoll A. (1981) 18*th Symp. (Int.) on Combustion*, 857.
24 Jones W. P. and Whitelaw J. H. (1984) 20*th Symp. (Int.) on Combustion*, 233.

C3.10 Problem

Combustion probes

C3.1. This problem investigates how temperatures may be deduced from the relative intensities of rotationally resolved LIF measurements, applies the technique to OH in a laser heated system and then discusses its extension to flames.

(i) The Einstein B coefficient, $B_{12}(v', J'; v'', J'')$ for a transition between rovibrational levels (v'', J'') and (v', J') in electronic states 1 and 2, respectively, is given, to a good approximation, by:

$$B_{12}(v', J'; v'', J'') = C_1 R_e^2(2, 1) q(v', v'') S(J', J'') / g_1(J'')$$

where C_1 is a constant, $R_e(2, 1)$ is the electronic transition moment, $q(v', v'')$ is the Franck–Condon factor, $S(J,', J'')$ is the rotational line strength and $g_1(J'')$ is the degeneracy (electronic and rotational) of the lower state. The rotational line strengths from a given J'' are subject to the sum rule:

$$\sum_{J'} S(J', J'') = (2J'' + 1).$$

Show that, provided fluorescence quenching is negligible, the total fluorescence intensities excited from a series of J'' states, in the same (v', v'') band, are given by:

$$I_f(J', J'') = C_2 S(J', J'') N(v'', J'')/(2J'' + 1)$$

where C_2 is a constant for a specific vibronic transition and $N(v'', J'')$ is the population of the (v'', J'') state. Hence show that the rotational temperature may be obtained by plotting $\ln\{I_f(J', J'')/S(J'', J'')\}$ vs the energy of the lower state.

(ii) The reactions of OH at high temperatures may be studied using laser heating to generate a large increase in temperature on a short timescale (see section B1.5). OH, obtained by passing an argon/water mixture through a microwave discharge, was injected into 65 mTorr SF_6 in 2 Torr Ar in a flow tube. The system was exposed to IR radiation from a pulsed CO_2 laser and the LIF spectrum of OH in the range 32360 to 32440 cm^{-1} was recorded at a zero delay time (Fig. C3.10a), when the OH was still at ambient temperature, and again at a delay time of 100 μs (Fig. C3.10b) when the components in the irradiated volume had equilibrated at a new high temperature.

The transitions shown all involve $OH[X^2\Pi (v = 0)]$, which corresponds to Hund's case (a) and has energy levels $F_1(J'')$, where $J'' = N'' + 1/2$ and $F_2(J'')$, where $J'' = N'' - 1/2$, and where $N'' = 1, 2, 3 \ldots$ The transitions shown in Fig. C3.10 are labelled according to the spin component involved in the upper and lower levels [e.g. P(21) is an $F'_2 - F''_1$ transition] and to the value of J''. The spin splitting in the $A^2\Sigma^+$ upper state is so small that the Q(11) and P(21) lines are blended and contributions from both lines must be considered. Despite the complications introduced by spin and orbital angular momentum, the relationship between I_f and N derived above still applies.

Table C3.1 lists energies and rotational line strengths. Using the heights of the lines in Fig. C3.10 as a measure of intensity, evaluate the rotational temperatures in $OH[X^2\Pi (v = 0)]$ at the two delay times.

(iii) Discuss any approximations implicit in the above approach and the difficulties in extending it to the measurement of temperature in an atmospheric pressure flame.

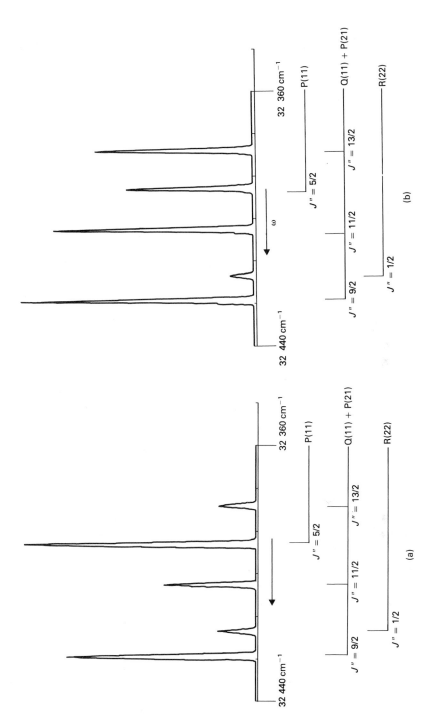

Fig. C3.10. Simulated traces of laser induced fluorescence from OH ($v = 0$) in a mixture of OH, SF$_6$ and Ar, in a flow tube, following pulsed CO$_2$ laser excitation of the SF$_6$: (a) zero delay, (b) 100 μs delay.

Table C3.1. Rotational line strengths and energies for OH X^2 ($v = 0$).

Branch	J''	$S(J', J'')$	$F(J'')/\text{cm}^{-1}$
P(11)	5/2	0.796	45.74
Q(11)	9/2	2.108	316.85
P(21)	9/2	0.303	316.85
Q(11)	11/2	2.638	505.35
P(21)	11/2	0.275	505.35
Q(11)	13/2	3.167	729.25
P(21)	13/2	0.249	729.25
R(22)	1/2	0.167	88.10

C3.11 Answer to problem

Combustion probes

C3.1. (i) If quenching is negligible, $A_{21} \gg k_q[\text{Q}]$, so that equation 5 becomes:

$$I_f = \{hv\rho\Omega V/4\pi\}B_{12}[\text{X}_1].$$

Adapting this equation to a molecular transition and ignoring the small change in frequency

$$I_f(v', J'; v'', J'') = C_3 R_e^2(2, 1)q(v', v'')S(J', J'')N(v'', J'')/g_1(J'')$$

where C_3 is a constant. For excitation in the same vibronic band, R_e^2, q and the electronic degeneracy are constant, so that

$$I_f(J', J'') = C_2 S(J', J'')N(v'', J'')/(2J'' + 1). \tag{1}$$

Provided a Boltzmann distribution applies in the v'' state,

$$N(v'', J'') = N(v'')(2J'' + 1)\ \exp\{-hcF(J'')/k_B T\}/Q_{rot} \tag{2}$$

where $hcF(J'')$ is the energy of the rotational level, $N(v'')$ is the population of the v'' state and Q_{rot} is the rotational partition function. Combining equations 1 and 2:

$$\ln\{I_f(J', J'')/S(J', J'')\} = \ln\{C_2 N(v'')/Q_{rot}\} - hcF(J'')/k_B T.$$

The first term on the right hand side is constant, at a given temperature, and T may be determined by plotting the left hand side vs $F(J'')$.

(ii)

Branch	J''	$I(\text{rel})$	$\ln(I/S)$	$I(\text{rel})$	$\ln(I/S)$
		zero delay		100 μs delay	
P(11)	5/2	63.2	4.37	8.0	2.31
Q(11)	9/2	51.5	3.06	14.8	1.81
P(21)					
Q(11)	11/2	25.3	2.16	12.7	1.47
P(21)					
Q(11)	13/2	10.0	1.07	10.0	1.07
P(21)					
R(22)	1/2	10.8	4.16	1.5	2.19

The intensities, $I(\text{rel})$, are determined relative to a value of 10.0 for the lowest frequency transition.

Figure C3.11 shows a plot of $\ln(I/S)$ vs $F(J'')$. The slope is equal to $k_\text{B}T/hc$:

temperature after a zero delay $= 300\,\text{K}$
temperature after a 100 μs delay $= 800\,\text{K}$.

(iii)

(a) The traces shown in Fig. C3.10 are simulations—care should be taken when recording experimental spectra to ensure that the signals have been properly normalised to allow for variations in laser intensity. An advantage in

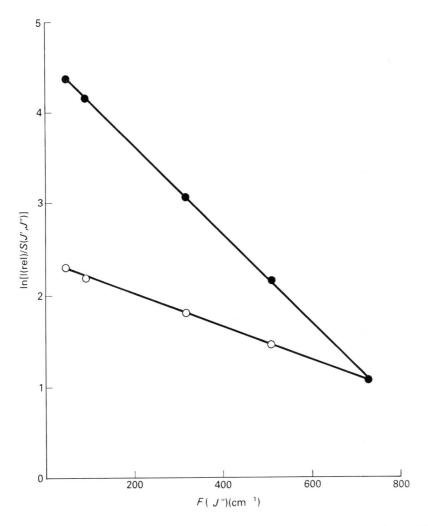

Fig. C3.11. Plot of $\ln[I(\text{rel})/S(J',J'')]$ vs $F(J'')$, ●, zero delay, ○, 100 μs delay. The slopes of the plots are equal to $hc/k_\text{B}T$ where T is the rotational temperature at the two delay times.

recording several lines rather than the minimum of two is that errors arising from artefacts of this sort are more likely to show up.

The theoretical development assumed the validity of the Born–Oppenheimer approximation. Chidsey and Crosley (1980) in *J. Quant. Spectrosc. Radiat. Transfer*, **23**, 187 have discussed two sources of error. The electronic transition moment depends on internuclear separation and has to be determined as a function of the r-centroid, the mean separation in the transition. In addition the Franck–Condon factor depends on J via the dependence of the effective potential energy curve on J.

(b) The major problem facing extension of the technique to flames is the importance of quenching and the consequent need to determine k_q and its dependence on J and to assess the importance of rotational relaxation in the upper state. A quantitative discussion of the use of LIF to determine flame temperatures and species concentrations may be found in the article by Daily J. W. (1980) in *ACS Symposium Series*, **134**, 61.

Chapter C4
Numerical Integration and Sensitivity Analysis

M. J. PILLING

C4.1 Introduction

Numerical integration of the coupled differential equations which describe a reaction system is becoming an increasingly important tool in chemical kinetics. It enables us (a) to model the kinetics of complex processes, without the need to invoke the steady-state approximation, (b) to simulate experimental systems employed, for example, in pulsed measurements of elementary reactions, in order to test the assumptions implicit in the adoption of a specific analytic rate law, (c) to analyse experimental decay data under conditions where no analytic solution is available and (d) to assess, when coupled with sensitivity analysis, which rate constants are of particular importance in a given complex system. In this chapter, we shall review the basis of numerical integration, as applied to chemical kinetics, and of sensitivity analysis, and give examples of applications in each of the four areas outlined above.

C4.2 Numerical integration

A complex chemical reaction consists of a number of elementary steps and the time dependences of the component chemical species are described by ordinary differential equations (o.d.e.s) of the type:

$$dy_i/dt = f_i(y_1, y_2, \ldots, y_n)$$

where y_i is the concentration of the ith species and the function f_i contains rate parameters and may be non-linear in the y_i. For example, for the reaction scheme:

$$Cl_2 \rightarrow 2Cl \qquad (1)$$
$$Cl + H_2 \rightarrow H + HCl \qquad (2)$$
$$H + Cl_2 \rightarrow Cl + HCl \qquad (3)$$
$$Cl + Cl \rightarrow Cl_2 \qquad (4)$$

we may write, setting $y_1 = [Cl_2]$, $y_2 = [H_2]$, $y_3 = [Cl]$, $y_4 = [H]$,

$$\frac{dy_1}{dt} = -k_1 y_1 - k_3 y_1 y_4 + k_4 y_3^2$$

etc.

In the following discussion it will make life easier if we use, as an example, a single differential equation $dy/dt = f[y(t)]$, with an initial condition that, at $t = 0$, $y = y_0$.

The time-honoured method for the numerical integration of o.d.e.s is the Runge–Kutta method,[1] in which the solution at time $t + \delta t$ is determined from that at time t by evaluating a function, ψ, which is an explicit function of f:

$$y(t + \delta t) = y(t) + \psi.$$

For example, with the fourth-order Runge–Kutta method,

$$\psi = \frac{1}{6}(\alpha_0 + 2\alpha_1 + 2\alpha_2 + \alpha_3), \tag{5}$$

where

$$\alpha_0 = f[y(t)]\delta t; \qquad \alpha_1 = f\left[y(t) + \frac{1}{2}\alpha_0\right]\delta t$$

$$\alpha_2 = f\left[y(t) + \frac{1}{2}\alpha_1\right]\delta t; \qquad \alpha_3 = f\left[y(t) + \alpha_2\right]\delta t.$$

The time step, δt, is made sufficiently small so that y does not change too drastically during the time period $t \rightarrow t + \delta t$, but not too small, otherwise the computer time used is excessive. y at time δt may then be determined from y_0 by evaluating ψ, that at time $2\delta t$ from $y(\delta t)$ and so on. A simple example of the application of the fourth-order Runge–Kutta method is given in the problem at the end of this chapter.

Explicit methods of this type run into difficulties when they are applied to kinetic problems with reactions occurring on very different time-scales. For example, the reaction scheme:

$$A \xrightarrow{k_1} B \xrightarrow{k_2} C$$

presents no problems if $k_1 \sim k_2$, but if $k_2 \gg k_1$, the step-size, δt, which is required to produce an accurate solution is determined by k_2 throughout the integration ($\delta t \ll k_2^{-1}$). If we are interested in the overall removal of A, then we must use step sizes which are many orders of magnitude less than the overall integration time, and the computation is very time-consuming. We can make approximations, for example we might assume a steady-state for B at times somewhat in excess of k_2^{-1}, but this approximation is less self-evidently valid for more complex reaction schemes.

The problem alluded to above is termed 'stiffness', which arises, in our context, when the different elementary reactions defining the o.d.e.s have very different time-constants. It is a general problem in the numerical integration of o.d.e.s which, interestingly, was first referred to in the context of chemical kinetics. It can be overcome by using so-called implicit methods,[1] in which the function ψ, required to evaluate $y(t + \delta t)$, is determined not only from the value of y at the time t, but also from that at time $t + \delta t$. A trivial example is the trapezium rule:

$$y(t + \delta t) = y(t) + \delta t \{f[y(t)] + f[y(t + \delta t)]\}/2.$$

The solutions cannot be determined directly, as with explicit methods, but only by iteration, except for the case where f is linear in y.

A highly successful algorithm for the solution of stiff o.d.e.s was proposed by Gear.[2] It employs a hybrid explicit/implicit, so-called predictor–corrector, method in which an explicit equation is used to predict $y(t + \delta t)$, which is then corrected using an implicit technique. The computational effort involved for each time step is significantly increased, but much larger time steps can be employed.[1,2]

The technique may be illustrated most succinctly by representing the coupled differential equations in matrix form:

$$\frac{dy}{dt} = f(y, t; k) \tag{6}$$

where the dependence of the concentration vector, y, on time and on the rate parameters, k, is explicitly recognised. The concentrations at zero time are subject to the condition:

$$y(t = 0) = y_0.$$

The Gear algorithm makes estimates of y, which we shall represent as Y, at any time $t_n > 0$, via the equation

$$Y(t_n) = \sum_{l=1}^{q_n} \alpha_l Y(t_{n-l}) + \delta t \beta f[Y(t_n), t_n; k] \tag{7}$$

with

$$Y(t = 0) = y_0.$$

The first term on the right hand side of the equation is a predictor, i.e. the concentrations at time t_n are predicted from those at the earlier times t_{n-1}, t_{n-2}, t_{n-3} etc. The second term is the implicit corrector which depends on the concentration at t_n itself. α_l and β are coefficients and q_n is the order of approximation, i.e. the number of earlier concentrations included in the calculation at time t_n. The program itself determines the step size, δt, and the order of approximation, q_n, in order to satisfy a predetermined accuracy.

The Runge–Kutta method is very easy to implement and should generally be employed as a first attempt. For users of NAG routines, a useful implementation (DO2ABF) is available which uses Merson's algorithm and which also provides a check on stiffness. For more complex, stiff problems, there is a NAG implementation of the Gear algorithm, but there are also several complete packages available which are particularly straightforward because they have been written for chemical problems and the set of o.d.e.s is determined from an input set of chemical equations. In the U.K., FACSIMILE,[3] coded in Fortran for IBM 370 computers, but generally convertible to other machines, is widely used. A more international list of programs is given by Deuflhard *et al.*[4]

C4.3 Sensitivity analysis

Many chemical systems are so complex that it is difficult to determine, by inspection, the interplay between competing groups of reactions and to discern which elementary reactions constitute the important steps. Numerical solutions are more efficient if the basis set of kinetic equations is kept to a minimum, but this basis set must be realistically defined. Similarly, an experimentalist needs to know which reactions in an overall mechanism need particularly precise rate constants, so that he can concentrate his efforts accordingly. Sensitivity analysis can provide answers to these questions by specifying quantitatively the sensitivity of the overall reaction, or of a given reactant, product or intermediate, at a specific time, to a particular rate constant.

The sensitivity coefficient of species y_i with respect to rate constant k_j is defined by

$$S_{ij} = \frac{\partial y_i}{\partial k_j}; \quad S_{ij}^r = \frac{\partial \ln y_i}{\partial \ln k_j} \tag{8}$$

where S_{ij}^r is the reduced sensitivity coefficient, which is particularly useful when comparing sensitivities.

Sensitivity coefficients can be calculated most simply by the 'brute force' method in which the rate constants are varied one by one, and the concentrations of the species recalculated anew for each set of rate parameters. Clearly such a technique can only be applied to simple reaction systems or where only a limited set of sensitivity coefficients is required.

In the so-called direct method an auxiliary set of equations is derived by differentiating equation 7 with respect to the rate parameters (see below). The kinetic and auxiliary equations are then coupled and solved together. The technique is difficult to apply and the solutions are often unstable. A widely used technique which has been developed to overcome this difficulty is the so-called Green's function method (GFM)[5] in which the Green's function for the auxiliary equations is determined and the S_{ij}^r coefficients evaluated by integration. A particularly efficient version is the GFM–AIM (analytically integrated Magnus) modification.

Implementation of the Green's function method may be achieved using the package CHEMSEN,[6] which also contains a stiff o.d.e. solver. The setting up of the differential equations from an input chemical mechanism is based on the program CHEMKIN which contains a thermodynamic data base, so that reverse reaction rate constants are automatically defined. The package is written in FORTRAN and is particularly suitable for VAX usage.

Dunker[7] has recently described an efficient and stable decoupled form of the direct method which is accurate and much easier to implement than the Green's function method, although no packages are currently available. The auxiliary equations are:

$$\{I - h\beta J[Y(t_n), t_n; k]\}S_j(t_n) = \sum_{l=1}^{q_n} \alpha_l S_j(t_{n-l}) + h\beta f'_j[Y(t_n), t_n; k], \qquad (9)$$

where $S_j = \partial Y/\partial k_j$, I is the identity matrix, J the Jacobian with elements $J_{mn} = \partial f_m/\partial y_n$ and $f'_j = \partial f/\partial k_j$. In Dunker's method, equations 7 and 9 are decoupled and solved separately, i.e. $Y(t_n)$ is determined from equation 7, as in the normal Gear algorithm, and then the set of sensitivity coefficients, $S_j(t_n)$, is evaluated from equation 9 from the previously determined values $S_j(t_{n-l})$. $Y(t_{n+1})$ is next evaluated, followed by $S_j(t_{n+1})$ etc. Dunker illustrates the application of the technique with respect to three stiff systems relevant to Section C of this book, namely the photochemical oxidation of hydrocarbons (Chapter C1), the oxidation of formaldehyde (Chapter C2) and the pyrolysis of ethane (Chapter C5).

C4.4 Modelling of a complex chemical reaction: $H_2 + O_2$

The hydrogen/oxygen reaction represents a good example of a complex reaction which benefits from a combined numerical integration/sensitivity analysis approach. The overall mechanism employed by Dougherty and Rabitz[8] was very detailed, containing 62 reactions. Figure C4.1 shows the explosion limits for a stoichiometric mixture of H_2 and O_2. Figures C4.2a,b show the computed behaviour of the reactants, intermediates and products in the slow reaction phase between the second and third limits and the explosive regime just beyond the second limit; the locations referred to are shown on Fig. C4.1. The equations are very stiff indeed, with time constant ratios in the range of 10^{10}–10^{11} and an efficient solver was required.

The mechanisms determined by the sensitivity analysis in the slow regime are very similar to those proposed previously from a steady-state analysis. The major features of the mechanism in the slow region near the second explosion limit are

$H_2 + O_2 \rightarrow 2OH$	chain initiation	(10)
$OH + H_2 \rightarrow H + H_2O$	propagation	(11)
$H + O_2 \rightarrow OH + O$	chain branching	(12)
$H + O_2 + M \rightarrow HO_2 + M$	chain terminating	(13)
$O + H_2 \rightarrow OH + H$	chain branching	(14)
$\left.\begin{array}{l}2HO_2 \rightarrow H_2O_2 + O_2 \\ HO_2 + H_2 \rightarrow H_2O_2 + H\end{array}\right\}$ H_2O_2 production		(15) (16)
$H_2O_2 \rightarrow 2OH$	chain restoring	(17)
$\left.\begin{array}{l}HO_2 \rightarrow \text{wall} \\ H_2O_2 \rightarrow \text{wall}\end{array}\right\}$ chain terminating.		(18) (19)

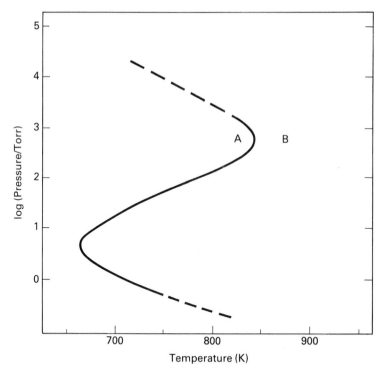

Fig. C4.1. Explosion peninsula for a stoichiometric H_2–O_2 mixture in a spherical KCl-coated vessel of 7.4 cm diameter. Points A and B mark the coordinates of the simulations in Fig. C4.2.

At these temperatures, $k_{13}[M] > k_{12}$ and the concentrations of the propagating radicals are kept low, whilst $[HO_2]$ and $[H_2O_2]$ gradually build up. As the temperature increases, k_{12}, which has a large associated activation energy, increases rapidly and the concentrations of atoms build up (Fig. C4.2b) at the expense of HO_2. The mechanism now becomes particularly sensitive to the H-atom concentration, with the moduli of the sensitivity coefficients for the reactions

$$H + H_2O \rightarrow H_2 + OH \tag{20}$$

$$H + H + M \rightarrow H_2 + M \tag{21}$$

$$H + HO_2 \rightarrow H_2O + O \tag{22}$$

$$H + HO_2 \rightarrow H_2 + O_2 \tag{23}$$

increasing rapidly. It is in this explosive regime, where the steady-state approximation is inapplicable, that the numerical technique becomes particularly valuable.

Finally, a cautionary note! It might be presumed, from the above discussion, that we can use the computer to sort out fully the chemistry for us. Nothing could be further from the truth, since much chemical knowledge and experience goes into the initial formulation of the model, especially for systems which are more chemically complex than the H_2/O_2 reaction, such as are found in the combustion

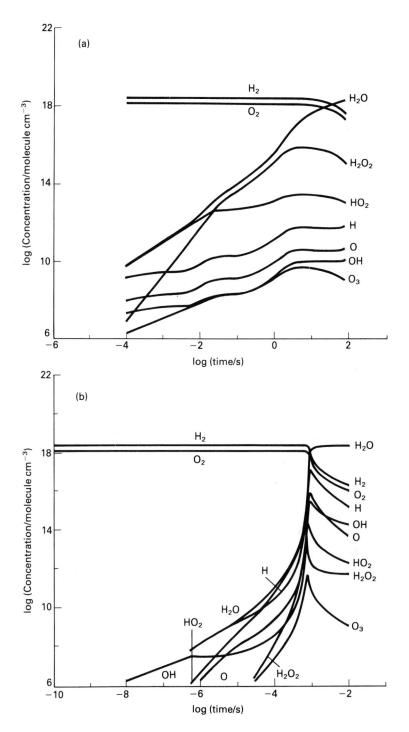

Fig. C4.2. Calculated concentration profiles for a stoichiometric H_2–O_2 mixture: (a) $[H_2]_{t=0} = 2.8 \times 10^{18}$ molecule cm^{-3}, $T = 825\,K$, non-explosive behaviour; (b) $[H_2]_{t=0} = 2.8 \times 10^{18}$ molecule cm^{-3}, $T = 875\,K$, explosion occurs. Reproduced, with permission, from reference 8.

of hydrocarbons. There is, in addition, the problem of incorporating realistic rate constants, where we rely heavily on data evaluations.

C4.5 Simulation of experimental systems: $H + C_2H_4$

In Chapter C5 we shall find that the reaction

$$H + C_2H_4 \rightarrow C_2H_5 \tag{24}$$

is very important in the cracking of hydrocarbons. It may be measured by the flash photolytic production of H atoms, followed by their time-resolved detection by resonance fluorescence.[9] Typical H-atom concentrations are $\sim 5 \times 10^{10}$–10^{12} molecule cm^{-3} and the time scale employed is ~ 0.1–10 ms, corresponding to a first-order decay constant in the range 10^2–10^4 s. The atom–radical reactions

$$H + C_2H_5 \Big\langle {}^{\displaystyle C_2H_6}_{\displaystyle 2CH_3} \tag{25a}$$
$$\tag{25b}$$

$$H + CH_3 \longrightarrow CH_4 \tag{26}$$

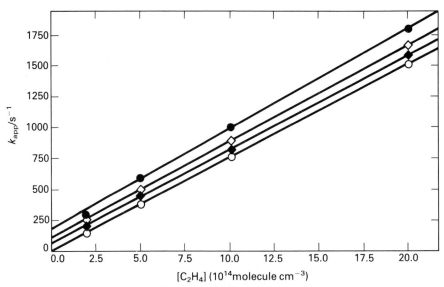

Fig. C4.3. Apparent pseudo-first-order rate constants, obtained by analysing simulated H-atom concentration profiles, calculated using FACSIMILE for reactions 24–26, with a first-order decay program.[9] The simulations refer to increasing initial H-atom concentrations (units 10^{11} molecule cm^{-3}): $\bigcirc$, ~ 0.0, $\blacklozenge$, 4.0, $\diamond$, 7.5, $\bullet$, 16. The lines refer to linear fits to the points for $[C_2H_4] \geqslant 3 \times 10^{14}$ molecule cm^{-3}. The slopes of these plots all lie within 6% of the value of k_1 input into the FACSIMILE program, demonstraing that k_1 can be determined from the slopes of plots of k_{app} vs. $[C_2H_4]$, provided the experiments are conducted at constant flash energy, $[C_2H_4] \geqslant 3 \times 10^{14}$ molecule cm^{-3} and $[H]_{t=0} \leqslant 10^{12}$ molecule cm^{-3}.

also occur, with rate constants of $\sim 10^{-10}\,\text{cm}^3\,\text{molecule}^{-1}\,\text{s}^{-1}$. At the transient concentrations in the experiments, these reactions can make a significant contribution (at the 1–10% level) and this contribution must be assessed. One approach is to simulate the decays using numerical integration of the o.d.e.s describing reactions 24–26. This generates a decay curve for the concentration of H atoms, which differs from the experimental one only in that it is noise free. The type of decay found experimentally may then be reproduced by putting normally distributed random noise on to the decay profile to match the experimental signal to noise ratio. The simulated profile may then be analysed, on the same timescale as that studied experimentally, using the simple first-order program employed to analyse the fluorescence decay curve. In the present case, it was found that the apparent pseudo-first-order rate constant was increased slightly by the atom/radical reactions, but that the absolute increase was independent of $[C_2H_4]$, provided a constant initial atom concentration was employed.[9] Thus the rate constant for reaction 24 could be determined by plotting the apparent first-order decay constant versus $[C_2H_4]$, whilst maintaining the laser pulse energy constant. Figure C4.3 shows examples of the apparent pseudo-first-order rate constants, plotted vs. $[C_2H_4]$, for various initial atom concentrations.

C4.6 Numerical analysis of radical decay profiles: $CH_3 + O_2$

Flash photolysis studies of the reaction:

$$CH_3 + O_2 \longrightarrow CH_3O_2 \tag{27}$$

using absorption spectroscopy have produced very imprecise rate constants because of contributions from the reactions:

$$CH_3 + CH_3 \rightarrow C_2H_6 \tag{28}$$

which can be catered for analytically and

$$CH_3 + CH_3O_2 \rightarrow 2CH_3O \tag{29}$$

which cannot. Modern signal averaging techniques enable the initial CH_3 to be reduced sufficiently that contributions from reactions 28 and 29 are negligible, so that k_{27} can be determined precisely.[10] Data at higher $[CH_3]$ ($t = 0$) can, on the other hand, be used to determine k_{29}. The FACSIMILE program incorporates a parameter fitting procedure, in which the experimental and simulated decay data are compared and, with k_{27} and k_{28} known, k_{29} (in the present context) is varied until the sum of squares of the residuals is minimised. The program works well provided the number of variable parameters is kept to a minimum—it should be remembered that there is comparatively little information in a decay profile!

C4.7 Sensitivity analysis of a potentially complex profile: $CH_3 + O_2$

The system discussed in the previous section is potentially much more complex, since the following reactions may also take part:[10]

$$2CH_3O_2 \rightarrow \text{inert products} \tag{30}$$

$$2CH_3O_2 \rightarrow 2CH_3O + O_2 \tag{31}$$

$$CH_3 + CH_3O \rightarrow \text{products} \tag{32}$$

$$CH_3O + CH_3O_2 \rightarrow \text{products} \tag{33}$$

$$2CH_3O \rightarrow \text{products} \tag{34}$$

$$CH_3O + O_2 \rightarrow \text{products.} \tag{35}$$

Not all of these rate constants are known accurately and, if a reaction contributed significantly to the CH_3 decay, then the analysis of k_{29} discussed above would be invalidated. The importance of reactions 30–35, with reasonable (or even widely varied) rate constants, can be assessed by evaluating the reduced sensitivity coefficients. Figure C4.4 demonstrates that the CH_3 decay is, indeed, governed primarily by reactions 27–29.

The numerical integration technique discussed in section C4.6 was applied to

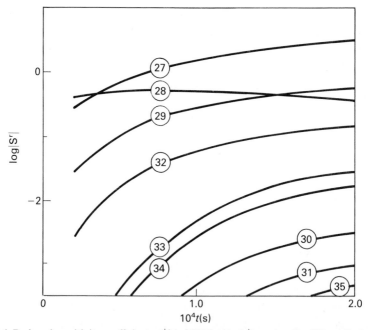

Fig. C4.4. Reduced sensitivity coefficients, ($|\partial \ln [CH_3]_t / d \ln k_j|$) vs. time for CH_3 radicals according to the reaction scheme 27–35, for $[CH_3]_{t=0} = 3.4 \times 10^{14}$ molecule cm^{-3}, $[O_2] = 3.7 \times 10^{16}$ molecule cm^{-3} and $T = 298$ K. The circled numbers refer to the reactions; the time-scale covers 99% of the methyl radical decay.[10]

an analysis of the build-up and decay of CH_3O_2 produced in reaction 27 and monitored via its absorption at 254 nm.[10] The build-up was satisfactorily accounted for by the model (reactions 27–35) with available literature values for the rate constants and with k_{27} and k_{29} as determined from the decay of $[CH_3]$. It was found, however, that the experimental decay of CH_3O_2 was far faster than the computed decay. An analysis of the sensitivity coefficients, coupled with an assessment of the precision of the experimental rate constants for reactions 30–35, demonstrated that the experimental decay could be satisfactorily modelled provided k_{33} was made significantly larger than the rather qualitative literature estimates. Direct experimental determination of k_{33} is needed, however, before this conclusion can be confirmed.

C4.8 Conclusions

Numerical integration is now widely employed by both modellers and experimentalists. A fuller appreciation of the range of possible applications, including modelling of spatially inhomogeneous systems, may be found in the general references given at the end of this chapter. A development which has still to come to full fruition in a chemical kinetic context is the construction of so-called Expert systems, in which the program itself constructs the model for, say, alkane pyrolysis, given a set of general rules. Even such programs, however, are unable to provide their own rate constants and experimental kineticists are unlikely to become redundant just yet.

Perhaps the most encouraging recent development has been the coming together of modellers and experimentalists so that the one may appreciate the difficulties and requirements of the other. This potential symbiosis, not yet fully realised, may be appreciated by reading the book edited by Ebert et al.[11] which reports the proceedings of a workshop, held in Heidelberg in 1980, on modelling of a wide variety of chemical kinetic systems. A further such workshop takes place in the summer of 1986 and promises to be a productive forum.

C4.9 Suggestions for further reading

Ebert K. H., Deuflhard P. and Jäger W. (Eds) (1981) *Modelling of Chemical Reaction Systems*, Springer Series in Chemical Physics, Vol. 18, Springer-Verlag, Berlin.
Rabitz H., Kramer M. and Dacol D. (1983) *Ann. Rev. Phys. Chem.*, **34**, 419.
Jones W. P. and Whitelaw J. H. (1984) *20th Symp. (Int.) on Combustion*, 233.

C4.10 References

1 Jones A. (1975) Specialist Periodical Report. *Reaction Kinetics*, (Ed. by P. G. Ashmore), Vol. 1, p. 291. The Chemical Society, London.
2 Gear C. W. (1971) *Commun. Association for Computing Machinery*, **14**, 176, 185.
3 Curtis A. R. (1979) *The FACSIMILE numerical integrator for stiff initial value problems.* Tech. Report, AERE-R 9352, Harwell.

4 Deuflhard P., Badet G. and Nowak U. (1981) *Modelling of Chemical Reactions Systems*, (Ed. by K. H. Ebert, P. Deuflhard and W. Jäger), Springer Series in Chemical Physics, Vol. 18, p. 38. Springer-Verlag, Berlin.
5 Hwang J.-T., Dougherty E. P., Rabitz S. and Rabitz H. (1978) *J. Chem. Phys.*, **69**, 5180.
6 Kramer M. A., Calo J. M., Rabitz H. and Kee R. J. (1982) Sandia National Labs., Rep. SAND 82-8231; Kramer M. A., Kee R. J. and Rabitz H. (1982) SAND 82-8230.
7 Dunker A. M. (1984) *J. Chem. Phys.*, **81**, 2385.
8 Dougherty E. P. and Rabitz H. (1980) *J. Chem. Phys.*, **72**, 6571.
9 Lightfoot P. D. and Pilling M. J., *J. Phys. Chem.*, to be published.
10 Pilling M. J. and Smith M. J. C. (1985) *J. Phys. Chem.*, **89**, 4713.
11 Ebert K. H., Deuflhard P. and Jäger W. (Eds), (1981) *Modelling of Chemical Reactions Systems*, Springer Series in Chemical Physics, Vol. 18, Springer-Verlag, Berlin.

Note added in proof:
    A further discussion of the direct method of calculating sensitivity coefficients may be found in Caracotsios M. and Stewart W. E. (1985)—*Computers Chem. Eng.*, **9**, 359. A program is available from Professor Stewart which includes an efficient implicit integrator with sensitivity analysis.

C4.11 Problem

Numerical integration and sensitivity analysis

C4.1. For the reaction scheme

$$R + O_2 \xrightarrow{1} RO_2$$

$$2R \xrightarrow{2} R_2$$

determine the concentration of R after $500\,\mu s$, given that $[R]_{t=0} = 10^{13}\,molecule\,cm^{-3}$, $[O_2] = 10^{15}\,molecule\,cm^{-3}$, $k_1 = 10^{-12}\,cm^3\,molecule^{-1}\,s^{-1}$ and $k_2 = 5 \times 10^{-11}\,cm^3\,molecule^{-1}\,s^{-1}$. Work in concentration units of $10^{13}\,molecule\,cm^{-3}$ and time units of $10^{-3}\,s$; use the Runge–Kutta method with $\delta t = 10^{-4}\,s$. Compare with the prediction of the analytical solution:

$$[R]_t = \left[\{[R]_{t=0}^{-1} + (2k_2/k_1[O_2])\}\exp(k_1[O_2]t) - \frac{2k_2}{k_1[O_2]} \right]^{-1}$$

(n.b. The rate of removal of R by reaction 2 is defined by $d[R]/dt = -2k_2[R]^2$.)

C4.12 Answer to problem

Numerical integration and sensitivity analysis

C4.1.
    Let $[CH_3] = y$ and $k_1[O_2] = k_1'$
    $$dy(t)/dt = -k_1'y - 2k_2 y^2.$$

Expressing y in units of 10^{13} molecule cm^{-3} and t in 10^{-3} s,

$k_1' = 1.0\,(\text{ms})^{-1}$

$k_2 = 0.5\,\text{cm}^3\,(10^{13}\,\text{molecule})^{-1}\,(\text{ms})^{-1}$.

Let $x = (y/10^{13})$, then $x_0 = 1.0$ and

$dx(t)/dt(t) = -x - 2 \times 0.5\,x^2$.

Runge–Kutta: $x(t + \delta t) = x(t) + \psi(t)$

$$x_{n+1} \quad = x_n + \psi_n$$

where

$\psi_n = (\alpha_0 + 2\alpha_1 + 2\alpha_2 + \alpha_3)/6$

$\alpha_0 = f(x_n)\delta t = (-x_n - x_n^2) \times \delta t$

$\alpha_1 = f(x_n + \alpha_0/2)\delta t$ etc. Setting $\delta t = 0.1$ ms,

For $n = 1$ $(t = 10^{-4}\text{s})$ $\alpha_0 = -(1.0 + 1.0) \times 0.1 = -0.2$

$\alpha_1 = -(0.9 + 0.81) \times 0.1 = -0.171$

$\alpha_2 = -(0.914 + 0.835) \times 0.1 = -0.175$

$\alpha_3 = -(0.825 + 0.68) \times 0.1 = -0.150$.

Thus $\psi_1 = -0.174$ and $x_1 = x_0 + \psi_1 = 0.826$.

Similarly, $\psi_2 = -0.133$, $x_2 = x_1 + \psi_2 = 0.693$

$\psi_3 = -0.107$, $x_3 = 0.586$

$\psi_4 = -0.084$, $x_4 = 0.502$

$\psi_5 = -0.069$, $x_5 = 0.433$.

Thus, at $t = 5 \times 10^{-4}$ s, $[CH_3] = 4.33 \times 10^{12}$ molecule cm^{-3}.

The analytical solution, expressed in reduced units, gives $x(0.5\,\text{ms}) = 0.435$ or $[CH_3] = 4.35 \times 10^{12}$ molecule cm^{-3}.

Chapter C5
Hydrocarbon Cracking

M. J. PILLING

C5.1 Introduction

Hydrocarbon cracking is an important process in the petrochemical industry, because of its use in the generation of a wide range of basic chemicals. In particular, it is used to generate ethene, which is a starting material in the manufacture of a variety of products. Detailed kinetic models are frequently employed in industry to try to understand and optimise the pyrolysis process and such models pose several interesting and fundamental kinetic problems. Many of the reactions involved (dissociation, recombination and addition) are pressure dependent and have not been widely studied at the operating temperatures (1100–1200 K) so that there are problems of extrapolation. The rate parameters for the central propagating step, H-abstraction, are subject to considerable uncertainties in this temperature range, with much literature discussion of the presence or absence of curvature in the Arrhenius plots. Some radicals establish local equilibrium, so that their thermodynamic parameters and, in particular, their heats of formation are of central importance, yet, here again, significant uncertainties exist. Finally, there is the question of which reactions are of greatest importance and cracking, especially of the higher hydrocarbons, is a fertile area for sensitivity analysis.

In order to simplify the discussion, we shall concentrate on ethane pyrolysis; ethane is an increasingly important feedstock and is the predominant source of ethene in the U.S., so that our discussions are not without relevance. At the end of the chapter we shall briefly discuss propane pyrolysis to demonstrate the extension of our ethane model to higher alkanes.

C5.2 Ethane pyrolysis

(i) Simple mechanism

An outline mechanism is shown in Table C5.1, whilst Fig. C5.1 shows a cycle diagram and is more helpful in appreciating the most important steps. The cycle diagram contains approximate reactive fluxes, F, through each reaction where,

Table C5.1. Simple mechanisms of C_2H_6 pyrolysis

Reaction type	Reaction	Reaction number	$k*$
Initiation	$C_2H_6 \rightarrow 2CH_3$	(1)	0.2
Propagation	$CH_3 + C_2H_6 \rightarrow CH_4 + C_2H_5$	(2)	4.7×10^{-15}
	$C_2H_5 \rightarrow C_2H_4 + H$	(3)	1.2×10^5
	$H + C_2H_4 \rightarrow C_2H_5$	(4)	2.9×10^{-12}
	$H + C_2H_6 \rightarrow H_2 + C_2H_5$	(5)	3.4×10^{-12}
	$CH_3 + H_2 \rightarrow CH_4 + H$	(6)	4.2×10^{-14}
Recombination	$CH_3 + CH_3 \rightarrow C_2H_6$	(7)	4.0×10^{-14}
	$CH_3 + C_2H_5 \rightarrow C_3H_8$ (and $CH_4 + C_2H_4$)	(8)	5.0×10^{-11}
	$C_2H_5 + C_2H_5 \rightarrow C_4H_{10}$ (and $C_2H_4 + C_2H_6$)	(9)	1.6×10^{-11}

*k refers to 1118 K and has units s^{-1} for reactions 1 and 3 and cm^3 molecule^{-1} s^{-1} for reactions 2, 4–9.

for example, the reactive flux for reaction 2 is defined as:

$$F_2 = k_2[CH_3][C_2H_6].$$

Commercial crackers operate to optimal (and therefore significant) conversions and so several of the channels refer to reaction of the radicals with the major products, C_2H_4 and H_2. A detailed analysis is required if we are to make quantitative comments on the mechanism but we can go a long way with a simple qualitative treatment.

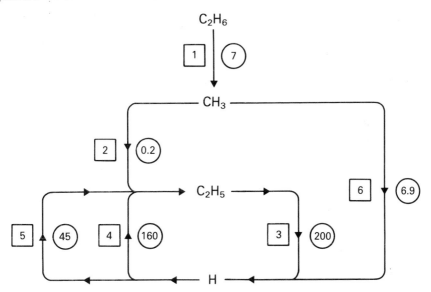

Fig. C5.1. Cycle diagram for ethane pyrolysis at 1118 K and 70% conversion. The initial ethane pressure is 38 Torr and the diluent gas is nitrogen. The numbers in squares refer to the reaction numbers and those in circles to the reactive flux in units of 10^{16} molecule cm^{-3} s^{-1}.

(a) Step 1, which is comparatively slow, is balanced by the total rate of recombination in the steady-state. The chains for ethane cracking are quite long and, to a reasonable approximation, only the propagation steps need be considered when discussing product formation.

(b) Abstraction from ethane by methyl radicals is not as fast as that by H, so that step 2 is comparatively slow and the methyl concentration is maintained at a high level in the early stages of pyrolysis, but is depressed as more H_2 is formed.

(c) Once C_2H_5 has been formed, we enter the rapid section of the chain reaction. The $C_2H_5 \rightleftharpoons H$ interconversion is by far the fastest part of the cycle and, at the high temperatures employed, these radicals are close to a local equilibrium, with $[C_2H_5] > [H]$. The overall rate of ethane consumption is largely determined by step 5.

(d) At ~ 1100 K, conversions of 85–90%, which are close to the equilibrium values, are employed. Further increases in temperature increase the overall rate of conversion, but push the equilibrium increasingly towards C_2H_2. Mechanistically, this means that the routes through C_2H_3, which will be discussed below, become more important.

(ii) Quantitative discussion of individual reactions

In a more quantitative discussion several of the reactions need to be examined in greater detail.

Reactions 1, 3, 4, 7, 8, and 9 are pressure dependent (they are all combination, addition or decomposition reactions). In particular, reactions 1, 3, 4 and 7 are below their high pressure limits at temperatures in the region of 1100 K and at pressures of 1–2 atm, although rate constants quoted for many of them refer to lower temperatures and to the high pressure limit. Thus realistic allowances should be made for fall-off, although this is very rarely done.

We illustrate the procedure which may be adopted by reference to reaction 4, which is of central importance. The high pressure limiting rate constant, k_4^∞, has been determined accurately over the temperature range 200–600 K,[1,2] whilst the fall-off behaviour has been studied for $300 \leqslant T/K \leqslant 600$.[3] The data have been analysed using the factorisation method discussed in section A4.10. The reaction occurs on a Type I potential (p. 102) and transition state parameters have been calculated using a combination of potential energy surface calculations and comparison with the experimental data.[4] Thus most of the parameters needed to calculate k_4 as a function of temperature and pressure (viz. k_4^∞, k_4^0 and S_k) are well established.[3] The experimental measurements were conducted in helium, and a value was taken for β, the collisional stabilisation efficiency (p. 104), of 0.28 at 300 K, based on experimental measurements. Values at other temperatures may be calculated from equation 15 in Chapter A4, assuming $\langle \Delta E \rangle (= 285 \, \text{cm}^{-1})$ to be temperature independent, whilst β values for other third body molecules may be

Table C5.2. Molecular parameters for $H + C_2H_4 \leftrightharpoons C_2H_5$[3,4]

	Ethyl radical	Ethylene	Activated complex
Vibrational frequencies	3112	3104.9	3036
(cm^{-1})	3033	3102.5	3009
	2987	3026.4	2995
	2920	3021	2944
	2842	1630	1504
	1462	1443.5	1446
	1440	1342.2	1218
	1427	1220	1185
	1366	1023	994
	1175	979.3	938
	1138	939.6	869
	948	826.0	821
	713		339
	540		369
External moments of inertia	24.5	20.35	
(amu Å^2)	22.8	16.84	
	4.9	3.46	
Internal moment of inertia	1.12		
(amu Å^2)			
Symmetry numbers	6	4	

$\Delta H_0^{\ominus} = -159\,\text{kJ mol}^{-1}$, $\langle \Delta E \rangle = 285\,\text{cm}^{-1}$,
$k_4^{\infty} = (4.39 \pm 0.56) \times 10^{-11} \exp[-(1087 \pm 36)/T]\,\text{cm}^3\,\text{molecule}^{-1}\,\text{s}^{-1}$
$k_4^0 = (1.39 \pm 0.76) \times 10^{-29} \exp[-(569 \pm 220)/T]\,\text{cm}^6\,\text{molecule}^{-2}\,\text{s}^{-1}$
(95.5% confidence limits)

estimated from experimental data or by calculation. Table C5.2 lists all the molecular parameters which are needed for calculations on the $H + C_2H_4 \leftrightharpoons C_2H_5$ system, and Fig. C5.2 shows a fall-off calculation for k_4 at 1000 K.

Reaction 3 has been studied by end product analysis methods e.g., following the mercury photosensitised decomposition or the pyrolysis of ethane.[5,6] The experimental rate constants for reaction 4 and those for reaction 3, have been shown to be compatible with the recently measured enthalpy changes for reaction (3,4) (see problem B4.1). Thus values for k_3 under cracker conditions may be calculated from those for k_4, (based, in turn, on the model discussed above) and the equilibrium constant. Such an approach is desirable, since k_3 and k_4 are then automatically linked via thermodynamic parameters and, as we have discussed above, the two radicals are close to a local equilibrium. $\Delta S_{3,4}^{\ominus}$ may be determined from the spectroscopic data given in Table C5.2 and is combined with $\Delta H_{298}^{\ominus}$ (3,4) and calculated specific heats to obtain values of $K_{3,4}$ at the required temperatures (see Table C5.3).

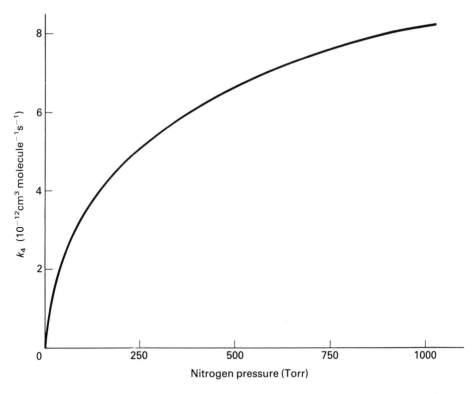

Fig. C5.2. Fall-off curve for $H + C_2H_4 \rightarrow C_2H_5$ at $1000\,K$ in a nitrogen diluent, $k_4^\infty = 1.48 \times 10^{-11}\,cm^3\,molecule^{-1}\,s^{-1}$.

Reactions 1 and 7

$$C_2H_6 \rightarrow 2CH_3 \tag{1}$$

$$CH_3 + CH_3 \rightarrow C_2H_6 \tag{7}$$

are also into the fall-off region at $1100\,K$ and $1\text{–}2\,atm$. Reaction 7 has been thoroughly studied over a range of temperatures[7] and $\Delta H_f^\ominus(CH_3)$ and $S^\ominus(CH_3)$ are also well known, so the procedure discussed above for reactions 3 and 4 could, in principle, be adopted again. However, reaction 7 has been less successfully modelled than has reaction 4; it occurs on a Type II potential surface (p. 102) so

Table C5.3. Equilibrium constants and thermodynamic parameters for $H + C_2H_4 \rightleftharpoons C_2H_5$

T/K	$\Delta H^\ominus/kJ\,mol^{-1}$	$\Delta S^\ominus/J\,mol^{-1}\,K^{-1}$	$10^{17}\,K_{3,4}/cm^3\,molecule^{-1}$
1053	-158.7	98.66	7.53
1093	-159.0	98.91	4.03
1118	-159.1	99.08	2.77
1148	-159.3	99.20	1.82

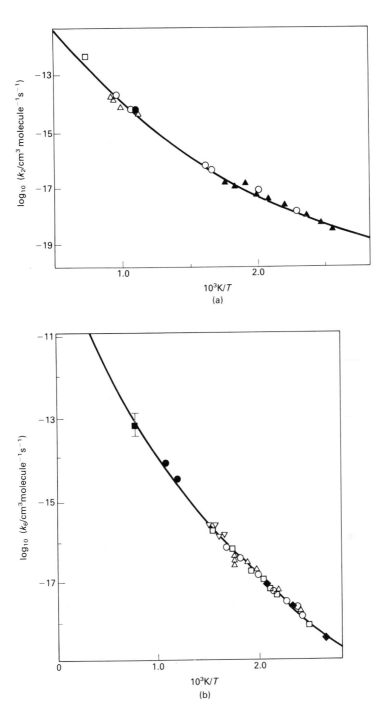

Fig. C5.3. Arrhenius plots for (a) $CH_3 + C_2H_6 \rightarrow CH_4 + C_2H_5$ and (b) $CH_3 + H_2 \rightarrow CH_4 + H$. Reproduced, with permission, from reference 8.

that the configuration of the activated complex depends on energy. k_7^∞ appears to decrease slightly with temperature and this property is difficult to reproduce in a model. Whilst the decrease is only modest, the inadequacy of the models casts some doubt on the extrapolations to cracker temperatures.

There is considerable uncertainty in the literature concerning the temperature dependence of many H-atom abstraction reactions in the region of interest. Figure C5.3 shows Arrhenius plots for $CH_3 + C_2H_6$ and for $CH_3 + H_2$. The lower temperature data for reaction 2 were obtained by end-product analysis of photolysed $(CD_3)_2CO$ in the presence of C_2H_6 and rely on comparative yields of CH_3CD_3 and C_2D_6, i.e. k_2 is determined relative to $(k_7)^{1/2}$. The higher temperature measurements, which lie well above the straight line obtained by extrapolating the lower temperature data, were obtained by a variety of techniques, including ethane pyrolysis in a flow reactor, and again rely on end product analysis measurements. In one study, k_2 was determined relative to $(k_8)^{1/2}$, which was assumed to be at its high pressure limit. The latter assumption may not be fully justified, but this approximation cannot account for the large discrepancy.

The deviation from straight line behaviour is common to many H-atom abstraction reactions and has attracted much theoretical interest. Furue and Pacey[8] recently examined the experimental data in detail and showed that, if the rate constant is expressed in the form $AT^n \exp(-E_{act}/RT)$, then, for reaction 2, $n \sim 12$, much higher than can be accommodated from the temperature dependence of partition functions (cf. section A3.6). For reaction 6, the curvature is less marked and n is about 5; calculations of partition functions, based on an *ab initio* surface, were unable to reproduce the experimental curvature unless tunnelling was invoked. The tunnelling models employed were quite crude and an understanding of the reactions would be improved by a more detailed theoretical investigation using the techniques discussed in section A1.5.

(iii) Extended mechanism

The simple mechanism is unable to explain many aspects of ethane pyrolysis at high temperatures. Figure C5.4 shows typical product yields for ethane cracking in a jet stirred reactor at 1118 K. Note particularly the yields of acetylene and butadiene. In addition, aromatic compounds are formed in low yields, and also 'coke' which can, eventually, lead to shut down of the reactor. Several of these products result from the formation of the vinyl radical by reaction 10, followed by reactions of C_2H_3:

$$H + C_2H_4 \rightarrow H_2 + C_2H_3 \tag{10}$$

$$C_2H_3 \rightarrow C_2H_2 + H \tag{11}$$

$$C_2H_3 + H_2 \rightarrow C_2H_4 + H \tag{12}$$

$$C_2H_3 + C_2H_4 \rightarrow C_4H_6 + H. \tag{13}$$

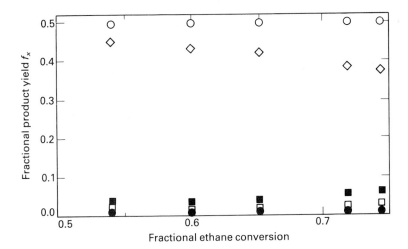

Fig. C5.4. Product yields for ethane cracking at 1118 K as a function of ethane conversion. The fractional product yield is defined as $f_x = [X]/([C_2H_6]_0 - [C_2H_6])$. $\bigcirc = H_2$; $\diamondsuit = C_2H_4$; $\blacksquare = CH_4$; $\square = C_2H_2$; $\bullet = C_4H_6$.

The rate constant for reaction 10 is very poorly characterised, with estimates at 1118 K ranging from 3×10^{-13}–3×10^{-11} cm^3 molecule^{-1}s^{-1} and the rates for vinyl radicals are generally poorly understood. The rate constant for reaction 14:

$$H + C_2H_2 \rightarrow C_2H_3 \qquad (14)$$

has been extensively studied at lower temperatures and procedures could be adopted, similar to those discussed for k_4 in section 5.2(ii), with k_{11} being calculated via the equilibrium constant, $K_{11,14}$. Estimates could also be made of the rate of the H-abstraction reaction, k_{10}, from other H-abstractions, using the Evans–Polanyi relationship,[9] in which activation energies for a similar group of atom transfer reactions are related to bond dissociation energies. The only problem is that there is a very large uncertainty in the heat of formation of the vinyl radical, with estimates for $\Delta H^{\ominus}_{f,298}$ ranging from 265 to over 300 kJ mol^{-1}. An uncertainty of 40 kJ mol^{-1} is reflected in calculated values for k_{11} differing by a factor of 70 at 1118 K! As a consequence, one of the major barriers to developing a full understanding of ethane pyrolysis under industrial conditions is the parlous state of our present knowledge of the vinyl radical.

That the vinyl radical can have a significant effect is demonstrated by the radical concentrations shown in Table C5.4, which were calculated for two reaction schemes: (a) the simple scheme discussed above and (b) a full and realistic scheme which includes C_2H_3 and which employs numerical integration (Chapter C4). The C_2H_3 rate constants were calculated using the lowest heat of formation (265 kJ mol^{-1}) which, admittedly, gives the greatest emphasis to C_2H_3, but which

Table C5.4. Radical concentrations. Ethane pyrolysis, 1118 K, residence time in reactor $= 0.6$ s

Radical	10^{-12} concentration/molecule cm^{-3}	
	Simple scheme	Extended scheme
H	3.9	0.55
CH_3	3.7	4.0
C_2H_5	27.6	5.0
C_2H_3	—	12
C_3H_5	—	25

has much support. Note the large concentrations of allyl which are formed and which result from the reactions

$$C_2H_3 + CH_3 \rightarrow C_3H_6 \tag{15}$$

$$H + C_3H_6 \rightarrow H_2 + C_3H_5 \tag{16}$$

It is a comparatively stable radical and undergoes radical molecule reactions very slowly.

(iv) Conclusions

It is clear from these comments that our attempts to understand and model even the cracking of ethane under industrial conditions are far from complete. Indeed one can really only say that the major problems have been identified. Many more experimental studies of sensitive elementary reactions are required, coupled with detailed and realistic extrapolations of pressure dependent rate constants to industrial conditions.

Attempts to model ethane cracking under industrial conditions provide a salutary tale of the dangers of adjusting rate parameters in order to match experimental data. Such a procedure is widely adopted and may be illustrated by the work of Sundaram and Froment.[10] Their kinetic transgressions were no worse than those of other modellers, indeed their model is widely used, but an examination of their approach demonstrates the sweeping assumptions that are often made. They constructed a 49 reaction scheme using some literature rate constants, but they also allowed rate constants for 23 of the reactions to float. The magnitude of the permitted variations can be gauged by reference to reaction 2, for which the proposed Arrhenius parameters predict a rate constant, at 1100 K, a factor of twenty greater than the experimental values shown on the curved Arrhenius plot in Fig. C5.3, well outside experimental error. Such an approach is clearly untenable. For reaction systems with significant uncertainties, such as those found for ethane cracking, the important function of modelling is to demonstrate, through numerical integration and sensitivity analysis, which reactions require further experimental or theoretical investigation. The justification of such a limited approach, and the dangers accompanying overindulgent parameter adjustment, are demonstrated by the inability of the model of Sundaram and Froment to reproduce experimental data outside their fitted temperature range.

C5.3 Propane cracking

Figure C5.5 shows a cycle diagram for propane cracking. It was constructed from the results of a numerical integration at 1147 K and at ~60% conversion. The diagram is quite complex and, rather than show the fluxes in numerical form, a diagramatic representation is adopted, with the thickness of the arrows correlating

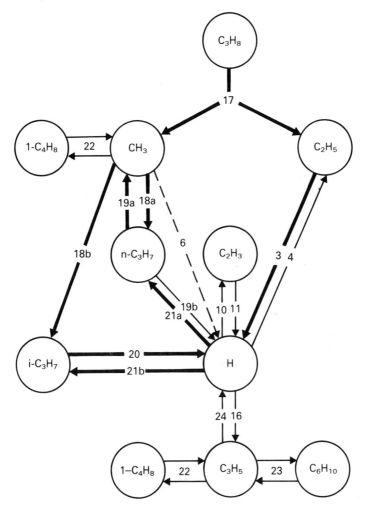

Fig. C5.5. Cycle diagram for propane pyrolysis at 1147 K and 60% conversion. The magnitudes of the reactive fluxes correlate roughly with the thickness of the arrow. The numbers refer to the reaction numbers.

roughly with the magnitudes of the fluxes. The higher temperature and the lower activation energy for the initiation reaction:

$$C_3H_8 \rightarrow CH_3 + C_2H_5 \qquad (17)$$

result in a higher total radical concentration. The initially generated radicals are channelled through to H, C_2H_5 by dissociation (reaction 3) and CH_3 by abstraction from C_3H_8 to generate propyl radicals, which then decompose:

$$CH_3 + C_3H_8 \rightarrow CH_4 + n\text{-}C_3H_7 \qquad (18a)$$
$$\rightarrow CH_4 + i\text{-}C_3H_7 \qquad (18b)$$
$$n\text{-}C_3H_7 \rightarrow C_2H_4 + CH_3 \qquad (19a)$$
$$\rightarrow C_3H_6 + H \qquad (19b)$$
$$i\text{-}C_3H_7 \rightarrow C_3H_6 + H. \qquad (20)$$

The major propagation steps involve abstraction by H from C_3H_8:

$$H + C_3H_8 \rightarrow H_2 + n\text{-}C_3H_7 \qquad (21a)$$
$$\rightarrow H_2 + i\text{-}C_3H_7. \qquad (21b)$$

Perhaps the most interesting feature of the mechanism is the generation of the allyl radical by hydrogen atom abstraction from C_3H_6 by H (reaction 16). As discussed earlier, allyl is a comparatively stable radical and undergoes radical/molecule reactions only slowly. It reacts primarily by radical/radical recombination:

$$CH_3 + C_3H_5 \rightarrow 1\text{-}C_4H_8 \qquad (22)$$
$$2C_3H_5 \rightarrow 1,5\text{-}C_6H_{10}. \qquad (23)$$

However, but-1-ene and hexa-1,5-diene both have weak C—C bonds and rapidly decompose to regenerate C_3H_5 via the reverse of reactions 22 and 23. [C_3H_5] is, in consequence, high and allyl represents a radical pool, which is slowly fed back into the main cycle via abstraction reactions, e.g.

$$C_3H_5 + H_2 \rightarrow C_3H_6 + H. \qquad (24)$$

The most uncertain aspects of propane pyrolysis are the generation of allyl via reaction 16 and the allyl abstraction reactions. By contrast with ethane pyrolysis, C_2H_3 is comparatively unimportant and the radical does not reach such high concentrations. Presumably, it is still a major intermediate in the generation of acetylene and of coke, although its effects on the concentrations of other radicals are of minor importance.

C5.4 Conclusions

The basic mechanism of hydrocarbon pyrolysis has been understood, in outline, for many years, since the pioneering work of Rice and Herzfeld in 1934.[11] Even this basic mechanism is difficult to model quantitatively, because of the pressure-dependence of decomposition, addition and recombination reactions and of the reasonably well characterised, although poorly understood, curvature in the Arrhenius plots for abstraction reactions. The major complications in the mechanism, however, concern the generation of the unsaturated radicals, C_2H_3 and C_3H_5, which become particularly important at high temperatures and high conversions. Experimental characterisation of the kinetics of reactions involving these radicals is clearly required.

Acknowledgement

The calculations on ethane were performed by Phillip Lightfoot to whom I am very grateful. They may be found in his doctoral thesis (Lightfoot P. D., *D. Phil. Thesis*, Oxford University, 1986).

C5.5 Suggestions for further reading

Albright J. and Crynes J. (Eds) (1982) *Pyrolysis: Theory and Industrial Practice*, Academic Press, New York.
Allara D. L. and Shaw R. (1980) *J. Phys. Chem. Ref. Data*, 9, 523.
Edelson D. and Allara D. L. (1980) *Int. J. Chem. Kinet.*, 12, 605.

C5.6 References

1 Lee J. H., Michael J. V., Payne W. A. and Stief L. J. (1978) *J. Chem. Phys.*, 68, 1817.
2 Sugawara K., Okazaki K. and Sato S. (1981) *Bull. Chem. Soc. Jpn*, 54, 2872.
3 Lightfoot P. D. and Pilling M. J., *J. Phys. Chem.*, to be published.
4 Hase W. L. and Schlegel H. B. (1982) *J. Phys. Chem.*, 86, 3901.
5 Loucks F. L. and Laidler K. J. (1967) *Can. J. Chem.*, 45, 2795; Lin M. C. and Back M. H. (1966) *Can. J. Chem.*, 44, 2357.
6 Trenwith A. B. (1986) *J. Chem. Soc., Faraday Trans. 2*, 82, 457.
7 Macpherson M.T., Pilling M. J. and Smith M. J. C. (1985) *J. Phys. Chem.*, 89, 2268.
8 Furue H. and Pacey P. D. (1986) *J. Phys. Chem.*, 90, 397.
9 Evans M. G. and Polanyi M. (1938) *Trans. Faraday Soc.*, 34, 11.
10 Sundaram K. M. and Froment G. F. (1978) *Ind. Eng. Chem., Fundam.*, 17, 174.
11 Rice F. O. and Herzfeld K. F. (1934) *J. Am. Chem. Soc.*, 56, 284.

C5.7 Problem

Hydrocarbon cracking

C5.1. At 1118 K, for the conditions shown in Fig. C5.1, assume that, because $[C_2H_5] > [CH_3]$, reaction 7 may be neglected. Use the simple scheme in Table C5.1 and the steady state approximation to show that

$$[H] \simeq (k_3[C_2H_5])/(k_4[C_2H_4] + k_5[C_2H_6])$$

and

$$[C_2H_5] = \frac{2k_1[C_2H_6]}{k_8[CH_3]} - \frac{(k_2[C_2H_6]+k_6[H_2])}{k_8}.$$

Obtain an expression for $[CH_3]$ containing no other radical concentrations (it is a cubic) and solve by trial and error. Hence obtain estimates for the radical concentrations.

Assuming that these radical concentrations are unaffected by reactions 10–13 in the extended scheme, show that

$$[C_2H_3] = k_{10}[H][C_2H_4]/(k_{11}+k_{12}[H_2]+k_{13}[C_2H_4])$$

and hence obtain expressions for the relative rates of C_2H_2 and C_4H_6 production versus C_2H_4 production, assuming the following rate constants for $k_{10}-k_{13}$:

$$k_{10} = 1.0 \times 10^{-12}\,\text{cm}^3\,\text{molecule}^{-1}\,\text{s}^{-1}, \quad k_{11} = 1.5 \times 10^3\,\text{s}^{-1},$$

$$k_{12} = 4.2 \times 10^{-14}\,\text{cm}^3\,\text{molecule}^{-1}\,\text{s}^{-1},$$

$$k_{13} = 6.7 \times 10^{-15}\,\text{cm}^3\,\text{molecule}^{-1}\,\text{s}^{-1}.$$

Conditions: 1118 K, 70% conversion. Take:

$$[C_2H_4] = [H_2] = 1.9 \times 10^{17}\,\text{cm}^{-3}, \quad [C_2H_6] = 9 \times 10^{16}\,\text{cm}^{-3},$$

$$[CH_4] = 4 \times 10^{16}\,\text{cm}^{-3}.$$

C5.8 Answer to problem

Hydrocarbon cracking

Let $a = [H]$, $b = [CH_3]$, $c = [C_2H_5]$, $E = [C_2H_6]$ and express the rates of bimolecular reactions involving stable molecules in terms of pseudo-first-order rate constants, k' e.g. $k_2' = k_2[C_2H_6]$. Set up steady-state equations:

$$da/dt = k_3c - k_4'a - k_5'a + k_6'b = 0 \tag{1}$$

$$db/dt = 2k_1E - k_2'b - k_6'b - k_8bc = 0 \tag{2}$$

$$dc/dt = k_2'b - k_3c + (k_4'+k_5')a - 2k_9c^2 - k_8bc. \tag{3}$$

From (1),

$$a = (k_3c + k_6'b)/(k_4'+k_5') \tag{4}$$

and from (2)

$$c = [2k_1E - (k_2'+k_6')b]/k_8b. \tag{5}$$

Adding (1) + (2) + (3) and dividing by 2,

$$k_1E = k_8bc + k_9c^2. \tag{6}$$

Substituting (5) into (6)

$$k_8^2(k_2'+k_6')b^3 - \{k_9(k_2'+k_6')^2 + k_2 Ek_8^2\}b^2 + 4k_9k_1 E(k_2'+k_6')b - 4k_9k_1^2 E^2 = 0$$

—the required cubic, whose solution is $b = [CH_3]$.

Inserting numerical values:

$$b^3 - 5.57 \times 10^{13}b^2 + 4.61 \times 10^{26}b - 9.86 \times 10^{38} = 0.$$

The roots can be found approximately by a graphical method. Only the lowest values give positive solutions for c from equation 5. More exact numerical solution gives $b = 3.68 \times 10^{12}\,\text{cm}^{-3}$, $c = 2.76 \times 10^{13}\,\text{cm}^{-3}$. These values show that $k_6'b$ may be neglected in equation 4, giving

$$a \simeq k_3 c/(k_4'+k_5') = 3.86 \times 10^{12}\,\text{cm}^{-3}:$$

$$[H] = 3.86 \times 10^{12}\,\text{cm}^{-3}, [CH_3] = 3.68 \times 10^{12}\,\text{cm}^{-3}, [C_2H_5] = 2.76 \times 10^{13}\,\text{cm}^{-3}.$$

Let $[C_2H_3] = v$:

$$dv/dt = k_{10}'a - (k_{11}+k_{12}'+k_{13}')v = 0$$

$$v = k_{10}'a/(k_{11}+k_{12}'+k_{13}')$$

$$[C_2H_3] = 6.8 \times 10^{13}\,\text{cm}^{-3}$$

$$r_1 = \frac{d[C_2H_4]}{dt} = k_3 c - k_4'a - k_{10}'a + k_{12}'v - k_{13}'v = 8.6 \times 10^{17}\,\text{cm}^{-3}\,\text{s}^{-1}$$

$$r_2 = \frac{d[C_2H_4]}{dt} = k_{11}v = 1.0 \times 10^{17}\,\text{cm}^{-3}\,\text{s}^{-1}$$

$$r_3 = \frac{d[C_4H_6]}{dt} = k_{13}'v = 8.8 \times 10^{16}\,\text{cm}^{-3}\,\text{s}^{-1}$$

$$\frac{r_2}{r_1} = 0.12,\quad \frac{r_3}{r_1} = 0.10.$$

Comments

Note the very high vinyl concentration. If this is correct, then our original derivation is invalid, because recombination involving C_2H_3 will be important. The rate constants for C_2H_3 have been evaluated assuming that the C_2H_3—H bond energy is similar to that for CH_3—H and that the vinyl rate constants are similar to those for CH_3. There is, however, a large uncertainty in the heat of formation of C_2H_3 and k_{11}–k_{13} may be significantly faster. This increase, coupled with a reduced k_{10}, would reduce $[C_2H_3]$.

Moral

This type of steady-state problem is becoming complex and a great deal of further work is required if we are to refine the model—it is better to use numerical simulation.

Chapter C6
Chemistry in Lasers

J. P. SIMONS

C6.1 Introduction

Laser action necessarily requires a population inversion between the radiatively coupled states. In a chemical laser the inversion is generated by the results of exothermic chemical reactions: laser action may operate directly on the reaction products themselves or indirectly, on energy levels in a diluent gas where the population inversion is created through collisional energy transfer. The understanding of the 'modus operandi' of a chemical laser (actual or potential) requires an understanding of the chemistry *in* lasers, and that embraces an understanding of the way in which energy is disposed in the products of elementary reactions, of the way in which it is redistributed through collisional energy transfer processes and of the state-to-state kinetics and spectroscopy of the lasing system.[1] Since, by definition, a chemical laser operates between quantum state selected levels, the chemistry in lasers is the quantum state-to-state chemistry of exothermic reactions. This is the prime emphasis of the final chapter of this book.

(i) A little history[1-5]

The writer of this chapter gained his PhD in 1958, two years before Maiman achieved the first successful demonstration of laser action. At that time Norrish and Thrush[2] were exploiting the flourishing technique of flash photolysis in their pioneering studies of vibrational energy disposal in exothermic reactions of the type

$$O(^3P) + ClO \rightarrow O_2(v) + Cl \tag{1a}$$

$$O(^3P) + NO_2 \rightarrow O_2(v) + NO \tag{1b}$$

$$O(^1D) + O_3 \rightarrow O_2(v) + O_2 \tag{1c}$$

$$H + O_3 \rightarrow OH(v) + O_2 \tag{1d}$$

and molecular photodissociation reactions, such as

$$ClNO + hv \rightarrow Cl + NO(v) \tag{2a}$$

$$(CN)_2 + hv \rightarrow 2\,CN(v). \tag{2b}$$

During the same period, Polyani[4] was pioneering the technique of infra-red chemiluminescence and applying it to measurements of vibrational energy disposal in hydrogen halides, generated in exothermic atom transfer reactions, such as

$$H + XY \rightarrow HX(v) + Y \tag{3a}$$

$$X + HY \rightarrow HX(v) + Y. \tag{3b}$$

These early experiments were of major significance in contributing to the intellectual climate that led rapidly to the development of the first chemical lasers, based first on exothermic atom transfer reactions

$$H + Cl_2 \rightarrow HCl(v) + Cl \tag{4a}$$

$$HCl(v) \rightarrow HCl(v-1) + h\nu(\lambda = 3.7 \ \mu m) \tag{4b}$$

and, shortly afterwards on molecular photodissociation

$$CF_3I + h\nu \xrightarrow{\ 250 \, nm\ } CF_3 + I(^2P_{1/2}) \tag{5a}$$

$$I(^2P_{1/2}) \rightarrow I(^2P_{3/2}) + h\nu(\lambda = 1.315 \ \mu m). \tag{5b}$$

Both were first demonstrated by Pimentel and Kasper, the viability of the HCl laser having been predicted by Polanyi. The two methods of initiation, discharge flow and pulsed photolysis were soon complemented by the TEA technique—transverse electrical atmospheric pressure discharge, used for example, in the CO infra-red laser generated in CS_2/O_2 mixtures; in the ultra-violet rare gas halide lasers generated in rare gas/halogen mixtures; and in the invaluable CO_2 infra-red laser. The basic kinetics of such systems are fairly well understood, but the detailed kinetics are a major exercise in computer modelling, involving the solution of a very large number of coupled rate equations for a wide range of reactive and inelastic state resolved collision processes under (self-evidently) non-thermal equilibrium conditions, possibly for both neutral and ionic species. The major pumping mechanism in the TEA excited rare gas (Rg) halide lasers, for example, is probably the ion combination

$$Rg^+ + X^- + M \rightarrow RgX^* + M \tag{6a}$$

rather than the neutral reaction

$$Rg(^3P_J) + X_2 \rightarrow RgX^* + X. \tag{6b}$$

The philosophy of developing computer codes parallels that outlined in earlier chapters for combustion, pyrolysis and atmospheric chemical reactions, but with the added problem of quantum state resolution and the contribution made by photon emission and absorption processes. Detailed rate coefficients for state resolved energy transfer processes from high (v, J) levels, for example, in a chemical laser operating on inverted vibrational–rotational population densities, are simply not available. In this chapter, we focus instead on the basic kinetics of chemical laser systems which has developed in parallel with the search for new laser systems.

C6.2 Laser equations and the measurement of detailed rate coefficients

The critical parameter in a laser medium is its gain coefficient, $\alpha(v)$—if there is to be a net amplification of the light intensity in a round trip of $2l$ through the laser medium, then $I/I_0 = \exp[\alpha(v)2l]$ must exceed unity, i.e. $\alpha(v) > 0$. Assuming a Doppler line profile, the gain at the centre of a transition from state $|2\rangle \rightarrow |1\rangle$ is

$$\alpha^0(v) = \left(\frac{\ln 2}{16\pi^3}\right)^{1/2} \frac{c^2 A_{21}}{v^2 \Delta v_D}\left(\frac{N_2}{g_2} - \frac{N_1}{g_1}\right) \tag{7}$$

where N_1, N_2, g_1 and g_2 are respectively the population densities and the degeneracies, Δv_D is the linewidth and A_{21} is the (spontaneous) Einstein coefficient. For a molecular transition $|2\rangle \rightarrow |1\rangle \equiv |v'J'\rangle \rightarrow |v''J''\rangle$ this equation becomes

$$\alpha^0(v) \propto \frac{v}{\Delta v_D}|R_{v'J' \rightarrow v''J''}|^2 \left\{\frac{N_{v'J'}}{2J'+1} - \frac{N_{v''J''}}{2J''+1}\right\} \tag{8}$$

where $R_{v'J' \rightarrow v''J''}$ is the vibrational–rotational transition moment. Provided $N_{v'J'}/(2J'+1) > N_{v''J''}/(2J''+1)$, the gain is positive and laser action is possible, in principle. In practice, the 'effective' gain is reduced by reflection losses at the mirrors and by losses within the cavity; the threshold gain for laser action, $\alpha_{\text{th}}(v)$ has to satisfy the balance

$$R_1 R_2 = \exp\{[\beta - \alpha_{\text{th}}(v)]2l\} \tag{9}$$

where R_1, R_2 are the mirror reflectivities and β is a coefficient for the cavity loss.

Under moderate pressure conditions, typically > 10 Torr, the rotational population distributions are likely to be equilibrated (though highly excited diatomic hydrides may be resistant to equilibration because of their high rotational spacings). Under such conditions, the gain coefficients on P and R branches of the transition $|v\rangle \rightarrow |v-1\rangle$ become

$$P \equiv (J-1) \rightarrow J: \ \alpha(v_P) \propto \frac{N_v}{N_{v-1}} B_v \exp(-B_v J(J-1)/kT)$$
$$- B_{v-1} \exp(-B_{v-1}J(J+1)/kT) \tag{10a}$$

$$R \equiv (J+1) \rightarrow J: \alpha(v_R) \propto \frac{N_v}{N_{v-1}} B_v \exp(-B_v J(J+2)/kT)$$
$$- B_{v-1} \exp(-B_{v-1}J(J+1)/kT). \tag{10b}$$

P branch lines clearly have the larger gain.

The gain equations can be 'approached' from the left or the right hand side. From the left they measure the amplification (attenuation) of the laser medium for a given (known) population ratio; from the right they provide a strategy for determining the population ratios and hence the relative values of the detailed rate constants for reaction in the observed levels. One obvious approach is to measure the 'small signal' gain on a narrow line probe laser successively tuned into the transitions in question. In practice this technique has not been strongly exploited because of the difficulty in obtaining IR lasers that can be tuned over the required frequency range, but it has been successfully used, for example to determine detailed rates for the population of $CO(v)$ through the (chemical laser) reaction

$$O(^3P) + CS \rightarrow CO(v < 15) + S(^3P). \tag{11}$$

Other approaches include measurements of the times for tuned laser action to reach threshold, following pulsed initiation (e.g. by flash photolysis) or adjustment of the rotational temperature by varying the diluent gas pressure, to achieve equal or zero gain on pairs of coupled states. Detailed discussions have been given by Berry.[5] Before closing this section, it should be noted that the determination of vibrational population distributions via laser gain measurements offers one significant bonus—the determination of $N_{v=0}/N_{v>0}$, a ratio that cannot be determined via IR chemiluminescence measurements.

C6.3 Information theory and the analysis of population inversions

(i) Surprisal analysis

Statistical mechanics defines the entropy of a system by the equation $S = k \ln W$, where W is the number of ways of realising the system. The statistical probability of any given 'realisation' would be $P = 1/W$, whence $S = -k \ln P$. Information theory defines the surprisal of a given outcome, $I(a)$, as

$$I(a) = -\ln[P(a)/P^0(a)] \tag{12}$$

where $P(a)$, $P^0(a)$ are the observed and anticipated 'prior' probabilities of that outcome, with $P^0(a)$ corresponding to the most entropic, i.e. least informative set of outcomes. Suppose the outcome were the detection of a reaction product in quantum state n, with the normalised probability $P(n)$; then the 'local' surprisal would be $I(n) = -\ln[P(n)/P^0(n)]$ and the 'global' surprisal of the distribution, averaged over all accessible states n (subject to energy conservation constraints), would be

$$I = -\sum_n P(n) \ln[P(n)/P^0(n)] \tag{13}$$

known as the information content. If $P(n) \equiv P^0(n)$ for all n then $I = 0$ and the distribution follows the statistical, democratic, *a priori* expectation. The difference between the observed and prior distributions may be expressed in terms of a molar entropy difference:

$$\Delta S = R \sum_n P(n) \ln [P(n)/P^0(n)] \tag{14}$$

where, for $I > 0$, ΔS is necessarily negative; ΔS is known as the 'entropy deficiency' and its magnitude measures the degree to which the experimental distribution departs from the most random one. Levine and co-workers[6-9] have developed a general principle which proposes that the observed distribution over all accessible quantum states will be that which maximises the entropy of the distributions (i.e. minimises the entropy deficiency) subject to satisfaction of all dynamical constraints—a high entropy deficiency implies a high degree of specificity in the selection of the final quantum states. If this specificity favours high energy states then one has the basis of a laser system.

In many exothermic atom transfer reactions, the measured surprisals are often (but by no means always) found to vary linearly with the fraction of the available energy associated with the populated quantum state. For example, if the vibrational energy of a product in level v is $E(v)$ and the total exoergicity into $E(v = 0)$ is E, the linear surprisal relationship would be

$$\ln [P(v)/P^0(v)] = - \lambda_1 f(v) - \lambda_0 \tag{15a}$$

where $f(v) = E(v)/E$, and

$$P(v) = P^0(v) \exp[- \lambda_1 f(v)]/\exp (\lambda_0). \tag{15b}$$

The form of the latter equation mirrors—but certainly does not reflect—the Boltzmann equation, with λ_1 representing a 'temperature-like' parameter and $\exp(\lambda_0)$ a partition function. The more negative the value of λ_1, the more likely is the reaction to be a basis for a potential chemical laser system.

If the vibrational energy disposal in an exothermic reaction

$$\text{e.g. } O(^3P) + CS \rightarrow CO(v) + S \tag{16}$$

follows a linear surprisal law [which it does(!) with $\lambda_1 = - 7.7$] then with $\langle f_v \rangle = \sum_v P(v) f(v)$

$$\lambda_1 = dI/d\langle f_v \rangle \neq 0. \tag{17}$$

The average, $\langle f_v \rangle$, the first moment of the vibrational population distribution, is known as an 'informative variable' or 'constraint'—in this case, where the surprisal is linear, it is the only constraint. Surprisals are not always linear however: if quadratic behaviour were found

$$\text{e.g. } I(v) = \lambda_0 + \lambda_1 f(v) + \lambda_2 f^2(v) \tag{18}$$

the constraints on the distribution would include both the first and the second moments, $\langle f_v \rangle$ and $\langle f_v^2 \rangle$. In general $I(n)$ is defined as

$$I(n) = \lambda_0 + \sum_i \lambda_i f_i(n). \tag{19}$$

(ii) The prior and the persuasion

The usual choice of the prior distributions, $P^0(n)$, are those which correspond to the most chaotic or 'statistical' distribution over the energetically accessible quantum states. Their values can be calculated by assuming that the probability of the outcome is proportional only to the number of ways in which it can be achieved, i.e. the number of energetically accessible product states, or, in classical terms, to the volume of the accessible phase space. Statistical prior distributions, based on density of states calculations for rigid rotor–harmonic oscillators, have been compiled for the common situation of vibrationally and rotationally excited diatomic molecules generated in reactions of the type

$$A + BC \rightarrow AB(v, J) + C \tag{20a}$$

or

$$A + BCD \rightarrow AB(v, J) + CD(v', J') \tag{20b}$$

and are listed by Levine and Kinsey.[6] (Problem 1 in Chapter A3 provides an example for a triatomic system, namely the reaction $O + CS \rightarrow CO(v) + S$.) A statistical assumption based on energy conservation alone is, of course, the simplest possible. It may be, however, that you think you know better and you would like to include more than simple energy conservation, for example, total angular momentum conservation. There may be a theoretical basis (e.g. a quantum scattering or, more likely, a classical trajectory study) for employing a prejudiced prior, or possibly for identifying an alternative (set of) dynamical constraint(s). If the theoretically based prior is $Q(n)$, then the function

$$H = \sum_n P(n) \ln [P(n)/Q(n)] \tag{21}$$

defines the 'persuasion' of the observed distribution. When $H = 0$ you are 'persuaded' that your prior analysis is in accord with experimental observations and that you have identified the constraints which limit the observed distributions—in the case of a chemical laser system, this would clearly include the specificity of energy disposal. This approach allows the information theoretic analysis to guide the analysis of the reaction dynamics, so adding to its practical utility in compacting known experimental data and in extrapolating to inaccessible regions of the product distribution [e.g. to estimate $P(v = 0)$ from infra-red chemiluminescence (IRCL) experiments].

C6.4 Reaction dynamics and the creation of population inversions

There are many chemical laser systems! For a recent survey see the bibliography, reference 3. Rather than attempt an exhaustive (and exhausting) exploration, one representative family of systems will be highlighted for analysis, namely those based on the hydrogen halides.

(i) Hydrogen halide systems—forward reactions generating population inversions

The archetype is the HF laser based on the exothermic chain sequence

$$F + H_2 \rightarrow HF(v \leqslant 3) + H \tag{22a}$$

$$H + F_2 \rightarrow HF(v \leqslant 9) + F. \tag{22b}$$

Both reactions have been studied by Polanyi and his coworkers via IRCL techniques and Fig. C6.1 shows a 'triangle' contour plot for the detailed energy disposal into vibration, rotation and (by energy conservation) translation in the products of reaction (22a); such plots are a convenient way of representing the pattern of the energy release (cf. Fig. A2.9). The contours for the detailed rate constants, $k(v, J|;T)$, peak along a high ridge where the energy is released principally into vibration. The mean fraction of energy so disposed is $\langle f_v \rangle = 0.65$; translation takes a fraction $\langle f_t \rangle = 0.27$, and only a small fraction is disposed into rotation, $\langle f_R \rangle = 0.08$. This partitioning is typical of a reaction proceeding over a repulsive potential energy surface where the dynamics are associated with 'mixed' energy release, i.e. rapid release of H—H repulsion while the H—F bond is still extended, favouring classical trajectories which 'cut the corner' (see Chapter A3). The low rotational excitation does not correlate with the reagent orbital angular momentum and is believed to arise from repulsion in the exit channel as the products separate from a slightly bent collision geometry. The vibrational populations, which peak at $v = 2$, display a linear surprisal with $\lambda_1 = -6.5$.

Reaction (22b) also generates an inverted vibrational population distribution peaking at $v = 6$ with $\langle f_v \rangle = 0.58$, but the surprisal plot is closer to quadratic than to linear. The dynamics are strongly influenced by the rapid motion of the light H atom relative to the heavy target molecule. This allows a near rectilinear trajectory over the reactive potential energy surface and brings the three atoms into close proximity before the (impulsive) release of repulsion in the exit channel. The energy disposal into vibration and translation can be qualitatively understood by comparing the 'suddenness' of the dissociation into products following the impact of the light H atom, with that which occurs when the molecular target suffers direct photodissociation following photon impact (cf. the discussion of the potential energy surface for the reaction $H + Cl_2 \rightarrow HCl + Cl$ in Chapter A1). Repulsion is suddenly 'switched on' through the introduction of an electron into the antibonding σ_u orbital. The impulsive release introduces a Franck–Condon or momentum

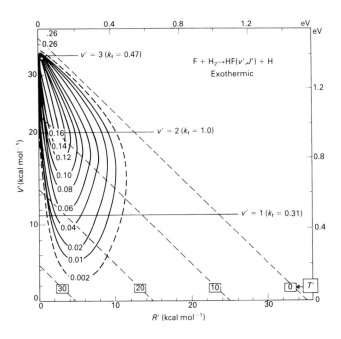

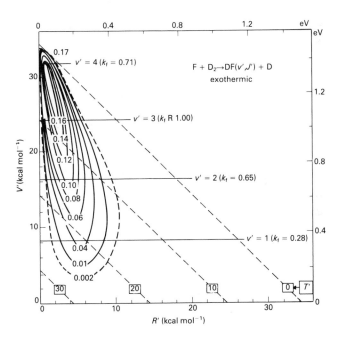

Fig. C6.1. Triangle contour plots summarising detailed rate constants, $k(v'J'|; T)$ for the reactions

$$F + H_2[D_2] \rightarrow FH[FD] + H[D]$$

at $T = 200$ K. (Reproduced, with permission, from Polanyi, J. C. and Tardy, D. C. (1969) *J. Chem. Phys.*, **51**, 5717.)

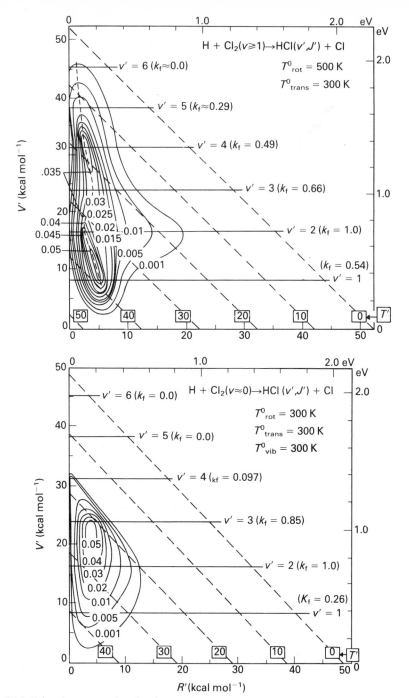

Fig. C6.2. Triangle contour plots for the reaction

$$H + Cl_2(v) \rightarrow HCl(v' J') + Cl$$

showing the consequences of vibrational excitation of the reagent Cl_2 (principally into $v = 1$). (Reproduced, with permission, from Ding, A. M. G., Kirsch, L. J., Perry, D. S., Polanyi, J. C. and Schreiber, J. L. (1973) *Faraday Discuss. Chem. Soc.*, **55**, 252.)

transfer constraint on the translational energy disposal, allowing the associated energy distributions to be modelled by a surprisal that is quadratic in the final momentum

$$I(E_t) = \lambda_t(E_t^{1/2} - E_{max}^{1/2})^2 + \lambda_0 \tag{23}$$

where $E_{max} \equiv E_t$ measured at the peak of the translation energy distribution. The Gaussian probability contour reflects that of the instantaneous F—F distance in $F_2 (v = 0)$ at the instant of H atom (or photon) impact.

The Franck–Condon control is particularly well revealed when the population distributions displayed in the reactions

$$H + Cl_2(v = 0) \rightarrow HCl(v' < 4, J') + Cl \tag{24a}$$

$$H + Cl_2(v \geqslant 1) \rightarrow HCl(v', J') + Cl \tag{24b}$$

are compared—see the triangle plots in Fig. C6.2. Both produce population inversions, neither give linear surprisals, but, most significantly, the reaction with $HCl(v > 1)$ gives a bimodal distribution. Furthermore, there is a propensity for $v \rightarrow v'$.* The sister reaction in the H_2/Cl_2 chain

$$Cl + H_2 \rightarrow HCl(v = 0) + H \tag{24c}$$

is slightly endoergic, with $\Delta H_0^{\ominus} = +4 \, kJ \, mol^{-1}$ which diminishes the utility of the overall reaction as an efficient chemical laser. The generation of $HCl(v = 0)$ also diminishes the population inversion $N(v = 2)/N(v = 1)$ because of the near-resonant $v \rightarrow v$ transfer,

$$HCl(v = 2) + HC!(v = 0) \rightarrow 2HCl(v = 1). \tag{25}$$

The triangle plot for the chemical laser reaction

$$H + ClI \rightarrow HCl(v' \leqslant 7, J') + I \tag{26}$$

is even more intriguing, see Fig. C6.3. At first glance, the bimodal distributions would seem to indicate branching into $I(^2P_{3/2})$ and $I(^2P_{1/2})$ but the absence of any emission from the spin–orbit excited atom is disillusioning. The smaller, low rotation component of the energy disposal contour resembles the behaviour in $H + Cl_2$ (see Fig. C6.2) but along the major ridge the energy disposal is quite different—it has been ascribed by Polanyi to a 'migratory' mechanism in which an initial attack at the I atom end of the target is followed by migration to the Cl atom. The large impact parameter necessary for this type of collision would generate the high rotational excitation observed in the product. The weaker component, associated with low rotational excitation, would then be generated through the low impact parameter, 'head-on' collisions.

*In the reaction $OH(v) + Cl \rightarrow O + HCl(v')$ this propensity provides a route to the generation of $HCl(v' \leqslant 11)$.

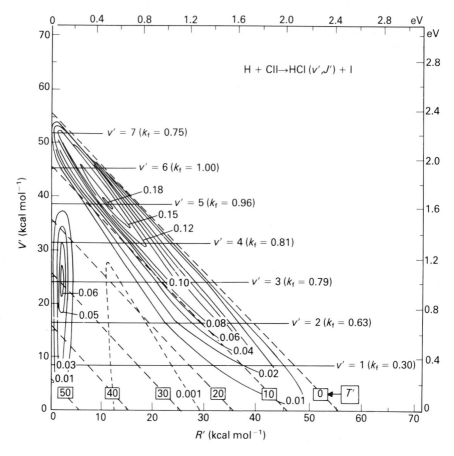

Fig. C6.3. Triangle contour plots for the reaction

$$H + ICl \rightarrow HCl + I$$

at $T = 300$ K. (Reproduced, with permission, from Nazar, M. A., Polanyi, J. C. and Skrlac, W. J. (1974) *Chem. Phys. Letters*, **29**, 473.)

(ii) Hydrogen halide systems—the backward reactions

Studies of the kind described so far led naturally to analysis of the reverse, endoergic reactions and the specificity of energy utilisation. Microscopic reversibility requires that vibrational excitation in the erstwhile products will promote the reverse endoergic reaction. For reactions occurring at pressures where the rotational and translational motions are thermally equilibrated, application of microscopic reversibility gives

$$\frac{k(\text{fwd})}{k'(\text{bwd})} = \frac{k(v'|v; T)}{k(v|v'; T)} = \frac{Q'_R}{Q_R}\left(\frac{\mu'}{\mu}\right)^{3/2} \exp\left\{-\frac{(E_{v'} - E_v)}{kT}\right\} \exp\left\{\frac{-\Delta E_0}{kT}\right\} \qquad (27)$$

where $Q_R(Q'_R)$ and $\mu(\mu')$ are, respectively, the rotational partition functions and reduced masses of the reagents (products). Equation 27 readily rearranges to an Arrhenius-like form.

$$\ln\left[\frac{k'(\text{bwd})}{k(\text{fwd})}\right] = \frac{E_{v'}}{kT} - \ln\left\{\frac{Q'_R}{Q_R}\left(\frac{\mu'}{\mu}\right)^{3/2} \cdot \exp\left\{-\frac{(\Delta E_0 - E_v)}{kT}\right\}\right\} \tag{28}$$

which clearly indicates the enhancement of the reverse endoergic reaction by vibrational excitation. An empirical relation which approximates the observed vibrationally enhanced rate constants is

$$\ln\left(\frac{k_{v'}}{k_{v'=0}}\right) = \frac{-(E_{\text{act}} - \alpha E_{v'})}{kT} \tag{29}$$

where α represents the fraction of the vibrational energy that is 'available' for surmounting the activation energy E_{act} of the endothermic reaction. Equation 28 was given theoretical respectability by Manz and Levine who used an information theoretic/phase space analysis to determine an *a priori* dependence in the range $E_{v'} < \Delta E_0$

$$k_{v'}^0(T) \propto \exp[-(\Delta E_0 - E_{v'})/kT]. \tag{30}$$

Experimental data for endoergic reactions of $HX(v')$ fit the form of equations 28 and 29 but do not correlate with the measured values of ΔE_0 or E_{act}.

(iii) Hydrogen halide systems—relaxation mechanisms

In modelling the kinetics of chemical laser systems based on $HX(v, J)$, one must include near resonant V–V exchange, V–R, T exchange (and R–R, T exchange when the experimental conditions do not ensure full thermal equilibration among the rotational levels). A fourth process which may be important, and which has been much studied—particularly by Wolfrum—is relaxation in atom–molecule collisions

$$H + HX(v, J) \rightarrow HX(v-1, J') + H \tag{31a}$$

$$X + HX(v, J) \rightarrow HX(v-1, J') + X. \tag{31b}$$

Consider, for example, the possible breakdown of reaction 31b for $X \equiv Cl$. Relaxation could proceed via inelastic, or reactive collisions

e.g. $$Cl + HCl(v=1, J) \rightarrow Cl + HCl(v=0, J' \gg J) \tag{32a}$$

$$Cl + HCl'(v=1, J) \rightarrow HCl(v=0, J') + Cl' \tag{32b}$$

$$Cl(^2P_{3/2}) + HCl(v=1, J) \rightarrow HCl(v=0, J') + Cl(^2P_{1/2}). \tag{32c}$$

Unravelling their relative rates involves a major experimental campaign—clearly it is not enough simply to monitor the rate of relaxation; the rate of formation of alternative products must also be monitored, e.g. via isotopic labelling of the two Cl atoms in reaction 32b. This has been achieved by isotopic selection of the $HCl'(v = 1, J)$ reagent in a discharge flow system, using narrow line HCl laser radiation, coupled with molecular beam sampling into a mass spectrometer, to 'weigh' the products. It is found that the total rate of H atom transfer is at least double the overall relaxation rate, indicating a major contribution from the vibrationally adiabatic reaction

$$Cl + HCl'(v = 1, J) \rightarrow HCl(v = 1, J') + Cl' \tag{32d}$$

a result which has implications for the potential energy surface, for 'vibrational bonding'—the development of bound transition states along the minimum energy path—and for quantum mechanical resonances (see Chapter A1).

A final point to note is that vibrational relaxation need not always be the enemy of the chemical laser designer. It can be put to practical use in generating population inversions in the acceptor, as for example in the $HF(v)$ sensitised CO_2 laser

$$HF(v) + CO_2(0, 0°, 0) \rightarrow HF(v - 1) + CO_2(0, 0°, 1). \tag{33}$$

C6.5 References and suggestions for further reading

1 The literature for this final chapter is a vast one: an excellent general list of leading references, as well as a reference in its own right, may be found in R. B. Bernstein's "Hinshelwood Lectures". These were published (in 1982) as *Chemical Dynamics via Molecular Beam and Laser Techniques*, Clarendon Press, Oxford.

Additional sources include:

Sections C6.1 and C6.2

2 Norrish R. G. W. (1958) Liverside Lecture, *Proc. Chem. Soc.*, 247.
3 Lin M. C., Umstead M. E. and Djeu N. (1983) *Ann. Rev. Phys. Chem.*, **34**, 557.
4 Polanyi J. C. and Schreiber J. L. (1974) In *Physical Chemistry, An Advanced Treatise*, (Ed. by H. Eyring, W. Henderson and W. Jost), Vol. 6, p. 383, Academic Press, New York.
5 Berry M. J. (1975) *Ann. Rev. Phys. Chem.*, **26**, 259.

Section C6.3

6 Levine R. D. and Kinsey J. L. (1979) In *Atom–Molecule Collision Theory—a Guide for the Experimentalist*, (Ed. by R. B. Bernstein), Plenum, New York.
7 Levine R. D. (1978) *Ann. Rev. Phys. Chem.*, **29**, 59.
8 Levine R. D. (1979) In *Maximum Entropy Formalism*, (Ed. by R. D. Levine and M. Tribus), MIT Press, Cambridge, Massachusetts.
9 Ben-Shaul A., Haas Y., Kompa K. L. and Levine R. D. (1981) *Lasers and Chemical Change*, Springer Series in Chemical Physics, Vol. 10, Springer-Verlag, Berlin.

Section C6.4

10 References 1 and 4 listed above.
11 Holmes B. E. and Setser D. W. (1979) In *Physical Chemistry of Fast Reactions*, II, (Ed. by I. W. M. Smith), p. 83. Plenum, New York.

Index